Introduction to the Pipeline Industry

Trainee Guide
Third Edition

PEARSON

Boston Columbus Indianapolis New York San Francisco Amsterdam
Cape Town Dubai London Madrid Milan Munich Paris Montreal Toronto Delhi
Mexico City Sao Paulo Sydney Hong Kong Seoul Singapore Taipei Tokyo

NCCER

President: Don Whyte
Vice President: Steve Greene
Chief Operations Officer: Katrina Kersch
Director of Product Development: Daniele Dixon
Pipeline Mechanical Project Managers: Patty Bird, Elizabeth Schlaupitz, Justin Street, Chris Wilson
Senior Development Manager: Mark Thomas
Senior Production Manager: Tim Davis
Quality Assurance Coordinator: Karyn Payne
Desktop Publishing Coordinator: James McKay
Permissions Specialist: Adrienne Payne
Production Specialist: Adrienne Payne
Editors: Graham Hack, Karyn Payne, Debie Hicks, Jackie Strawbridge, Adrienne Payne

Writing and development services provided by Topaz Publications, Liverpool, NY

Lead Writers/Project Managers: Don Congdon, Veronica Westfall
Desktop Publisher: Joanne Hart
Art Director: Alison Richmond
Permissions Editors: Andrea LaBarge, Briana Morgan
Writers: Don Congdon, Tony Clay, Thomas Burke

Pearson

Director of Alliance/Partnership Management: Andrew Taylor
Editorial Assistant: Collin Lamothe
Program Manager: Alexandrina B. Wolf
Digital Studio Project Managers: Heather Darby, Tanika Henderson, Jose Carchi
Director of Marketing: Leigh Ann Simms
Senior Marketing Manager: Brian Hoehl
Composition: NCCER
Printer/Binder: RR Donnelley
Cover Printer: RR Donnelley
Text Fonts: Palatino and Univers

Credits and acknowledgments for content borrowed from other sources and reproduced, with permission, in this textbook appear at the end of each module.

10 9 8 7 6 5 4 3 2 1

PEARSON

ISBN-13: 978-0-13-480566-5
ISBN-10: 0-13-480566-6

Preface

To the Trainee

NCCER is pleased to release the 3rd edition of Pipeline Training and Qualification. The curriculum has been redesigned to offer fundamental and comprehensive training for workers in the pipeline industry as part of an integrated program that provides career path training, as well as covered task (CT) training via modules that align to the American Petroleum Institute's *API Recommended Practice for Pipeline Operator Qualification (OQ) 1161*, 3rd edition (2014).

The modules in this book provide training and education that enable a worker to operate safely and effectively within the pipeline construction and maintenance industry. They explain the basic functions and purposes of pipelines and facilities, identify the characteristics and hazards of common pipeline products and lay the foundation of understanding that supports the more specific covered task training. Training for some advanced topics that are not covered tasks is included as well.

CT modules are available as traditional printed levels and as individual module eBooks. Modules are completed by successfully passing the proctored Module Examination, and completing the Performance Profile under the supervision of an NCCER-certified evaluator. These completions are tracked within the NCCER Registry for training delivered through an Accredited Training Sponsor. Covered task module completions in the Registry are also documented in ISNetworld.

We invite you to visit the NCCER website at **www.nccer.org** for information on the latest product releases and training, as well as online versions of the *Cornerstone* magazine and Pearson's NCCER product catalog.

Your feedback is welcome. You may email your comments to **curriculum@nccer.org** or send general comments and inquiries to **info@nccer.org**.

NCCER Standardized Curricula

NCCER is a not-for-profit 501(c)(3) education foundation established in 1996 by the world's largest and most progressive construction companies and national construction associations. It was founded to address the severe workforce shortage facing the industry and to develop a standardized training process and curricula. Today, NCCER is supported by hundreds of leading construction and maintenance companies, manufacturers, and national associations. The NCCER Standardized Curricula was developed by NCCER in partnership with Pearson, the world's largest educational publisher.

Some features of the NCCER Standardized Curricula are as follows:

- An industry-proven record of success
- Curricula developed by the industry for the industry
- National standardization providing portability of learned job skills and educational credits
- Compliance with the Office of Apprenticeship requirements for related classroom training (*CFR 29:29*)
- Well-illustrated, up-to-date, and practical information

NCCER also maintains the NCCER Registry, which provides transcripts, certificates, and wallet cards to individuals who have successfully completed a level of training within a craft in NCCER's Curricula. *Training programs must be delivered by an NCCER Accredited Training Sponsor in order to receive these credentials.*

Additional Pipeline Resources

The following additional Pipeline Resources are available from NCCER's website at **www.nccer.org/pipeline**:

1. A complete copy of NCCER's Covered Task (CT) list
2. Access to:
   a. An online bookstore to order print and digital versions of NCCER's pipeline curricula.
   b. Instructor resources (Lesson Plans and PowerPoints)
   c. Exams for Operator Qualification
   d. A site to custom build curriculum with pipeline and/or other NCCER-developed modules

Pipeline Career Pathway Training

Entry-level Pipeline Craft Trainee → **Pipeline Fundamentals** → **Level 1**

The newly established pipeline career pathway represented on this page provides the recognition pipeline professionals deserve through standardized training and industry-recognized credentials.

Levels of training have been established by combining training modules into specific focus areas within the industry. This allows trainees to progress through a standardized program and earn a professional credential.

In addition, Levels Two and Three are comprised of covered task training modules that can lead to operator qualifications. *Pipeline Corrosion Control is the only exception which actually provides covered task training starting in Level One.*

The blue boxes denote the professional credential that can be earned with successful completion of each program.

A trained workforce is a safe and qualified workforce!

Pipeline Fundamentals:

Core Curriculum

Introduction to the Pipeline Industry

and

a.) Field Abnormal Operating Conditions
Module: AOCFG-17

or

b.) Control Center Abnormal Operating Conditions
Module: AOCCC_17

Level 1:

Field and Control Center Operations

Pipeline Maintenance and Mechanical

Electrical and Instrumentation

Pipeline Corrosion Control (CT)

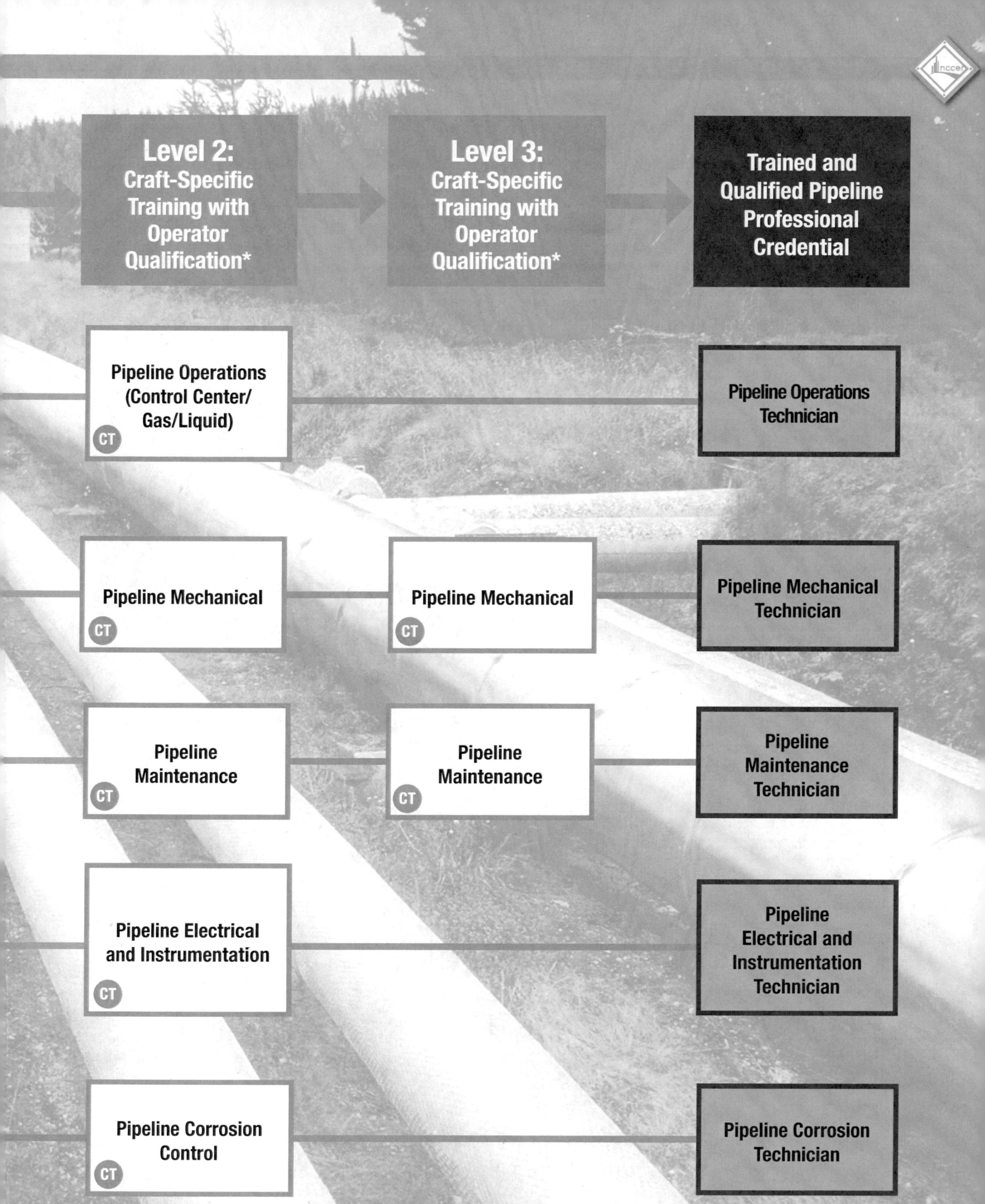

CT *Each of these titles have corresponding operator qualifications. The Covered Task list is available online at www.nccer.org/pipeline-program*

Special Features

In an effort to provide a comprehensive and user-friendly training resource, this curriculum showcases several informative features. Whether you are a visual or hands-on learner, these features are intended to enhance your knowledge of the construction industry as you progress in your training. Some of the features you may find in the curriculum are explained below.

Introduction

This introductory page, found at the beginning of each module, lists the module Objectives, Performance Tasks and Trade Terms. The Objectives list the knowledge you will acquire after successfully completing the module. The Performance Tasks give you an opportunity to apply your knowledge to real-world tasks. The Trade Terms are industry specific vocabulary that you will learn as you study this module.

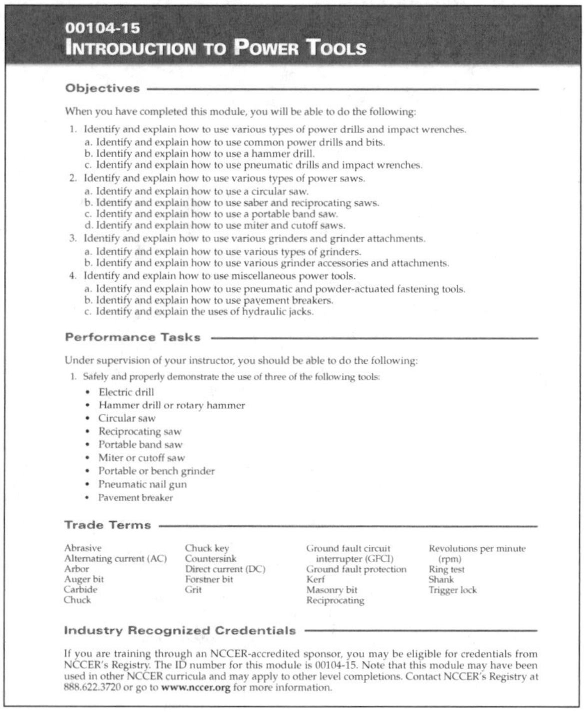

00104-15
INTRODUCTION TO POWER TOOLS

Objectives

When you have completed this module, you will be able to do the following:

1. Identify and explain how to use various types of power drills and impact wrenches.
   a. Identify and explain how to use common power drills and bits.
   b. Identify and explain how to use a hammer drill.
   c. Identify and explain how to use pneumatic drills and impact wrenches.
2. Identify and explain how to use various types of power saws.
   a. Identify and explain how to use a circular saw.
   b. Identify and explain how to use saber and reciprocating saws.
   c. Identify and explain how to use a portable band saw.
   d. Identify and explain how to use miter and cutoff saws.
3. Identify and explain how to use various grinders and grinder attachments.
   a. Identify and explain how to use various types of grinders.
   b. Identify and explain how to use various grinder accessories and attachments.
4. Identify and explain how to use miscellaneous power tools.
   a. Identify and explain how to use pneumatic and powder-actuated fastening tools.
   b. Identify and explain how to use pavement breakers.
   c. Identify and explain the uses of hydraulic jacks.

Performance Tasks

Under supervision of your instructor, you should be able to do the following:

1. Safely and properly demonstrate the use of three of the following tools:
   - Electric drill
   - Hammer drill or rotary hammer
   - Circular saw
   - Reciprocating saw
   - Portable band saw
   - Miter or cutoff saw
   - Portable or bench grinder
   - Pneumatic nail gun
   - Pavement breaker

Trade Terms

Abrasive	Chuck key	Ground fault circuit interrupter (GFCI)	Revolutions per minute (rpm)
Alternating current (AC)	Countersink	Ground fault protection	Ring test
Arbor	Direct current (DC)	Kerf	Shank
Auger bit	Forstner bit	Masonry bit	Trigger lock
Carbide	Grit	Reciprocating	
Chuck			

Industry Recognized Credentials

If you are training through an NCCER-accredited sponsor, you may be eligible for credentials from NCCER's Registry. The ID number for this module is 00104-15. Note that this module may have been used in other NCCER curricula and may apply to other level completions. Contact NCCER's Registry at 888.622.3720 or go to **www.nccer.org** for more information.

Trade Features

Trade features present technical tips and professional practices based on real-life scenarios similar to those you might encounter on the job site.

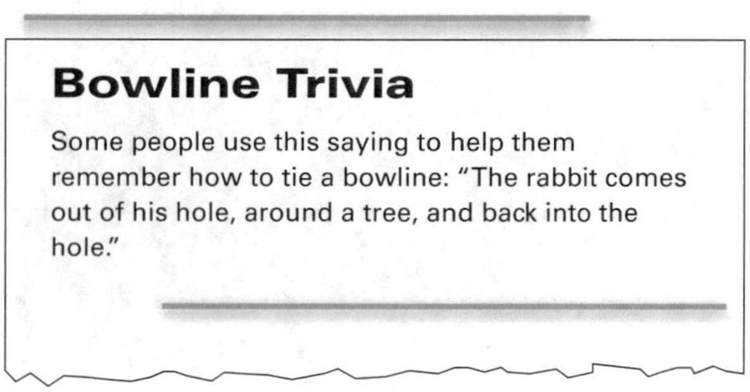

Figures and Tables

Photographs, drawings, diagrams, and tables are used throughout each module to illustrate important concepts and provide clarity for complex instructions. Text references to figures and tables are emphasized with *italic* type.

Notes, Cautions, and Warnings

Safety features are set off from the main text in highlighted boxes and categorized according to the potential danger involved. Notes simply provide additional information. Cautions flag a hazardous issue that could cause damage to materials or equipment. Warnings stress a potentially dangerous situation that could result in injury or death to workers.

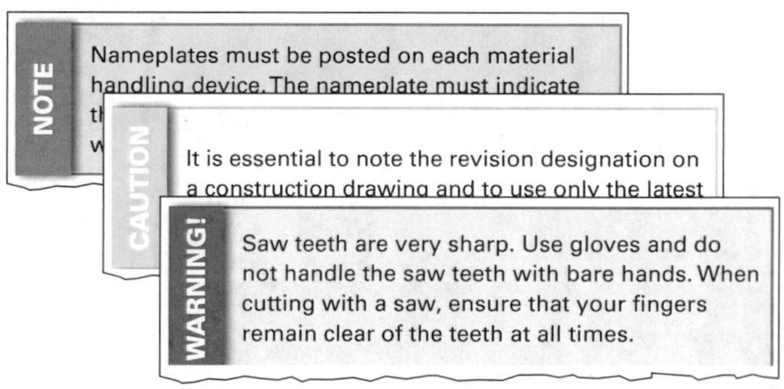

Case History

Case History features emphasize the importance of safety by citing examples of the costly (and often devastating) consequences of ignoring best practices or OSHA regulations.

Case History

Requesting an Outage

An electrical contractor requested an outage when asked to install two bolt-in, 240V breakers in panels in a data processing room. It was denied due to the 24/7 worldwide information processing hosted by the facility. The contractor agreed to proceed only if the client would sign a letter agreeing not to hold them responsible if an event occurred that damaged computers or resulted in loss of data. No member of upper management would accept liability for this possibility, and the outage was scheduled.

The Bottom Line: If you can communicate the liability associated with an electrical event, you can influence management's decision to work energized.

Going Green

Going Green features present steps being taken within the construction industry to protect the environment and save energy, emphasizing choices that can be made on the job to preserve the health of the planet.

GOING GREEN

Reducing Your Carbon Footprint

Many companies are taking part in the paperless movement. They reduce their environmental impact by reducing the amount of paper they use. Using email helps to reduce the amount of paper used,

Did You Know

Did You Know features introduce historical tidbits or interesting and sometimes surprising facts about the trade.

Did You Know?

Safety First

Safety training is required for all activities. Never operate tools, machinery, or equipment without prior training. Always refer to the manufacturer's instructions.

Step-by-Step Instructions

Step-by-step instructions are used throughout to guide you through technical procedures and tasks from start to finish. These steps show you how to perform a task safely and efficiently.

Perform the following steps to erect this system area scaffold:

Step 1 Gather and inspect all scaffold equipment for the scaffold arrangement.

Step 2 Place appropriate mudsills in their approximate locations.

Step 3 Attach the screw jacks to the mudsills.

Trade Terms

Each module presents a list of Trade Terms that are discussed within the text and defined in the Glossary at the end of the module. These terms are presented in the text with **bold, blue** type upon their first occurrence. To make searches for key information easier, a comprehensive Glossary of Trade Terms from all modules is located at the back of this book.

During a rigging operation, the **load** being lifted or moved must be connected to the apparatus, such as a crane, that will provide the power for movement. The connector—the link between the load and the apparatus—is often a sling made of synthetic, chain, or **wire rope** materials. This section focuses on three types of slings:

Section Review

Each section of the module wraps up with a list of Additional Resources for further study and Section Review questions designed to test your knowledge of the Objectives for that section.

Additional Resources

Materials Handling Handbook, The American Society of Mechanical Engineers (ASME) and The International Material Management Society (IMMS), Raymond A. Kulwiec, Editor-in-Chief. 1985. New York, NY: Wiley-Interscience.

Manufacturing Facilities Design & Material Handling, Matthew P. Stevens, Fred E. Meyers. 2013. West Lafayette, IN: Purdue University Press.

1.0.0 Section Review

1. For material handling tasks, it is just as important to be mentally fit as it is to be _____.
   a. physically fit
   b. physically aggressive
   c. closely supervised
   d. over 200 pounds

2. Which of the following is a type of knot that is often used to join the ends of two ropes in non-critical, low-strain applications?
   a. Bowline
   b. Clove hitch
   c. Half hitch
   d. Square knot

Review Questions

The end-of-module Review Questions can be used to measure and reinforce your knowledge of the module's content.

Review Questions

1. Identification tags for slings must include the _____.
   a. type of protective pads to use
   b. type of damage sustained during use
   c. color of the tattle-tail
   d. manufacturer's name or trademark

2. The type of wire rope core that is susceptible to heat damage at relatively low temperatures is the _____.
   a. fiber core
   b. strand core
   c. independent wire rope core
   d. metallic link supporting core

3. Synthetic slings must be inspected _____.
   a. once every month
   b. visually at the start of each work week
   c. before every use
   d. once wear or damage becomes apparent

4. An alloy steel chain sling must be removed from service if there is evidence that _____.
   a. the sling has been used in different hitch configurations
   b. replacement links have been used to repair the chain
   c. the sling has been used for more than one year
   d. strands in the supporting core have weakened

5. A piece of rigging hardware used to couple the end of a wire rope to eye fittings, hooks, or other connections is a(n) _____.
   a. eyebolt
   b. hitch
   c. shackle
   d. U-bolt

6. A lifting clamp is most likely to be used to move loads such as _____.
   a. steel plates
   b. piping bundles
   c. concrete blocks
   d. plastic tubing

7. Chain hoists are able to lift heavy loads by utilizing a _____.
   a. rope and pulley system
   b. rigger's strength
   c. stationary counterweight
   d. gear system

8. Before attempting to lift a load with a chain hoist, make sure that the _____.
   a. hoist is secured to a come-along
   b. load is properly balanced
   c. tag lines are properly anchored
   d. tackle is connected to its power source

9. A hitch configuration that allows slings to be connected to the same load without using a spreader beam is a _____.
   a. double-wrap hitch
   b. choker hitch
   c. bridle hitch
   d. basket hitch

10. To make the emergency stop signal that is used by riggers, extend both arms _____.
    a. horizontally with palms down and quickly move both arms back and forth
    b. directly in front and then move both arms up and down repeatedly
    c. vertically above the head and wave both arms back and forth
    d. horizontally with clenched fists and move both arms up and down

00106-15 Introduction to Basic Rigging Module Six 31

NCCER Standardized Curricula

NCCER's training programs comprise more than 80 construction, maintenance, pipeline, and utility areas and include skills assessments, safety training, and management education.

Boilermaking
Cabinetmaking
Carpentry
Concrete Finishing
Construction Craft Laborer
Construction Technology
Core Curriculum: Introductory Craft Skills
Drywall
Electrical
Electronic Systems Technician
Heating, Ventilating, and Air Conditioning
Heavy Equipment Operations
Highway/Heavy Construction
Hydroblasting
Industrial Coating and Lining Application Specialist
Industrial Maintenance Electrical and Instrumentation Technician
Industrial Maintenance Mechanic
Instrumentation
Insulating
Ironworking
Masonry
Millwright
Mobile Crane Operations
Painting
Painting, Industrial
Pipefitting
Pipelayer
Plumbing
Reinforcing Ironwork
Rigging
Scaffolding
Sheet Metal
Signal Person
Site Layout
Sprinkler Fitting
Tower Crane Operator
Welding

Maritime

Maritime Industry Fundamentals
Maritime Pipefitting
Maritime Structural Fitter

Green/Sustainable Construction

Building Auditor
Fundamentals of Weatherization
Introduction to Weatherization
Sustainable Construction Supervisor
Weatherization Crew Chief
Weatherization Technician
Your Role in the Green Environment

Energy

Alternative Energy
Introduction to the Power Industry
Introduction to Solar Photovoltaics
Introduction to Wind Energy
Power Industry Fundamentals
Power Generation Maintenance Electrician
Power Generation I&C Maintenance Technician
Power Generation Maintenance Mechanic
Power Line Worker
Power Line Worker: Distribution
Power Line Worker: Substation
Power Line Worker: Transmission
Solar Photovoltaic Systems Installer
Wind Turbine Maintenance Technician

Pipeline

Pipeline Fundamentals
Pipeline Corrosion Control
Pipeline Electrical and Instrumentation
Pipeline Maintenance
Pipeline Mechanical
Pipeline Operations (Control Center, Gas, Liquid)

Safety

Field Safety
Safety Orientation
Safety Technology

Supplemental Titles

Applied Construction Math
Tools for Success

Management

Construction Workforce Development Professional
Fundamentals of Crew Leadership
Mentoring for Craft Professionals
Project Management
Project Supervision

Spanish Titles

Acabado de concreto: nivel uno (*Concrete Finishing Level One*)
Aislamiento: nivel uno (*Insulating Level One*)
Albañilería: nivel uno (*Masonry Level One*)
Andamios (*Scaffolding*)
Carpintería: Formas para carpintería, nivel tres (*Carpentry: Carpentry Forms, Level Three*)
Currículo básico: habilidades introductorias del oficio (*Core Curriculum: Introductory Craft Skills*)
Electricidad: nivel uno (*Electrical Level One*)
Herrería: nivel uno (*Ironworking Level One*)
Herrería de refuerzo: nivel uno (*Reinforcing Ironwork Level One*)
Instalación de rociadores: nivel uno (*Sprinkler Fitting Level One*)
Instalación de tuberías: nivel uno (*Pipefitting Level One*)
Instrumentación: nivel uno, nivel dos, nivel tres, nivel cuatro (*Instrumentation Levels One through Four*)
Orientación de seguridad (*Safety Orientation*)
Paneles de yeso: nivel uno (*Drywall Level One*)
Seguridad de campo (*Field Safety*)

Acknowledgments

This curriculum was revised as a result of the farsightedness and leadership of the following sponsors:

American Petroleum Institute
Central NDT
Construction Craft Academy
CP Masters
EMS USA, Inc.
Enbridge Pipelines, Inc.
Enhance Co.
Enterprise
Flint Energy Services, Inc.
Jomax Construction
Kinder Morgan, Inc.
L.E. Bell Construction Co. Inc.
Marathon Pipe Line
Midwestern Contractors
OneOK
Plains All-American Pipeline
Praxair Services
Protech EIS
Safety Advantage LLC
Southland Safety
Sunoco Logistics, L.P.
Sunoco Pipeline
Thorco Holdings, LLC
URS/Flint Energy

This curriculum would not exist were it not for the dedication and unselfish energy of those volunteers who served on the Authoring Team. A sincere thanks is extended to the following:

Gary Affolter
Carl Allebach
Travis Ashby
Steve Barker
Art Birx
James Carroll
Leslie Carroll
Gordon Carson
Bill Emery
Paul Fontenot II
Paula Gates-Rippe
Trobi Hall
Sid Harrell
David Hoffman
Kimberly-Joy Harris
Les Harrison
Lisa Hartnup
Benjamin Johnson
Dan Lake
Tom Magness
Wallace McCracken
Tim Mills
Steve Montoy
Michael Nordstrom
Tom Pickthall
Donnie Pittsinger
William Rado
David Reeder
Tom Reeder
Randall Rogers
Dan Rohret
Dan Ruane
DeWayne Schapansky
Mark Seitz
Blaine Shellhammer
Lynn Simpson
Cameron Strother
Charles Whiten
Rick Wilenzick
Michael Williams
Erin Worthington
Donavon Wright
Will Wright

NCCER Partners

American Council for Construction Education
American Fire Sprinkler Association
Associated Builders and Contractors, Inc.
Associated General Contractors of America
Association for Career and Technical Education
Association for Skilled and Technical Sciences
Construction Industry Institute
Construction Users Roundtable
Design Build Institute of America
GSSC – Gulf States Shipbuilders Consortium
ISN
Manufacturing Institute
Mason Contractors Association of America
Merit Contractors Association of Canada
NACE International
National Association of Women in Construction
National Insulation Association
National Technical Honor Society
National Utility Contractors Association
NAWIC Education Foundation
North American Crane Bureau
North American Technician Excellence
Pearson
Prov
SkillsUSA®
Steel Erectors Association of America
U.S. Army Corps of Engineers
University of Florida, M. E. Rinker Sr., School of Construction Management
Women Construction Owners & Executives, USA

NCCER Business Partners

Contents

Module 66101-02

Introduction to the Pipeline Industry

Introduction to the Pipeline Operator Qualification Rule and NCCER

It could be said that petroleum products keep America running. Petroleum products like oil and natural gas are the principal sources of energy in the United States and are used in many manufacturing and chemical processes. Petroleum products are primarily transported by pipeline. There are more than a million miles of liquid and natural gas pipeline in this country, ranging in diameter from two inches to four feet. These pipelines bring crude oil and natural gas from production fields to processing plants and refineries—and ultimately to the consumer.

Pipeline Operator Qualification

Petroleum products are considered hazardous, so pipeline personnel must take extra precautions and follow specific procedures when working with them. To help ensure the safe handling and transportation of petroleum products, the United States Department of Transportation (DOT) mandates safe practices as federal regulations.

One such regulation, the Operator Qualification (OQ) rule, requires pipeline operators to develop a written qualification plan to evaluate their personnel and their contract personnel (see *49 CFR Part 192 Subpart N—Gas*, and *49 CFR Part 195 Subpart G—Liquid*). The evaluation hinges on the operator's ability to perform certain covered tasks and to be able to recognize and respond to abnormal operating conditions that may be encountered while performing these covered tasks.

Covered Tasks

A pipeline performance task is deemed covered if the activity passes the following four-part test:

1. Is it performed on a pipeline facility?
2. Is it an operations or maintenance task?
3. Is it performed as a requirement of *49 CFR Part 195 (Liquid)* or *49 CFR Part 192 (Gas)*?
4. Does it affect the operation or integrity of the pipeline facility?

If the answer to all four questions is yes, then the activity is considered covered and must figure into the evaluation process of pipeline personnel.

Consortium on Operator Qualification

The Consortium on Operator Qualification (COOQ) became the guiding force for the identification and development of covered tasks for the liquid pipeline industry. Their work product, API publication 1161, *Guidance Document for the Qualification of Liquid Pipeline Personnel*, formed the basis for the resulting Pipeliner Training and Assessment Program.

The covered task list created in API publication 1161 has now evolved to include gas, as well as construction covered tasks. NCCER maintains a version of the covered task list, which can be accessed at **www.nccer.org**.

NCCER

With the guidance and support of the Consortium on Operator Qualification, NCCER was able to design and implement the Pipeliner Training and Assessment Program.

NCCER is a not-for-profit 501(c)(3) education foundation established in 1995 by the world's largest and most progressive construction and maintenance contractors and national contractor associations. NCCER was founded to address the severe workforce shortage facing the construction and maintenance industries and to develop a standardized training process and curricula. Today, NCCER is supported by hundreds of leading construction and maintenance companies, manufacturers, and national associations and developed more than 70 construction and maintenance titles.

NCCER also maintains a Registry, which contains the records of trainees and instructors who have received NCCER training and certification. All completion information for every trainee is sent to NCCER and filed within the Registry. The Registry can then confirm training and skills for workers as they move from company to company, state to state, or even within their own company.

NCCER Standardized Curricula

NCCER standardized curricula are developed using the expertise of industry subject matter experts (SMEs). The use of SMEs as an integral part of every new development or revision ensures that the curriculum is current, industry-driven,

and indicative of industry practices. The development and/or revision process is facilitated from start to finish by NCCER project managers, who work closely with the SMEs and a select team of technical writer(s) on the development and presentation of the curriculum.

Pipeline: Five Training Paths

The pipeline curriculum focuses on essential skills and competencies within five skill areas:

- Corrosion Control
- Electrical and Instrumentation
- Maintenance
- Mechanical
- Operations (Field and Control Center)

SMEs from the pipeline industry have determined the skills and competencies needed based on API's covered tasks and the consensus of the group/industry. Specific company information is generally not incorporated into the curriculum unless the SMEs determine that the information is the industry standard.

Training Modules and Covered Tasks

The smallest unit of instruction in the curriculum is a *module*. Each curriculum is composed of modules (the equivalent of stand-alone chapters) that are logically sequenced into complete *levels* of instruction. Every module is task-oriented and requires the successful completion of a written (knowledge-based) examination and a performance (hands-on) test. Modules may be individually selected and/or combined for efficient task training. Modules in the pipeline curricula, to the extent possible, align with industry-developed covered tasks. While not all modules relate to covered tasks, all covered tasks relate to at least one module.

Benefits of Training to the Pipeline Marketplace

Applying NCCER's standardized training process allows pipeline owners and contractors in the United States to establish the benchmark for acceptable skill and competency levels within the pipeline industry. This process will assist training organizations in developing and/or improving standardized pipeline training programs that effectively match their resources and operating conditions while responding to regulatory demands and to the needs of the industry. The process also recognizes training organizations that meet the minimum standards of competence, warranting regulatory, public, and professional confidence. This, in turn, fosters national unity within the pipeline industry for the mutual benefit of individual technicians, owners, contractors, and ultimately pipeline-users and society.

Additional Resources

To review the Pipeliner Training and Assessment Program Implementation and Resource Guide and the latest version of pipeline covered tasks, visit NCCER's website at **www.nccer.org,** or contact:

NCCER
13614 Progress Boulevard
Alachua, FL 32615
(888) 622-3720
Website: **www.nccer.org**

Additional sources of information on the pipeline operator qualification regulation include:

DOT – Office of Pipeline Safety
www.ops.dot.gov

Transportation Safety Institute
www.tsi.dot.gov

COURSE MAP

This course map shows each curriculum path in the Pipeliner Training program. The suggested training order begins on the left with the Pipeline Core modules and proceeds to the right. Which curriculum path you follow depends on your areas of specialty and interest. On a given curriculum path, skill levels increase as you advance to the right on the course map, although some are single-level curricula. The local Training Program Sponsor may adjust the training order.

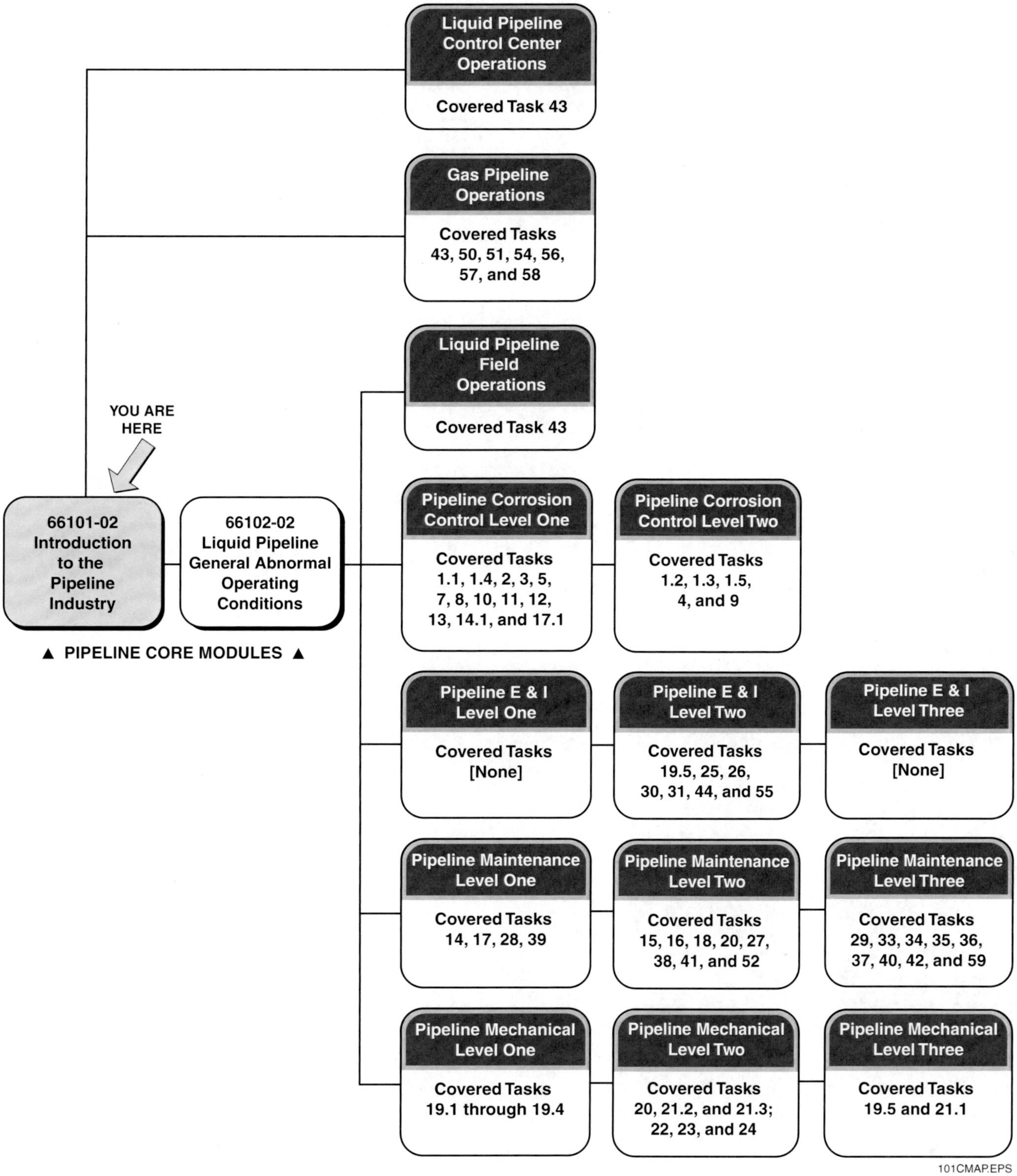

MODULE 66101-02 CONTENTS

Figures

Tables

MODULE 66101-02

Introduction to the Pipeline Industry

Objectives

When you have completed this module, you will be able to do the following:

1. Explain the basic functions and purposes of pipelines and facilities and identify the characteristics and hazards of common pipeline products.
2. Identify maps and drawings used to depict pipelines and facilities.
3. Explain the roles of control personnel and equipment in the overall operation of a pipeline.
4. Explain liquid pipeline hydraulics and gas pipeline pneumatics.
5. Explain the types and purposes of pipeline equipment.
6. Explain pipeline electrical power systems and corrosion control.
7. Review operations, maintenance, and emergency procedures and perform documentation required for pipeline operations.

Prerequisites

Before you begin this module, it is recommended that you successfully complete the following: Core Curriculum.

1.0.0 ◆ INTRODUCTION

Pipelines are used to transport **petroleum**, petroleum **products**, liquid carbon dioxide, and other liquids and gases over great distances. In spite of the costs of constructing pipelines, they are a relatively economical means of transport. A pipeline system consists of pipes, pump stations, compressor stations, and other facilities to support the safe, continuous movement of liquids and gases. Pipelines carry petroleum from the source, the oil and gas fields, to the **refineries**, to and from gas processing plants, and eventually to consumers. Pipes are most often buried in the ground, but they are also laid under water and sometimes erected above the ground.

Today, a network of gas and liquid pipelines crisscrosses the United States. The first pipeline on a continental scale was completed in the 1930s, but the real boom in cross-country lines came during World War II when the Big Inch and the Little Big Inch were constructed to provide fuel for wartime industry and the military. The Big Inch began operating in 1943, was 24 inches in diameter, and stretched 1,341 miles from Longview, Texas, to Phoenixville, Pennsylvania, and Linden, New Jersey. This made it the largest **crude oil** pipeline in the world. The Little Inch, opened in 1944, was 20 inches in diameter and 1,475 miles long. It carried **refined products** from Beaumont, Texas, to Linden, New Jersey.

The most famous crude oil pipeline in the U.S. crosses 800 miles of permanently frozen ground in Alaska. The Trans-Alaska Pipeline, which began operations in 1977, is 48 inches in diameter and carries crude oil from the Prudhoe Bay field to the shipping terminal at Valdez, Alaska. Along the way, it crosses 34 major rivers and streams and three major mountain ranges, going from sea level to as high as 4,800 feet and back to sea level. Protection of the delicate tundra, permafrost, wilderness environment, and wildlife in its path were and still are critical concerns.

All pipeline operators are concerned with safety and environmental impact, particularly those carrying petroleum products that are hazardous to the environment as well as to the people who live near the pipeline. Good construction practices, good maintenance and operations practices, continuous monitoring, and rapid response are necessary to prevent or to minimize the effects of pipeline leaks. Operations of a pipeline must be conducted in a fashion that will ensure the safety of employees and the general public and protect the environment.

The petroleum pipeline system can be divided into the following functions:

- Origination
- Transport
- Storage
- Delivery

1.1.0 Origination

All petroleum products produced or shipped originate from the domestic and foreign crude oil or natural gas from the oil and gas fields. Exploration teams from petroleum companies use geological and geophysical methods to find possible deposits of oil and gas deep under the layers of the earth. Drilling confirms and accesses the find; the well is completed and production begins.

1.1.1 Exploration

Oil and gas are formed in the earth from decaying organic matter that was buried long ago. The combination of time, pressure, and heat causes the matter to undergo a chemical change into hydrocarbons. These fluids and gases then tend to move, or migrate, upward in the earth until they either escape at the surface or are trapped in the earth's layers. It is this trapped fluid and gas that is sought by exploration.

The oil/gas-bearing formation is basically the layer of earth in which the hydrocarbon deposit is trapped until it is drilled. The formation must have the porosity to hold the fluid and gas and have the structure to keep it from moving further upward. Traps can have a variety of structures, and some of the most typical are the anticline trap, stratigraphic trap, fault trap, and salt dome trap, as shown in *Figure 1*.

1.1.2 Drilling and Completion

When a possible deposit is indicated, a well must be drilled to try to find it. Many wells have been drilled without finding the sought-after oil and gas. There may not be enough to warrant production costs or it cannot be drawn out economically. However, when a workable deposit is found, the well is completed for production. Completion includes casing the well with concrete, perforating the casing with

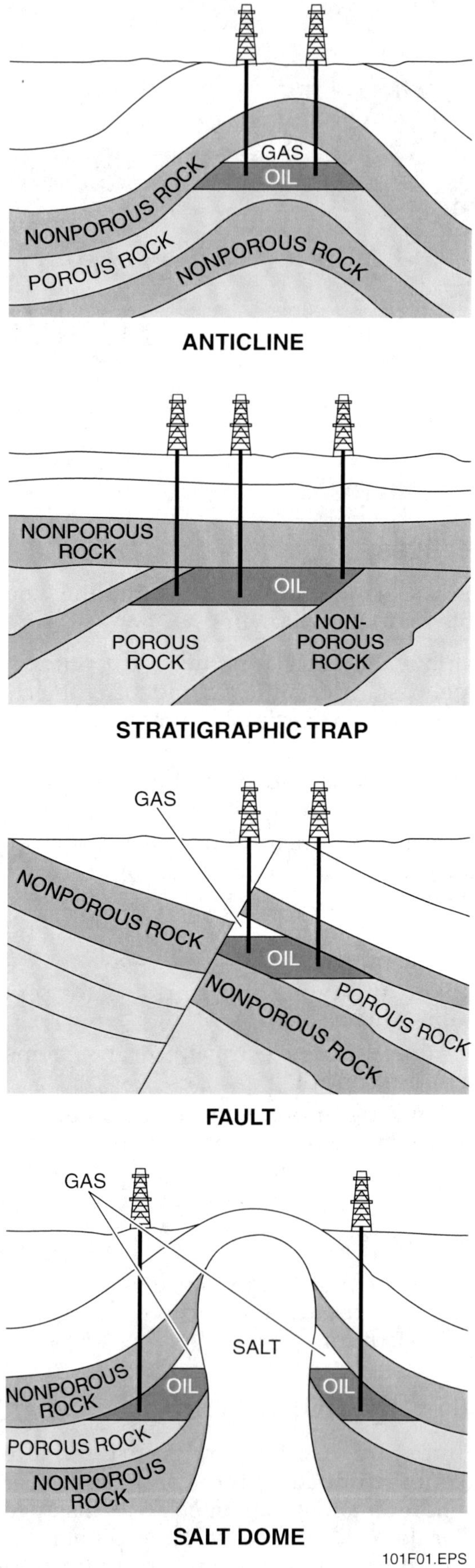

Figure 1 ◆ Types of traps.

holes for the product to flow through tubing within the casing, and installing pumps or valves on the top of the hole to control the flow. Tests of the producing well may show a large enough deposit to warrant other wells being drilled and completed in the area. All of these wells comprise a field. It is from these wells that the oil or gas is gathered to transport through the pipeline.

Many crude oil reservoirs have a gas cap. The oil produced from these reservoirs often contains various amounts of associated gas. Some formations, however, trap only gas. This trapped gas forms a gas or gas-condensate reservoir, which produces only gas or gas with varying amounts of liquid gas condensate. Drilling gas wells is very similar to drilling and completing crude oil wells. The primary difference between crude oil and gas wells is in the type of surface facility equipment needed to process the oil or gas after it is produced.

1.1.3 Production

Different types of pumps or valves, such as a Christmas tree and a beam pump, are used to cap the well depending on the properties of the product being withdrawn and the formation producing it (*Figure 2*). Oil and gas are often found in the same formation along with some amount of water. The amount of each may vary in relation to each other, and this may affect the production of the well. Since gas is lighter than oil, and oil is lighter than water, they are layered in that order: gas on the top, then oil, then water on the bottom.

Because the formation had to be closed off in some way for the product to accumulate, pressure is usually also trapped in the formation. When it is drilled, that pressure can push the oil, gas, and water into the hole with some force. When it is controlled, this pressure is good because it can push the oil and gas up to the surface with little or no pumping. Generally a newer well will have more pressure than an older one, so in time, pumping power may be needed to pull the product out. However, some wells start with only a little pressure and must be pumped from the beginning. Control of the flow is important since producing a well too fast can affect the life of the well. These conditions, along with other properties, will determine how the well is capped and produced. A Christmas tree of valves may be installed to control the flow, and a beam pump may be used to pull the product out. Gas wells are completed with only a Christmas tree.

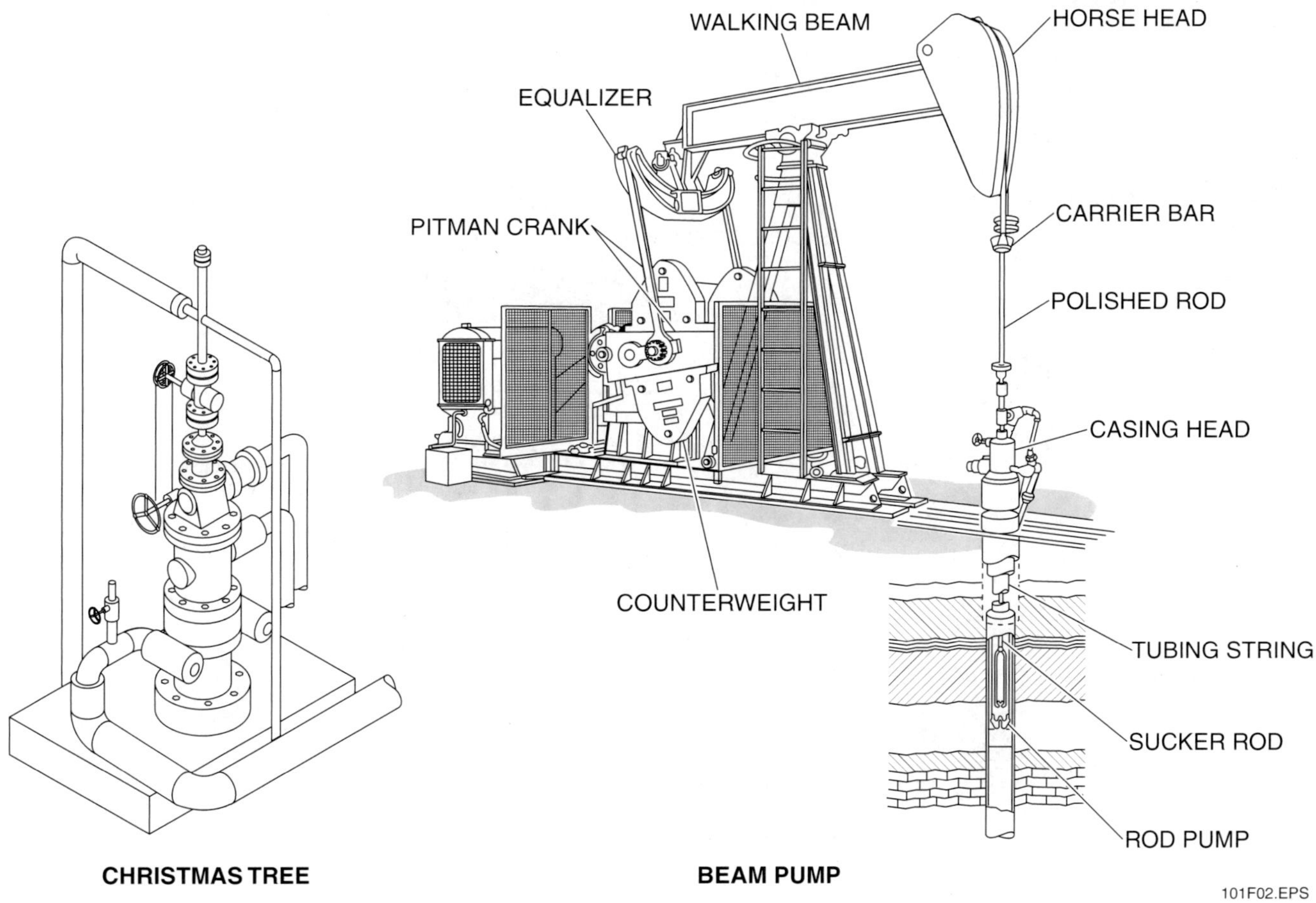

Figure 2 ◆ Christmas tree and beam pump.

1.2.0 Transport

Lines from the oil and gas fields are called *gathering lines* or *flow lines*. These gathering lines bring the oil and gas together into a facility for separation and processing. The vapor gas must travel in a different pipeline from the crude oil, so the gas and oil are separated in huge high-pressure **tanks**, called *gas/oil separators*. *Figure 3* shows the high-pressure separators.

From the separation or gas processing facility, the crude oil and gas is directed into a large **main line**, or trunk line, for the longest part of the journey to a refinery or a shipping terminal. The gas may be stored in tanks or routed through gas pipelines to market. Gas main lines transport gas to a gas distribution center for distribution to consumers, industrial facilities, or storage caverns. From the refinery, a large main line, or trunk line, will carry products across country with distribution lines branching off the main line along the way. A large distribution line may carry the liquid to a central distribution point, such as a regional or city station, from which smaller lines or vehicles carry it to consumers. Pipelines are constructed to meet *ASME Piping Standard, B31.4-1998, Pipeline Transportation Systems for Liquid Hydrocarbons, and other Liquids*. *Figure 4* shows the transportation chain.

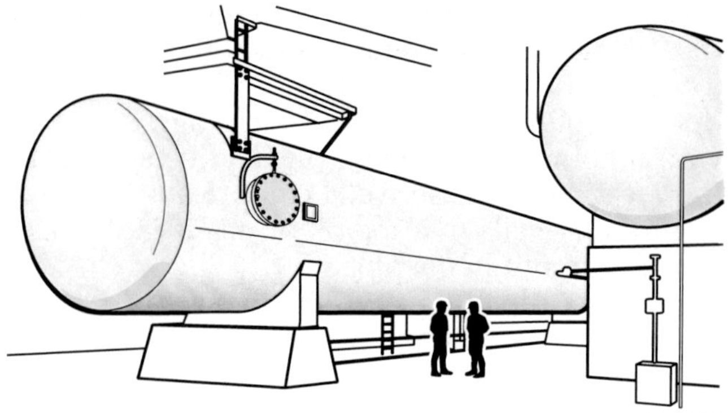

Figure 3 ◆ High-pressure separators.

1.2.1 Pipe

The size of pipe used in pipelines varies from 2-inch diameter pipe for gathering lines or small distribution lines, to 48-inch pipe for main lines or trunk lines. Most modern pipe is made of steel, either seamless or with a welded longitudinal seam, that is produced in lengths that are 40-60 feet long. These are often welded into 80-foot sections before being delivered to the construction site. Steel pipe is vulnerable to **corrosion**, which can cause damage to the pipe and reduce wall thickness, compromising the integrity of the pipe. Pipe coating and wrapping are the primary

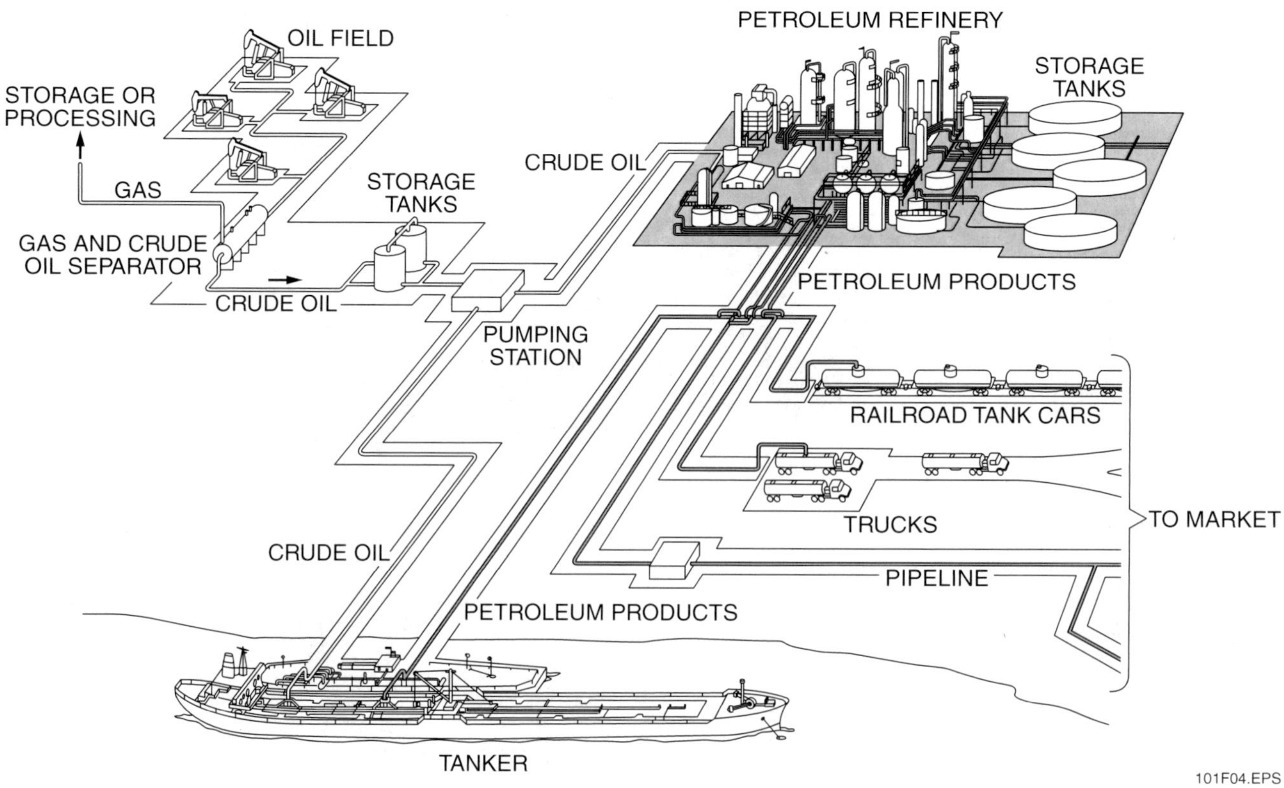

Figure 4 ◆ Transportation chain.

means of protecting the pipe. **Cathodic protection** is used after the pipe is buried to inhibit a corrosive environment.

1.2.2 Construction

At the time of construction, permits are obtained for a specified right-of-way for the pipeline. Earthmoving equipment clears the land, an access road is built along the route, and trenches are excavated for the pipe. Trenches require a minimum depth of coverage as stated in *Pipeline Safety Regulations, Part 195.248* for liquid and *Part 192.327* for gas.

Pipe is hauled to the right-of-way and welded in strings by teams of welders. After welding, radiographic inspection is used to check the welds. Sometimes, pipe must be bent by bending machines to fit the terrain or to go around obstacles. To inhibit corrosion, the pipe must be coated and wrapped before burial. This can be done at the mill or at the site.

The trench is dug and then prepared for the pipe with a layer of protective bedding. The pipe is strung or laid in the trench by sideboom tractors. Long sections are tied-in or welded together to form a continuous conduit. The trench is then filled with backfill to cover the pipe (see *Figure 5*). The pipe must be **hydrostatically tested** or checked for leaks. This involves filling the line with water and pressurizing it to see if the system will maintain the pressure intended.

When piping crosses lakes and rivers, it may be laid and weighted with concrete or steel anchors to keep it on the bottom. *Figure 6* shows an add-on weight installation. In most rivers, the crossings are drilled under the river bed without disturbing the bottom.

Pipelines are usually buried except at pump and compressor stations, some valve stations, and terminals. However, pipe is sometimes erected aboveground by using H-shaped supports, called bents, to hold the pipe. These bents are located at specified intervals along the line. The uprights on the bents are vertical support members (VSMs). In areas like Alaska, the VSMs have to be self-cooling to avoid heating the permafrost. This is not the normal installation, but it is necessary in frozen environments. *Figure 7* shows a bent with self-cooling VSMs.

Supports have sliding plates, or shoes, to allow the pipe to contract or expand with weather or to move with **seismic activity**. The shoes allow the pipe to move as much as 20 feet back and forth and 10 feet up and down. Aboveground lines are also constructed in a zigzag pattern instead of a straight line to allow for movement. *Figure 8* shows a common elevated pipeline section.

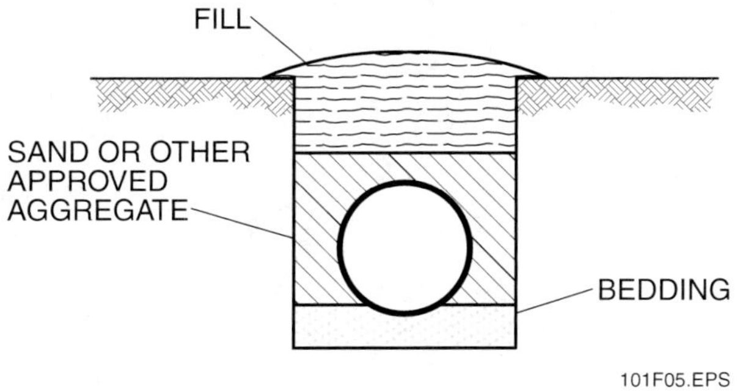

Figure 5 ◆ Pipe burial.

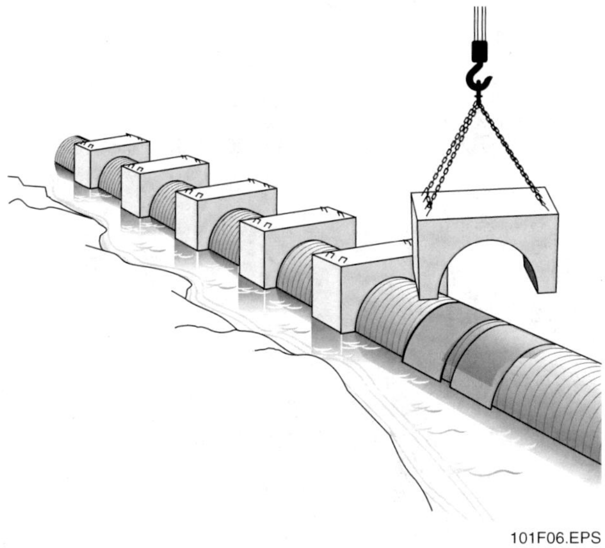

Figure 6 ◆ Add-on weight installation.

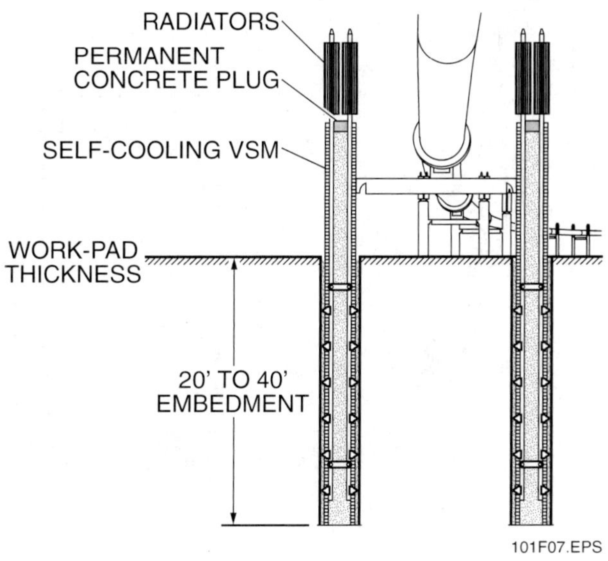

Figure 7 ◆ Bent with self-cooling VSMs.

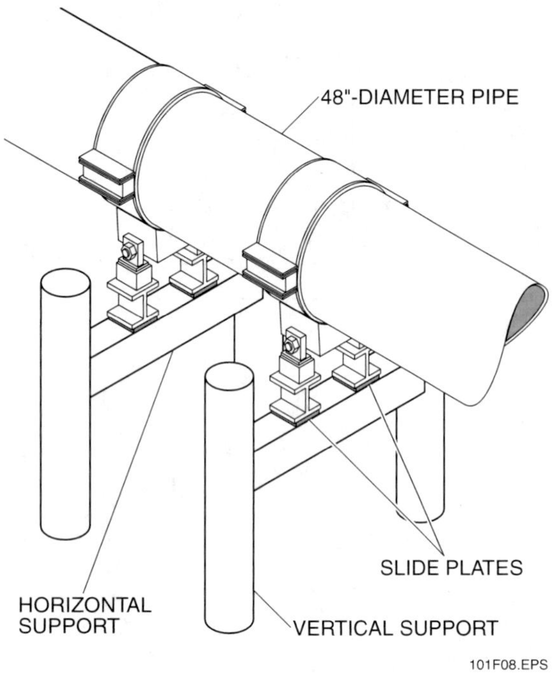

Figure 8 ◆ Common elevated pipeline section.

Pump and compressor stations are spaced strategically along the pipeline at intervals that are dependent on ground elevation, terrain, and other pipeline design factors to maintain the desired flow rate in the line. The pressure that causes the fluid or gas to flow is lost mostly through friction of the product but also by changes in elevation. This pressure must be changed by pumps and compressors to maintain flow. Drag-reducing agents (DRA) can also be added to the product in the line to reduce the friction loss.

A pump or compressor station facility includes pumps or compressors for moving great volumes of liquid or gas, methods of controlling the flow and pressure, means of measuring and accounting for the commodities being moved, storage facilities, and other devices for quality control. *Figure 9* shows an example of a liquid pipeline pump station layout. *Figure 10* shows an example of a gas pipeline compressor station layout.

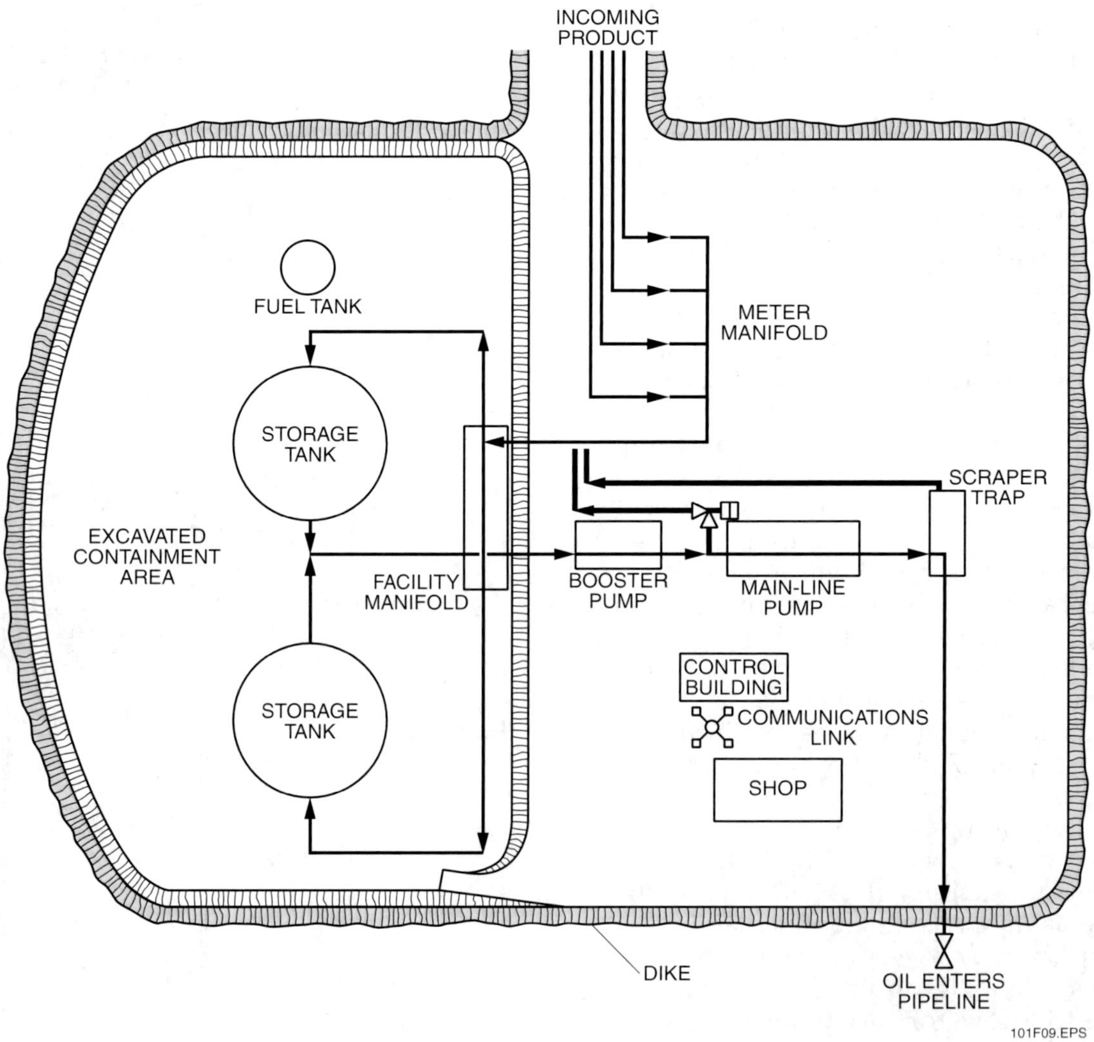

Figure 9 ◆ Example of a pump station layout.

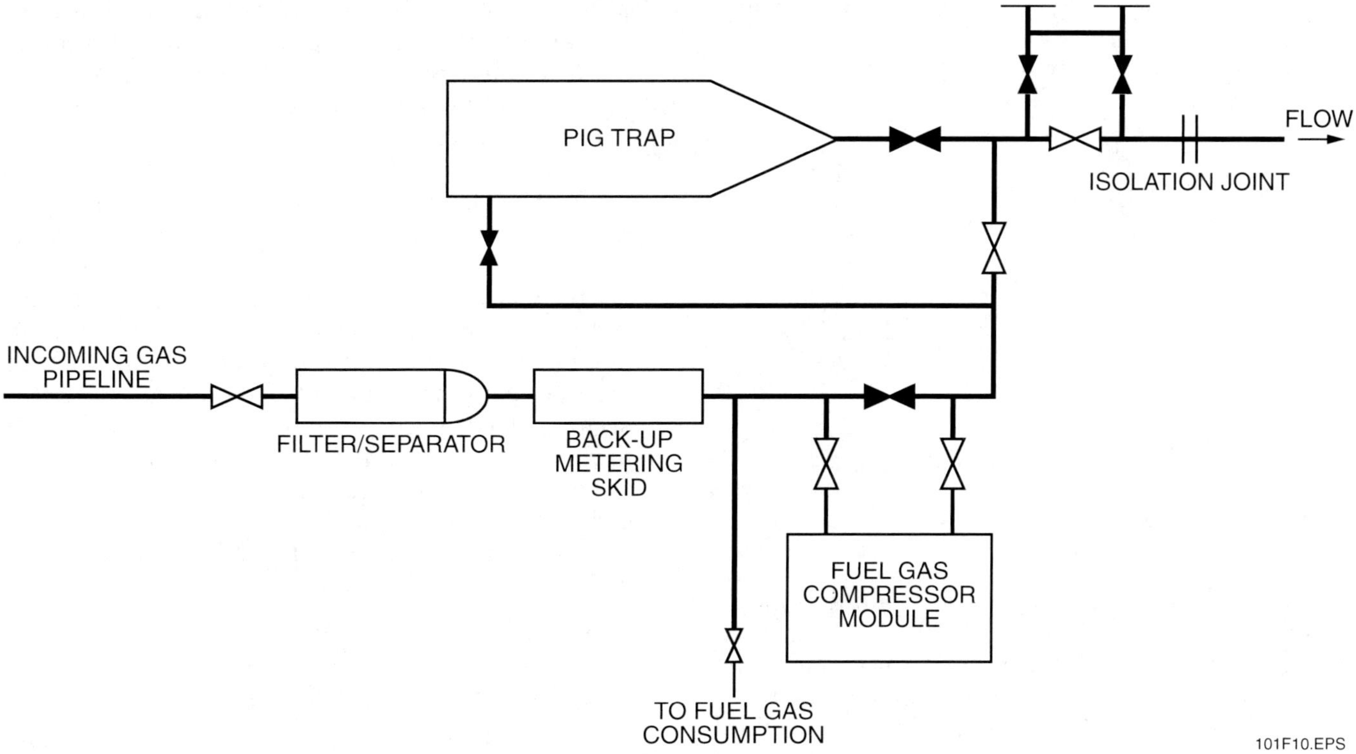

Figure 10 ◆ Example of a gas pipeline compressor station layout.

1.2.3 Operation and Maintenance of the Pipeline

The pipeline is monitored closely and inspected for potential equipment failure and line integrity. Crews identify problems and dig up the pipe to perform repairs. Small **anomalies** may be repaired temporarily while large anomalies may require replacing the pipe. In this case, a bypass line may be constructed so that flow continues while the damaged line is repaired.

The inside of the pipe must also be cleaned periodically to remove wax, scale, and debris. A scraper **pig** is inserted into the line at one point and flows with the liquid or gas to another point, where it is removed. The pig scrapes the inside walls using the hydraulic force of the flow behind it. An instrument pig, also called a smart pig, is inserted into the line periodically to check for anomalies.

1.2.4 Petroleum Products

Different petroleum products can be shipped consecutively from the point of origin in the same pipeline. This is called batching. To maintain quality, batches of each product are arranged in the line and pumped under pressure so that little or no mixing occurs during transit.

1.3.0 Storage

Storage facilities may be located along the pipeline route, at the origination sites, at pump stations, and at terminals. Storage is necessary to meet seasonal needs and other demand fluctuations. Storage is also necessary for the efficient operation of production equipment, pipelines, tankers, refineries, and for emergency use. Large volumes of natural gas are typically stored in underground storage fields throughout the United States.

The trend towards giant tankers created a need for large storage facilities at the loading and unloading ends of tanker runs. A large crude oil terminal typically has several tanks, each with capacities up to half a million **barrels**. A small pump station might have only one tank with a capacity of about 50,000 barrels. Stations along a product line may need separate tanks for storage of each product coming through. These are sometimes called relief tanks or surge tanks. *Figure 11* shows some storage tanks.

1.4.0 Delivery

A wide range of delivery methods is used for gas and petroleum products depending on their destination, intended use, and transport properties.

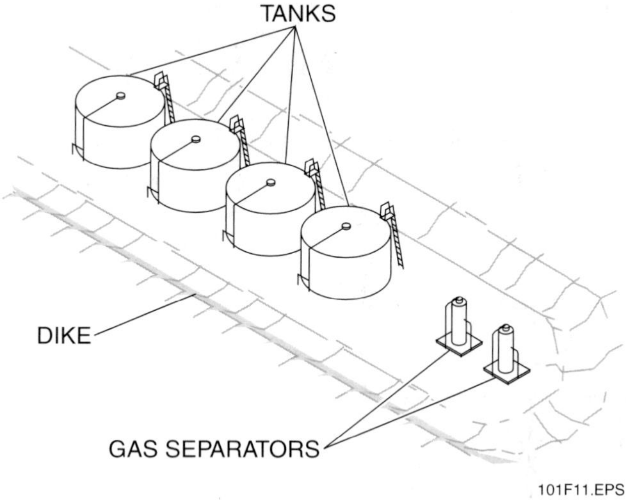

Figure 11 ◆ Storage tanks.

Petroleum is carried chiefly by pipeline, tanker, barge, tank truck, and railroad tank car. Nearly all product moves through pipelines for at least part of the route. Some products are delivered to city or regional stations by pipeline. From there they may be piped or delivered to local consumers or distribution sites. Gasoline, diesel, and kerosene are delivered by tank truck to gas stations for purchase by users. Industrial products may be shipped to companies by pipeline, truck, or railroad. Supertankers are used to move product on ocean routes, while barges are used on rivers and canals. While liquid petroleum products are typically measured in barrels, gas is typically measured using cubic feet. Depending on the volume being measured, gas quantities can be measured using billion cubic feet (BCF), thousand cubic feet (MCF), or hundred cubic feet (CCF). *Table 1* shows typical capacities that can be moved by different means of transport.

Table 1 Means of Transport

Transport	Capacity
Tanker Truck	200 barrels
Railroad Tanker	1,500 barrels
Barge	15,000 barrels
Supertanker	1 million barrels+
Large Liquid Pipeline	2 million barrels+/day
Large Gas Pipeline	1 BCF/day

101T01.EPS

2.0.0 ◆ PIPELINE PRODUCTS

A pipeline can carry a wide range of petroleum and gas products. Each of these products has its own characteristics and rules for handling. Each of these products also presents its own set of hazards for the pipeline workers, for the public, and for the environment. An understanding of the products in the pipeline, the hazards involved, and protective measures to be taken is essential to safe and secure operations.

Petroleum products are produced from crude oil in three basic stages. The first stage is separation and consists of separating the oil into its various fractions, or parts. The main process in this stage is fractional distillation, which separates light, medium, and heavy fractions. The second stage is conversion, in which less useful fractions are converted into more valuable ones. The third stage is treatment, which improves the quality and performance of petroleum products. *Figure 12* shows a diagram of the refinery process.

Products in a typical pipeline are discussed in terms of their characteristics and associated hazards. These products may include:

- Natural hydrocarbons
  - Crude oil
  - Liquefied petroleum gas
  - Wellhead natural gas
- Refined products
  - Gasoline
  - Diesel fuel
  - Fuel oil
  - Jet fuel
  - Consumer-ready natural gas
- Special products
  - Chlorine gas
  - Hydrogen chloride gas
  - Propylene
  - Acetone
  - Benzene
  - Ethane and ethylene
  - Anhydrous ammonia
  - Methane gas
  - Carbon dioxide

2.1.0 Natural Hydrocarbons

Natural hydrocarbons include crude oil, liquefied petroleum gas, and natural gas.

2.1.1 Crude Oil

Crude oil is unrefined petroleum—the source of the myriad of petroleum products. Its specific gravity ranges from 9 to 55 as compared with the specific gravity of water, which is 10. A light crude is one with a higher gravity, and a heavy crude has a lower gravity.

Because the petroleum may have different ingredients or fractions, its color can range from yellow

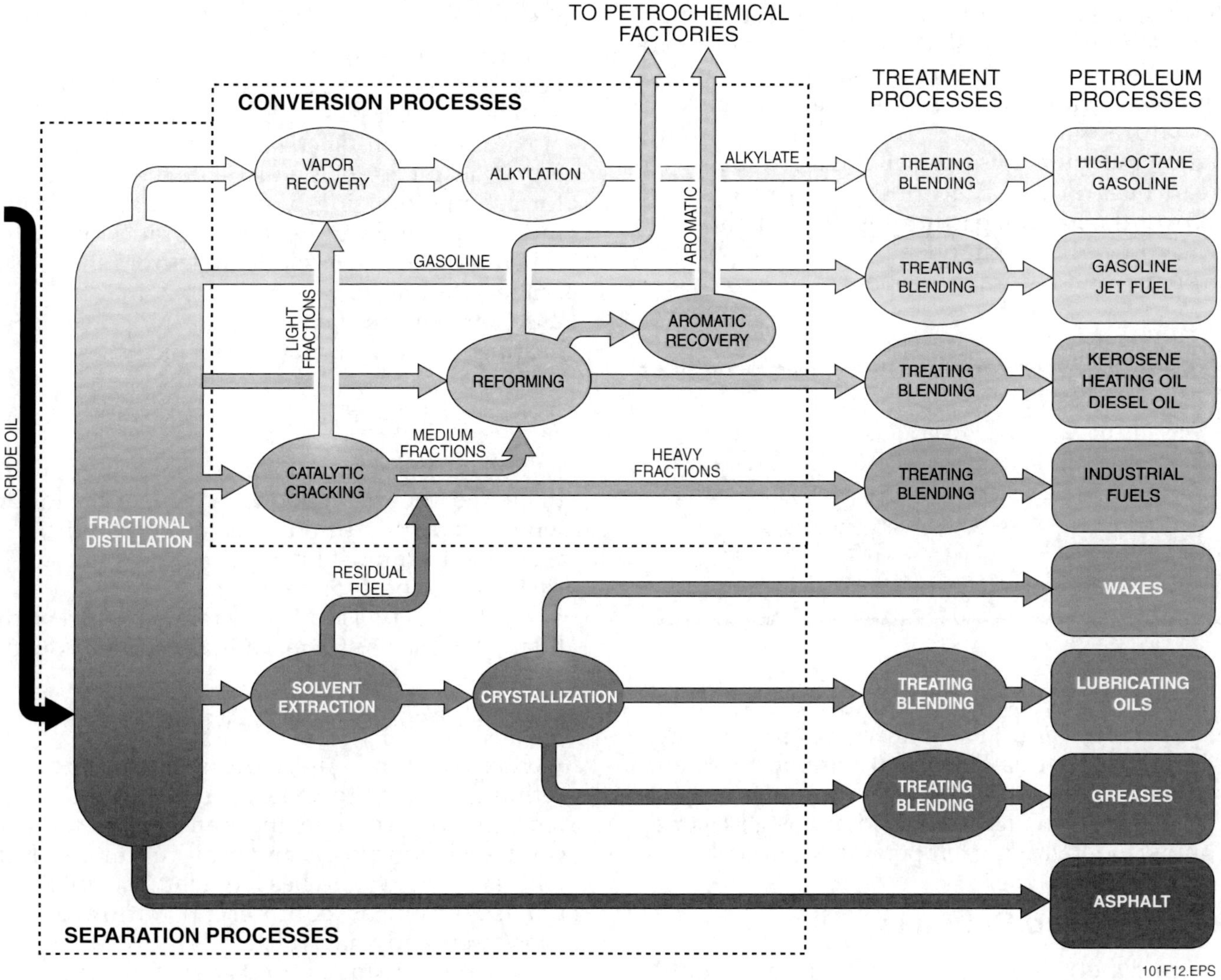

Figure 12 ◆ Diagram of the refinery process.

to black. It may have a paraffin, asphalt, or mixed base. Crude oil also contains substances other than hydrocarbons. These impurities include metallic compounds and sulfur and may make up as much as 10% of the oil. Oil containing a large amount of sulfur compound is called a sour crude. A sweet crude contains little or no sulfur.

2.1.2 Liquefied Petroleum Gas

Liquefied petroleum gas (LPG) is a mixture of heavier, gaseous, paraffinic hydrocarbons, principally butane and propane. These gases, easily liquefied at moderate pressures, may be transported as liquids and converted to gases on release of the pressure. Thus, LPG is a portable source of thermal energy that finds wide application in areas in which it is impractical to distribute natural gas. LPG is used in industry for cutting and welding metals and on farms for operating various kinds of machinery. The principal sources are natural and refinery gas, from which the liquefied petroleum gases are separated by fractionation.

Butane/butylene mix, a form of LPG, is extremely flammable. It releases vapors at well below ambient temperatures and readily forms flammable mixtures with air. If ignited, it will burn in open air or explode in confined spaces. Its vapors are heavier than air and may travel long distances to a point of ignition and then flash back. Alkane/chlorine gas mixtures have produced explosions. Closed systems will form a white frost at the point of the release. Liquid spills will vaporize, forming cold, dense vapor clouds that should be avoided even with respiratory equipment.

2.1.3 Wellhead Natural Gas

Natural gas is the name given to the naturally occurring hydrocarbon mixtures found in many underground wells. Natural gas is an unrefined product that is a naturally occurring gaseous mixture of hydrocarbons that may also contain nitrogen, helium, carbon dioxide, hydrogen sulfide, and water vapor. Because the content or makeup of natural gas can vary from well to well and region to region, it is normally referred to as wellhead natural gas. Natural gas is processed to recover the **natural gas liquids (NGLs)**, the heavier hydrocarbon components that include butane, propane, and natural gasoline.

> **WARNING!**
> Exposure to petroleum products can cause eye, skin, and respiratory tract irritation. Always wear proper personal protective equipment (PPE) and refer to company procedures and the product material safety data sheet (MSDS) for specific details.

Natural gas is highly compressible, highly expandable, and has a low specific gravity. Although gaseous at normal temperatures and pressures, the gases making up the mixture are variable in form and may be found as gases or as liquids under varying temperatures and pressures.

2.2.0 Refined Products

Refined petroleum and gas products include gasoline, jet fuel, diesel fuel, heating oil, and kerosene. Other refined products may include propane, butane, consumer-ready natural gas, and blending components.

2.2.1 Gasoline

Gasoline is a volatile, flammable liquid hydrocarbon refined from crude oil and used universally as a fuel for internal-combustion, spark-ignition engines. Natural gasoline, or casinghead gasoline, is the liquid hydrocarbon recovered from wet natural gas. Gasoline is classified as regular, premium, and aviation grades according to how smoothly it burns in an engine.

2.2.2 Diesel Fuel

Diesel fuel is a light hydrocarbon mixture for diesel engines, similar to furnace fuel oil. It has a boiling range just above that of kerosene. Diesel fuel requires less refining than gasoline.

2.2.3 Fuel Oil

Refined fuel oil comes in two grades. The specifications for both are defined by *American Society for Testing and Materials (ASTM) Specification D 396.*

No. 1 distillate, such as kerosene, is a light fuel. No. 2 fuel oil diesel is a distillate fuel oil prepared for use in atomizing-type burners or for smaller industrial burner units. Lesser grades of fuel oil, called *residual fuel oils*, are made from the topped crude of refinery operations. Residual fuel oil is used for the production of electric power, space heating, and various industrial purposes.

2.2.4 Jet Fuel

Jet fuel is a highly refined kerosene petroleum distillate. Jet fuel has good flow properties in low-temperature conditions, and it burns efficiently and cleanly. It must be delivered pure and free of rust, water, and sediment.

Military jet fuel is a highly refined kerosene distillate that meets stricter military specifications.

2.2.5 Consumer-Ready Natural Gas

Wellhead natural gas is normally run through a refining process to remove any undesired contents, remove contents that can be used for other commercial purposes, and achieve a more consistent safe product and heat content. The final product is known as commercial natural gas or consumer-ready natural gas.

Consumer-ready natural gas is still a mixture of different gases but is made up primarily of methane. Other gases, such as ethane, propane, nitrogen, and carbon dioxide can also be found in trace quantities. The composition of the consumer-ready natural gas remains fairly consistent over time. The heat value range of typical consumer-ready natural gas is 1,020 to 1,045 Btus per cubic foot.

2.3.0 Special Products

Special products include other by-products that are produced and carried through pipelines.

2.3.1 Chlorine Gas

Chlorine gas is a product used for water purification, chemical processes, and the manufacture of plastic. Chlorine gas is a strong oxidizing agent; it can react explosively with organic products and it is poisonous. Liquid and gas chlorine under pressure can be hazardous. Chlorine is nonflammable and nonexplosive, but like oxygen, it can support combustion of certain substances.

It can react violently or form explosive compounds with many chemicals, such as acetylene, turpentine, ether, ammonia gas, and hydrogen, and finely divided metals.

2.3.2 Hydrogen Chloride Gas

Hydrogen chloride gas is a fuming, poisonous gas used to make hydrochloric acid, an important industrial chemical used in many chemical compounds, in metallurgy, in food processing, and as a cleaning agent. Hydrogen chloride is nonflammable and noncombustible. However, if dissolved in water, it produces hydrochloric acid, which attacks most metals.

2.3.3 Hydrogen Sulfide

Hydrogen sulfide is a flammable, colorless, gaseous compound of hydrogen and sulfur, which in small amounts has the odor of rotten eggs. It is formed naturally by the decomposition of organic animal and vegetable matter by bacterial action. Hydrogen sulfide is found in crude oil, natural gas, sewers, and swamps and is created by a number of industrial processes. It is extremely corrosive and poisonous. Hydrogen sulfide causes damage to the eyes, skin, breathing passages, and lungs, attacks and paralyzes the central nervous system, especially the parts that control the lungs and heart. In large amounts, it deadens the sense of smell. Hydrogen sulfide is an extremely dangerous compound, and exposure is deadly.

2.3.4 Propylene

Propylene is used in plastics manufacturing. It is extremely flammable and will readily ignite at ambient temperatures. The flammable vapors, being heavier than air, may travel long distances along the ground to a source of ignition and then flash back. Propylene presents the danger of rocketing cylinders, ruptured containers, and explosion. Reignition or explosion can occur if the flames are extinguished without stopping the source flow. An electrostatic charge may build up during handling if the equipment is not grounded.

2.3.5 Acetone

Acetone is a clear, colorless liquid with a sweet, pungent odor that is used as a solvent. It is extremely flammable and may cause flash fire. It is incompatible with a variety of materials, including inorganic acids, caustic, amines, alkanolamines, oxygen, halogens, aldehydes, ammonia, oxidizing agents, and chlorinated compounds. Contact with these should be avoided during transport, handling, and storage.

2.3.6 Benzene

Benzene is an extremely flammable liquid, readily igniting at ambient temperatures. The vapor forms an explosive mixture with air. Flammable vapors, being heavier than air, may travel long distances to a point of ignition and flash back. Incomplete combustion can produce carbon monoxide.

2.3.7 Ethane and Ethylene

Ethane and ethylene are simple asphyxiants. They are dangerous when exposed to heat or flame and can react violently with oxidizing materials. Both are flammable over a wide range in air, with an even broader range in pure oxygen.

2.3.8 Anhydrous Ammonia

Anhydrous ammonia is the liquid form of pure ammonia gas. It is used as a fertilizer because of its high nitrogen content. It is also used as an industrial refrigerant in cold-storage and ice manufacturing plants, since it absorbs a large amount of heat when going from a liquid back to a gas. Anhydrous ammonia is made by compressing dry ammonia gas.

Ammonia is a colorless alkaline gas made up of one part nitrogen and three parts hydrogen. It is lighter than air and has a sharp, stinging odor. Ammonia can be safely inhaled if it is greatly diluted in air, but concentrated ammonia gas can cause suffocation and death. It does not burn in air, but it burns in oxygen with a weak yellow flame. Ammonia is made by combining hydrogen and nitrogen, and as a by-product in the production of coal and coke gas.

2.3.9 Methane Gas

Methane is a member of the alkaline or paraffin series of hydrocarbons. It is called marsh gas because it forms by anaerobic bacterial decomposition of vegetable matter in swampy land. Coal miners know it as *firedamp* because it is combustible when mixed with air. It is a major component of natural gas and of coal gas, and forms in large amounts in sewage disposal processes.

In addition to its use as a fuel, methane is important as a source of organic chemicals and of hydrogen. In the presence of steam and catalysts, it yields carbon monoxide and hydrogen, which can be converted to liquid hydrocarbons or to methanol and other alcohols. The incomplete combustion of methane with air produces carbon black, which is used in rubber and ink. Chlorination of methane yields various chlorides and chloroform.

2.3.10 Carbon Dioxide

Carbon dioxide (CO_2) is a colorless, odorless, tasteless gas which is about 15 times heavier than air. Under normal conditions, it is stable, inert, and nontoxic. The decay of all organic matter and the burning of hydrocarbon fuels produces carbon dioxide. It is obtained commercially from four sources: naturally occurring underground CO_2 wells, fermentation, combustion of carbonaceous fuels, and as a by-product of chemical processing.

Carbon dioxide gas may be liquid or solid based on compression and lowered temperatures. It may be used as a refrigerant, inerting medium, chemical reactant, neutralizing agent for alkalies, and pressurizing agent. Liquid carbon dioxide is an effective fire-extinguishing agent because it rapidly reduces the temperature of burning materials below the ignition point. The gas, being heavy and inert, displaces air, which is the source of the fire's oxygen, and blankets the flames. Oil tankers, barges, storage vessels, and pipelines are quickly and safely made inert with carbon dioxide to allow for welding and other repairs.

2.4.0 Protective Measures

Material safety data sheets (MSDSs) are good sources of information on these products and must be made available to all employees of the company. There is an MSDS for each hazardous substance, listing its properties and hazards, including a hazard rating for health, fire, reactivity, and other hazards. MSDSs also detail safe handling, personal protective equipment requirements, and first aid/emergency procedures.

All products pose hazards to the environment if spilled or disposed of improperly. Areas of possible spills should have containment structures and/or procedures. These include dikes around storage and transfer facilities and containment booms around tankers being filled. The marine environment is particularly sensitive to oil spills because the tides and currents spread the oil and complex marine life is so vulnerable. *Figure 13* shows temporary containment of a spill from a tank.

Oil may be harmful on land, too. It can seep into the groundwater or streams where it contaminates plant and animal life and eventually human

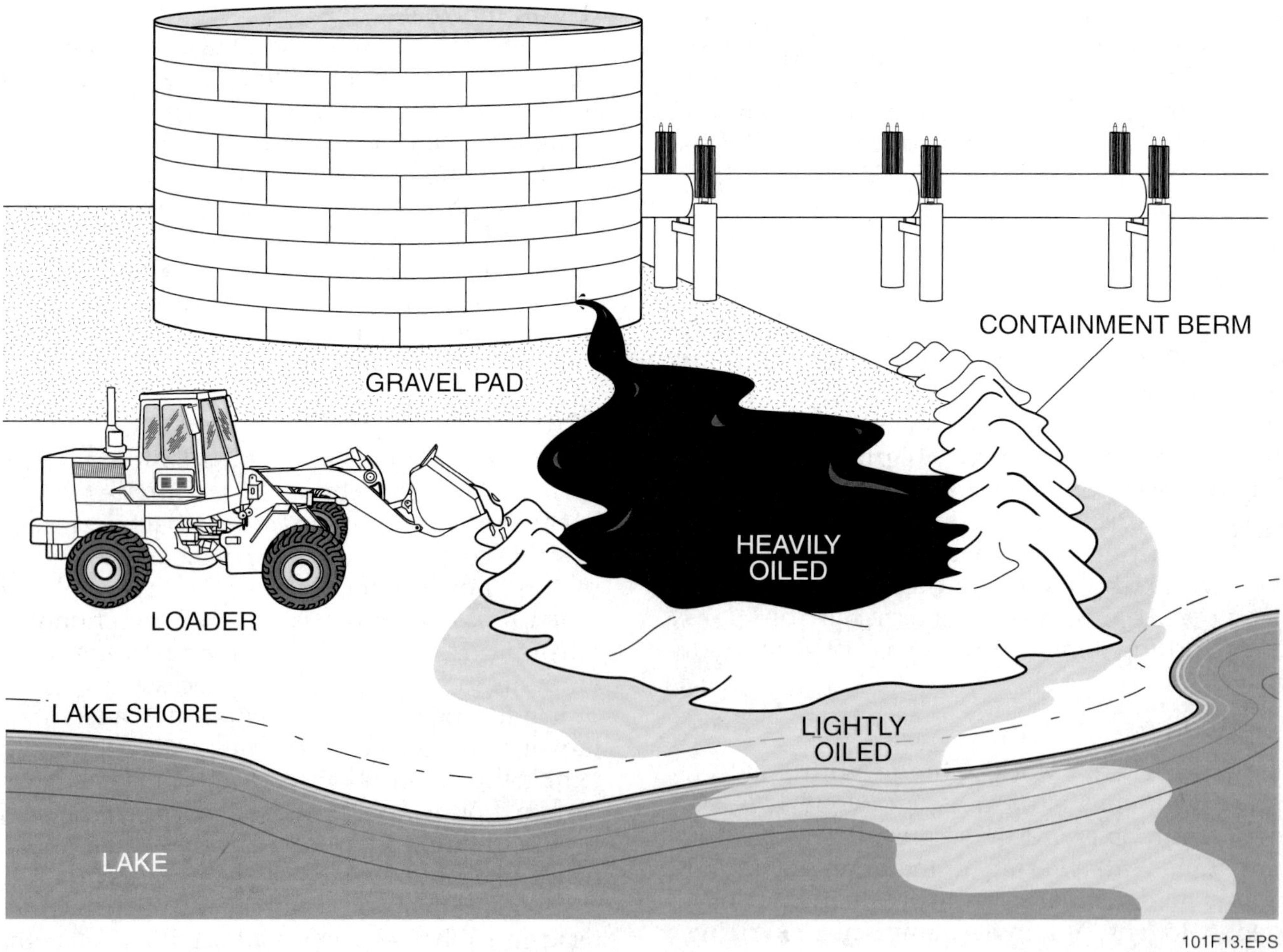

Figure 13 ◆ Temporary containment of a tank spill.

life. Handling and disposal of hazardous chemicals are addressed in company procedures and MSDSs. Detailed procedures exist for handling all petroleum products, and the observance of these procedures is strictly regulated and enforced by federal, state, and local agencies.

3.0.0 ◆ MAPS AND DRAWINGS

Maps and drawings are an important means of communicating information about pipeline equipment, flow, and processes. Some of these are piping and instrumentation diagrams (P&IDs), strip maps, blueprints, and alignment sheets.

3.1.0 Piping and Instrumentation Diagrams (P&IDs)

A P&ID depicts the functional layout of a piping system and its components, valves, and instrumentation as clearly and accurately as possible. It is accurate to the extent that all components are connected to each other as shown in relation to flow-path orientation. A P&ID does not represent the actual physical layout of equipment; for example, a valve that may appear to be right at the discharge of a pump can actually be located some distance away or even on another floor. Many times, a P&ID will use a broken line around groups of equipment to indicate that they are all in the same building (see *Figure 14*).

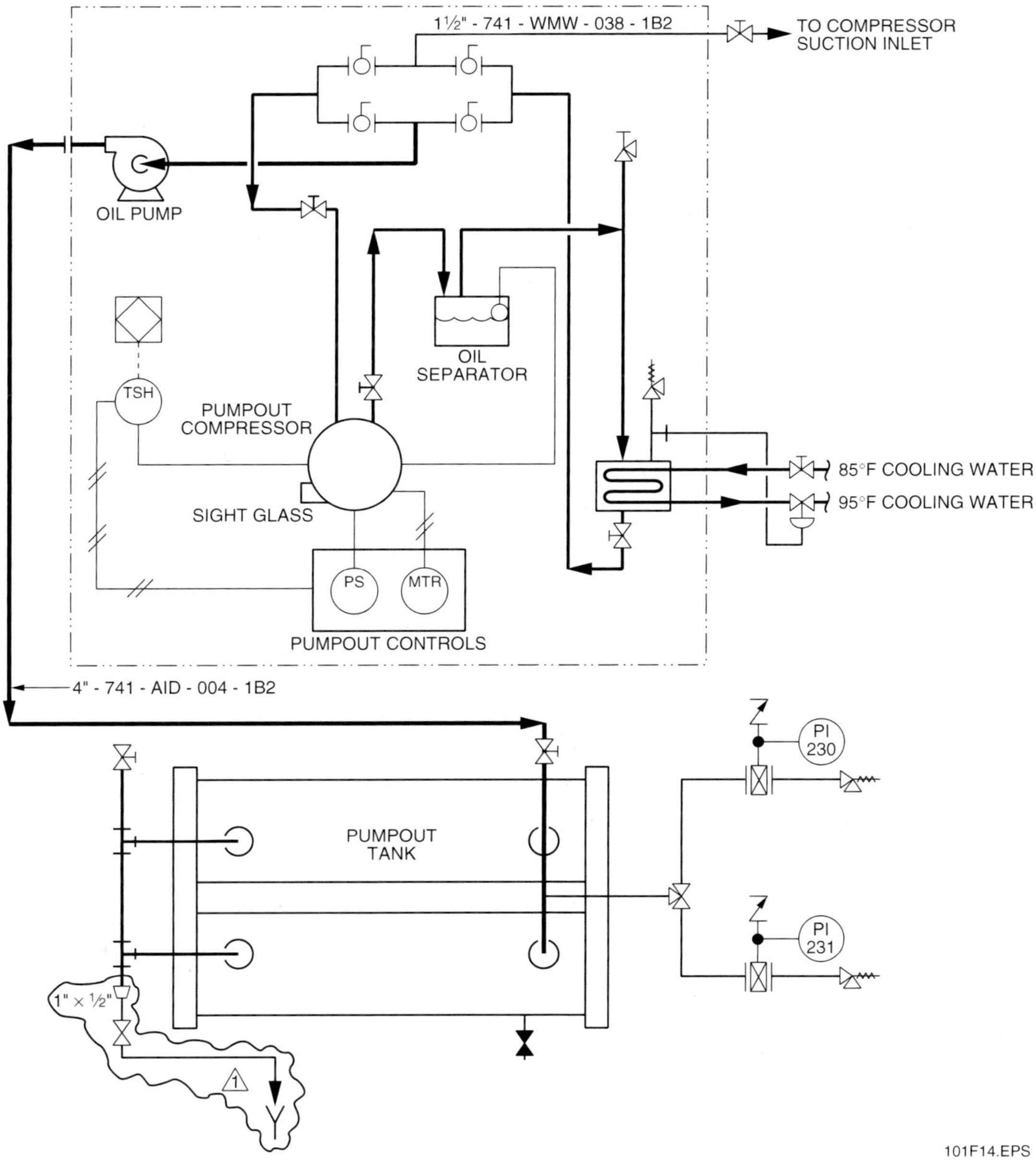

Figure 14 ◆ Sample P&ID.

on the same sheet so that they line up along the route. There may be views shown of the ground from an aerial, or plan view, and side, or elevation view, which shows the ground profile along a specific line. The pipeline will be superimposed on all views. Strip maps help to coordinate surface features with topography as well as with other pipeline data at any point on the route.

Alignment sheets are various types of maps, drawings, and diagrams that show different features along the pipeline route. The alignment sheets will show the relationship of the pipeline to true north, geographic features, political boundaries, fences, rivers, property lines, and other pipelines. *Figure 17* shows one type of alignment sheet.

Strip maps and alignment sheets vary greatly from company to company. These names may be used interchangeably or may mean very different things.

4.0.0 ◆ OPERATIONS

Controlling the fast-paced, continuous, and complex operation of the pipeline is a critical job done largely by the scheduler, controller, and dispatcher with the help of other personnel and systems like Supervisory Control and Data Acquisition (SCADA), programmable logic controllers (PLCs), and remote terminal units (RTUs). All control functions are documented by filling out forms or printing out reports.

4.1.0 Scheduling

Before a shipment of product reaches the pipeline and comes under the control of the dispatching office, it has to be accumulated and scheduled with the pipeline scheduler. The scheduler at a pipeline facility typically performs the following functions:

- Takes nominations from shippers
- Creates pumping orders for pipelines to meet shippers' needs
- Validates and reviews operating information for schedule adjustments
- Receives delivery information in time for shipments
- Validates all shipping information for billing
- Utilizes power optimization to save costs and maximize throughput efficiency

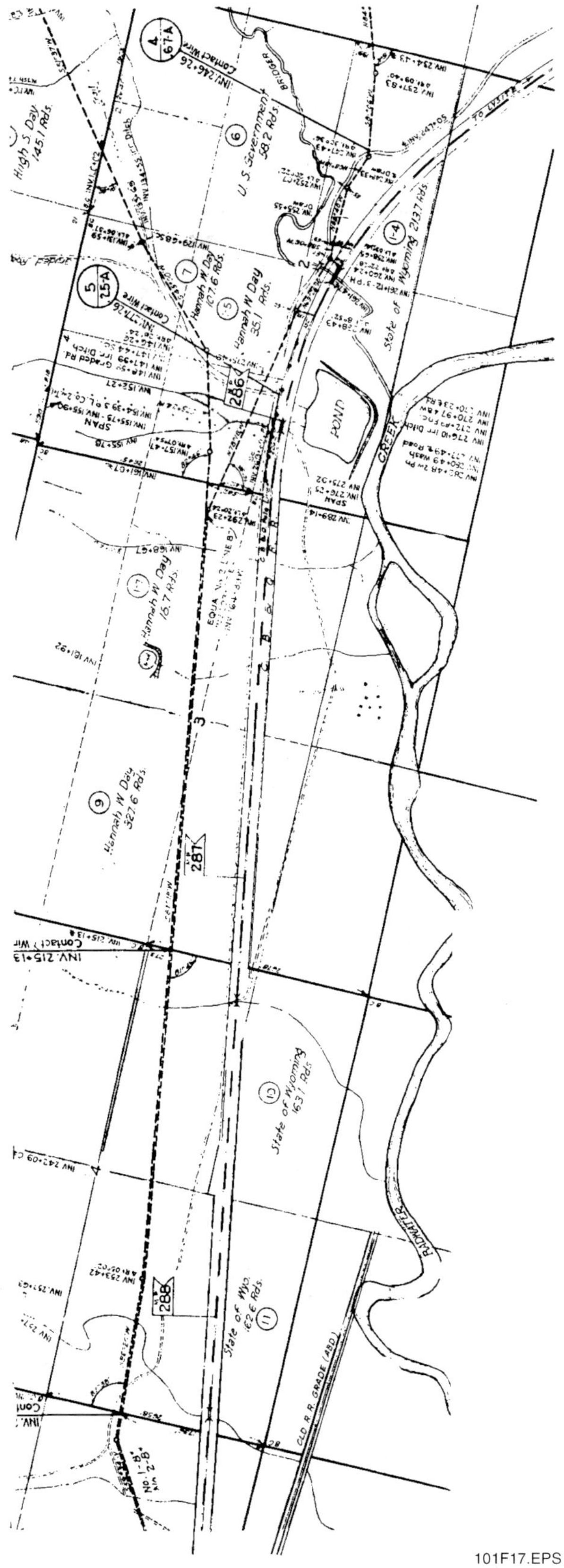

101F17.EPS

Figure 17 ◆ One type of alignment sheet.

4.2.0 Pipeline Operations

The pipeline is monitored closely and inspected for potential equipment failures and line integrity. These problems can arise from bad welds, corrosion spots, faulty valves, third party damage, or shifting of the pipe. When aboveground, the pipe may also be damaged by accidents or sabotage. In areas with faults or history of earthquakes, seismic activity must also be closely monitored. Flow can also be interrupted by obstructions in the pipe, such as stuck pigging devices, foreign objects, and debris.

Computers and automation are used to control and monitor the pipeline operation. Communication is maintained between all stations regarding the status of the operation. Controllers have the ability to shut down the pipeline flow if a problem is detected and to pinpoint the location of the problem. *Figure 18* shows a typical control center layout.

4.3.0 Control Systems

SCADA is the software used to monitor and control the flow of products through the pipeline every day. Wireless telemetry and SCADA systems provide remote alarm, control, and data collection for many types of pipeline operations.

SCADA is installed with the main computer system and hooked up through a server to RTUs and PLCs in the field. Sensors along the pipelines precisely measure operating conditions, flow rates, and the status of control devices, such as flow and pressure regulators and valves. The RTUs collect vital information in the field and transmit the data via telephone, microwave, or satellite circuits to the control center.

4.4.0 Communications

Communications are the systems that keep the pipeline operating. All along the pipeline, everyone must have open communications at all times because everything that occurs on the pipeline affects the entire system.

A variety of communications equipment and devices may be used in a pipeline company's operations. These are set up for continuity during normal or emergency conditions. It is important that communications be reliable and always available. Most companies have redundant, or backup, systems to ensure that this is true. Some of the equipment used includes:

- Dedicated satellites or microwave-based radio communications systems that provide linkage between the control room personnel and various pump stations and valve locations
- Hardwire communications that link the control room and all offices and all remotely monitored and operated facilities, such as pump stations and RTUs
- Portable cellular phones or vehicle phones provided to all key and supervisory personnel
- Locally activated pagers or nationwide pagers provided to key personnel for use in routine operating or emergency events

101F18.EPS

Figure 18 ◆ Typical control center layout.

- Fixed repeaters installed at points within the system that provide internal long distance and local radio communications

In the event of a power loss, backup systems are provided, which may include standby generators for powering communications equipment, batteries, or uninterruptible power systems (UPS). Portable self-contained mobile repeaters may be used at an incident scene and utilized to augment existing radio systems or replace any that may be inoperable. A radio link (base station) may be used with portable units to allow for emergency communications between facilities in the event that telephone and/or microwave transmissions become inoperable. Redundant communications and power systems should be installed at key facilities to allow for control in the event of loss of primary systems.

5.0.0 ◆ FLUIDS AND SYSTEM HYDRAULICS

The operation of any liquid pipeline is built on the principles of hydraulics, or the science of the behavior of fluids at rest or in motion. These principles are in turn built on the properties and characteristics of fluids.

5.1.0 Fluids

A fluid is a substance that flows, and some flow more easily than others. A fluid is also a substance that changes form with a slight force or pressure; it is elastic, however—meaning it will try to return to its previous form when the pressure is removed. Some fluids are more elastic than others. Elasticity is a measure of the fluid's resistance to change in shape.

Fluids include all liquids and gases. Water at ordinary temperature is a liquid fluid; at high temperatures, a gas fluid. Air is a gas fluid. All fluids can be greatly affected by changes in temperature and pressure.

No fluid is perfect and frictionless, and different fluids have different levels of friction, or resistance to flow. This resistance to flow is called *viscosity*.

Some fluids are thicker than others. This is called *density*. Density can be a large factor in the viscosity of a fluid. It is a measure of how tightly packed the fluid molecules are or how much of the fluid's matter exists within a given unit of volume.

Measuring density in a liquid is readily done with a hydrometer because they are not usually compressible. Measuring density in a gas is more difficult because gases can easily be compressed. Compressibility is the ability to be squeezed, or forced into a smaller volume.

Density helps determine the compressibility of a fluid. When a fluid is compressed, it becomes denser. The specific gravity of a substance is derived from its density. Specific gravity is the ratio of the density or mass of a given volume of the substance to the density of the same volume of a standard substance, such as water. The water and the liquid substance must be at the same temperature. For gases, the gas must be at the same temperature and pressure.

American Petroleum Institute (API) gravity is a unit of measure developed specifically for liquid petroleum products. The heavier the substance, the lower the number of its API gravity. Temperature will affect API gravity and should always be corrected to 60°F.

The properties of fluids that are used to describe the fluids are density, viscosity, compressibility, inertia, and elasticity. *Table 2* shows some fluid properties.

Table 2 Fluid Properties

Fluid	API Gravity @ 60°F	Specific Gravity @ 60°F	Vapor Pressure @ 100°F	Absolute Viscosity	Compressibility @ 1000 psi @ 60°F
EP Mix 80/20	214	0.41	550 psi	0.11 cP	7.3%
NGLS	146	0.51	325 psi	0.15 cP	3.8%
Raw Feed Propane	146	0.51	150 psi	0.15 cP	3.8%
Butane	112	0.58	88 psi	0.18 cP	2.0%
Anhydrous Ammonia	97	0.62	200 psi	0.25 cP	1.6%
Gasoline	63	0.66	2 psi	0.70 cP	0.8%
Jet Fuel #1	42	0.82	2 psi	1.4 cP	0.8%
Diesel Fuel #2	33	0.86	1 psi	4.7 cP	0.5%
Crude Oil	50 and below	0.78–0.98	> 1 psi	6–400 cP	> 1.0%
Water	10	1		1.12 cP	0.3%

101T02.EPS

Other properties of fluids, particularly those found in pipelines, include volatility and flammability. Volatility is the tendency of a fluid to readily vaporize. Sudden vaporization can cause a violent reaction or explosion. Flammability is a measure of how readily a fluid catches fire when in contact with heat or sparks. The volatility and flammability of specific pipeline products is described in another section of this module.

5.2.0 Basic Hydraulics

Some laws of hydraulics apply to gases as well as liquids but only under certain conditions. In general, the following sections pertain only to liquid fluids.

Hydraulics is usually divided into two branches: hydrostatics for liquids at rest and hydrodynamics for liquids in motion.

5.2.1 Hydrostatics

Hydrostatic principles are fundamental to the design of many machines and instruments. The design of hydraulic presses and the operation of instruments, such as manometers and barometers, depend on these principles.

A simple hydraulic press consists of a liquid-filled cylinder that contains two pistons, one smaller than the other. When a force is applied to the smaller piston, the force travels through the liquid to the larger piston, producing a force larger than the initial one. The increase in force is proportional to the ratio of the larger piston to the smaller one. *Figure 19* shows a simple hydraulic press.

The laws of hydrostatics are the following:

1. The amount of pressure caused by the weight of a column of liquid is determined by the height of the column.
2. An object placed in a liquid loses an amount of weight equal to the weight of the liquid it displaces.
3. A fluid in a container transmits pressure equally in all directions.

5.2.2 Hydrodynamics

Hydrodynamic principles are basic to hydraulic engineering, to the design of certain machines, and to pipeline engineering. Hydrodynamics define the plans for water systems, canals, and irrigation systems, as well as airplane design and hydraulic machines like water turbines.

Pipelines for water and other fluids have to take into account the pipe size and the fluid's friction on the pipe to determine fluid velocity through the pipe. Liquid flow can be steady or unsteady. An unsteady flow results from changes in the velocity, temperature, or pressure of a liquid or its movements around obstacles. The three basic laws of hydrodynamics apply only to steady-flowing liquids. The laws of hydrodynamics are the following:

1. The velocity of a liquid flowing through a pipe increases as the area of the pipe decreases and vice versa.
2. The pressure of a liquid increases as its velocity decreases and vice versa.
3. The velocity with which a liquid flows through an opening in a container equals the velocity of a body falling from the surface of the liquid to the opening.

Figure 20 shows liquid velocity and pressure relationships.

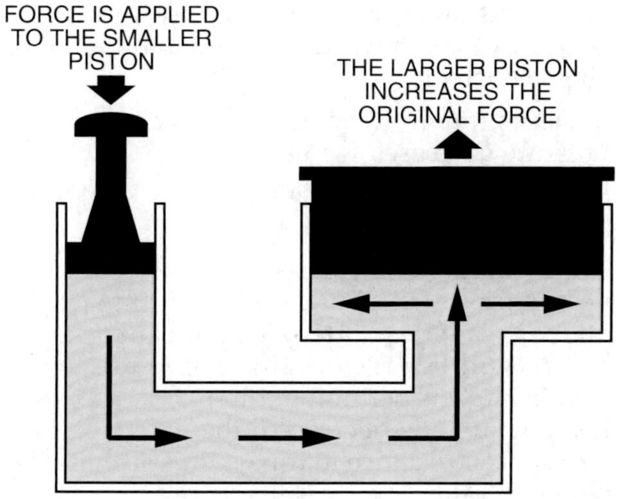

Figure 19 ◆ Simple hydraulic press.

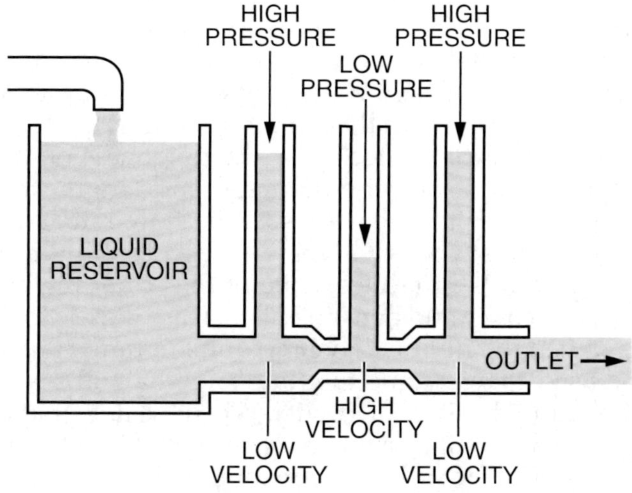

Figure 20 ◆ Liquid velocity and pressure relationships.

5.3.0 Pipeline Hydraulics

Fluid properties are major factors which affect the design of the pipeline. The primary factors in pipeline design are:

- Fluid characteristics
- Desired flow rate
- Maximum operating pressure (MOP)
- Maximum allowable operating pressure (MAOP) (for gas)
- Elevation profile/terrain
- Fittings/valves
- Installation costs
- Environmental concerns

Pressure is referenced in several ways in the pipeline industry. The maximum operating pressure (MOP) is defined as the maximum internal pressure allowed liquid during the operation of a pipeline. The term MOP is typically used more in the liquid pipeline industry and the regulations are referenced in *49 CFR 195*. The MAOP is defined as the maximum internal pressure permitted during the operation of a gas pipeline. The term MAOP is typically used more in the gas pipeline industry and the regulations can be referenced in *49 CFR 192*.

Although fluids are defined by their ability to flow, something has to start the flow and overcome their inertia and friction. A force, such as difference in pressure, gravity, or an external force physically pushing it, can do this. Usually, this is done by pumps. Stopping the flow once started will require another force, such as resistance (friction), or a stationary object, such as a closed valve.

The resistance or friction of the fluid against the pipe will be constantly trying to slow it down. Most pumps work by providing a difference in pressure and must continue providing that force all along the line to overcome the friction. There are drastic differences in the amount of pressure required to deliver the proper volume of different types of products. But the pressure in the system must be kept under the maximum limits.

Friction is also affected by the roughness of the pipe interior, the pipe diameter and length, the product density and viscosity, the flow rate, and obstacles, such as valves and bends. Pipeline design can help reduce some of these, but higher fluid viscosity and density will increase the friction.

When the fluid is traveling uphill, pumps must be used to overcome the fluid's inertia and resistance and the force of gravity. When the flow in the pipe is downhill, gravity may be all that is needed.

Another factor in the flow of the pipeline is the temperature. The higher the temperature, the lower the viscosity and vice versa. When the product is hot, it flows better. Sometimes heaters are included in the pipeline equipment to keep the product at the optimum temperature.

6.0.0 ◆ GASES AND SYSTEM PNEUMATICS

Pneumatics, as it relates to the pipeline industry, is the study of the storage, compression, and transmission of gases. The compression of gas has one primary goal, to deliver the gas at a pressure higher than the original pressure. The original pressure may vary from very low absolute pressures to several thousand pounds. The pressure rise may vary from a few ounces to many thousand pounds, and the volumes handled may vary from a few cubic feet per minute to hundreds or thousands of cubic feet.

There are three variables that have to be calculated and controlled when compressing and moving gas in a pipeline. These variables are pressure (p), volume (v), and temperature (k). These three characteristics of gases have a close relationship with one another and are explained in the ideal gas laws.

6.1.0 The Ideal Gas Laws

Pressure, volume, and temperature of gases have a close relationship with one another. By keeping one of them constant, the other two will have either a direct or inverse proportion. The relationships between pressure, volume, and temperature are explained using the ideal gas laws. These ideal gas laws are Boyle's law, Amonton's (Gay-Lussac's) law, and Charles' law.

6.1.1 Boyle's Law

Boyle's law states that when temperature is kept constant and volume is decreased, pressure will increase. His law gives the relationship between pressure and volume if temperature and amount are held constant.

- If the volume of a container is increased, the pressure decreases.
- If the volume of a container is decreased, the pressure increases.

If the volume is increased, the gas molecules have farther to travel and they will impact the container walls less often. This means the gas pressure will be less because there are fewer molecule impacts per unit of time.

If the volume is decreased, the gas molecules have a shorter distance to travel, thus striking the walls more often. This results in pressure being increased because there are more molecule impacts per unit of time.

6.1.2 Amonton's (Gay-Lussac's) Law

Amonton's law states that when volume is kept constant and temperature is increased, pressure also increases. Amonton's law was discovered by Joseph Louis Gay-Lussac in the early 1800s. His law gives the relationship between pressure and temperature when volume and amount are held constant.

- If the temperature of a container is increased, the pressure increases.
- If the temperature of a container is decreased, the pressure decreases.

If the temperature is increased, the gas molecules will move faster and they will impact the container walls more often. This means the gas pressure inside the container will increase. Gay-Lussac's law is a direct mathematical relationship. This means there are two connected values and when one goes up, the other also increases.

6.1.3 Charles' Law

Charles' law states that when pressure is kept constant and volume is increased, temperature also increases. The pressure that a gas exerts on the walls of its container is determined by the momentum of the atoms and molecules of the gas, which in turn is determined by the temperature. As the temperature increases, the atoms and molecules move faster, and so exert a greater pressure on the walls.

This law gives the relationship between volume and temperature if pressure and amount are held constant.

- If the volume of a container is increased, the temperature increases.
- If the volume of a container is decreased, the temperature decreases.

If the temperature is increased, the gas molecules will move faster and they will impact the container walls more often. This means the gas pressure inside the container will increase, but only for an instant. The greater pressure on the inside of the container walls will push them outward, thus increasing the volume. When this happens, the gas molecules will now have farther to go, thereby lowering the number of impacts and dropping the pressure back to its constant value.

6.2.0 Measuring Energy Gases

Quantities of natural gas are measured in volume units. A cubic foot of natural gas at a temperature of 60°F and an atmospheric pressure of 14.7 pounds per square inch gauge (psig) is the common unit of measure. Gas production from wells and supplies to power plants are measured in thousands or millions of cubic feet (MCF and MMCF). Resources and reserves are calculated in billions of cubic feet (Bcf) or trillions of cubic feet (Tcf). To put this in perspective, a trillion cubic feet is enough to fill a cube with sides 2 miles long.

The amount of energy that is obtained from the burning of a unit volume of natural gas is measured in British thermal units (Btus). One Btu is the amount of heat required to raise the temperature of 1 pound of water from 60° to 61°F at normal atmospheric pressure (14.7 psig). At sea level, it would take about 75 Btus to heat water enough for a cup of tea. A cubic foot of natural gas on the average gives off 1,000 Btus, but the range of values is 500 to 1,500 Btus. Therefore, 1 cubic foot of some natural gas may make only 7 cups of tea, while another would make as many as 20 cups of tea.

Energy content of natural gas varies because natural gas accumulations vary in the amount and types of energy gases they contain: the more noncombustible gases in a natural gas, the lower the Btu value. In addition, how much of any energy gas that is present in a natural gas accumulation—the mix of combustible gases—also influences the Btu value of natural gas. The more carbon atoms in a hydrocarbon gas, the higher its Btu value. To illustrate, methane typically represents more than 80% of energy gases. Methane contains one carbon atom per molecule; burning 1 cubic foot of methane gives off 1,012 Btus. Butane, possessing four carbon atoms, has a Btu value more than three times larger than that of methane. Molecular hydrogen, on the other hand, though combustible, contains no carbon atoms; its Btu value is three times smaller than that of methane.

6.3.0 Commercial Natural Gas Combustible Limits

Commercial natural gas is very flammable when sufficient amounts of oxygen (the oxygen present in air is more than adequate) are present. Mixtures of natural gas and air are explosive only when the mixture is less than 15% or more than 4% natural gas. If the natural gas is less than 4%, it will be too lean a mixture to burn, and over 15%, too rich a mixture to burn. The fact that only a small amount of gas is necessary to create an explosive mixture makes even small leaks in confined spaces very serious.

As mentioned earlier, an odorant is added to commercial natural gas so that it can be easily smelled if there should be a leak. However, it is possible for odorants to be filtered out or partially filtered out of commercial natural gas. This could happen when the natural gas passes through a filtering substance (soil or other). The presence of natural gas should be determined by testing the surrounding air with available instruments such as a combustible gas indicator or another suitable instrument.

6.4.0 Commercial Natural Gas Specific Gravity (Weight)

Specific gravity is a term used in the comparison of the weight of a gas compared to the weight of the same amount of air measured under the same temperature and pressure. Some gases may or may not be heavier than air. Air is judged to be the standard weight of one (1) with commercial natural gas weighing only .6 or 6/10 when compared to air. This means that natural gas weighs less than air so natural gas will rise up or actually float from the ground and fade away into the atmosphere when a leak occurs. Also, commercial natural gas will dissipate rapidly unless confined in a building, vault, or other structure. However, propane has a specific gravity of 1.5, is heavier than air, and will sink or fall to low places in buildings or in the ground. Butane is even heavier, weighing 2.0 or twice the weight of air, and will cling to the ground. It can be extremely dangerous when leaks occur.

Commercial natural gas is lighter than air and will normally rise and travel the path of least resistance. This means that a leak might travel uphill underground until it reaches a blockage and then travel back down the hill because that is the easiest path to follow.

7.0.0 ◆ PIPELINE EQUIPMENT

The pipeline system includes all the pipes, equipment, stations, terminals, and other facilities necessary for operation. The major pipeline equipment consists of:

- Pumps and motors
- Compressors
- Valves
- Actuators/operators
- Tanks
- Piping
- **Manifolds**
- Traps
- Measurement/sampling equipment

7.1.0 Pumps and Motors

Pumps provide the pressure and force to move product. Various types of pumps may be used in the pipeline, including centrifugal, positive displacement, hydraulic, and rotary pumps. Each is used in applications requiring different capacities or other pump characteristics. Power is supplied from many different sources, such as turbines, hydraulics, diesel engines, and electric motors. One pump station may use positive-displacement pumps driven by diesel engines, whereas another station may use centrifugal pumps driven by electric motors.

7.1.1 Centrifugal Pumps

Centrifugal pumps work by rotating and forcing product outward as product flows into the center. Revolving blades or vanes create the suction that pulls product into the pump. The product takes up the space between the vanes and rotates with them. It is then pushed outward until it leaves through the discharge, or exit, pipe. *Figure 21* shows how centrifugal pumps work.

7.1.2 Positive-Displacement Pumps

In a typical positive-displacement pump (see *Figure 22*), a piston slides back and forth in a cylinder. With each movement, it traps part of the product to be moved. This product is then moved toward the discharge end of the pump, where the pressure rises as the product is squeezed by the piston. The pressure forces the product out. There are several different types of positive-displacement pumps, including lift, or suction, pumps and force pumps.

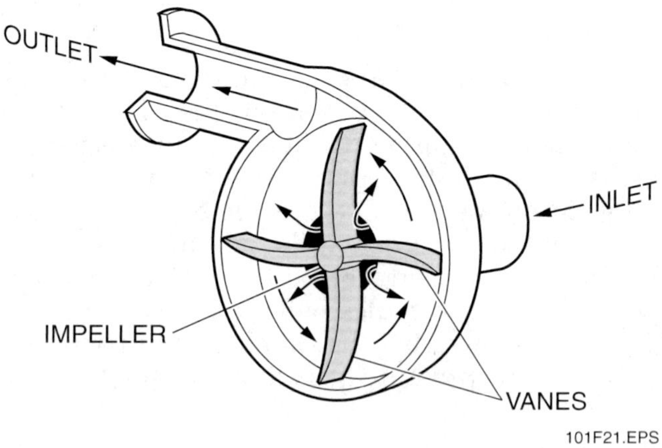

Figure 21 ◆ How the centrifugal pump works.

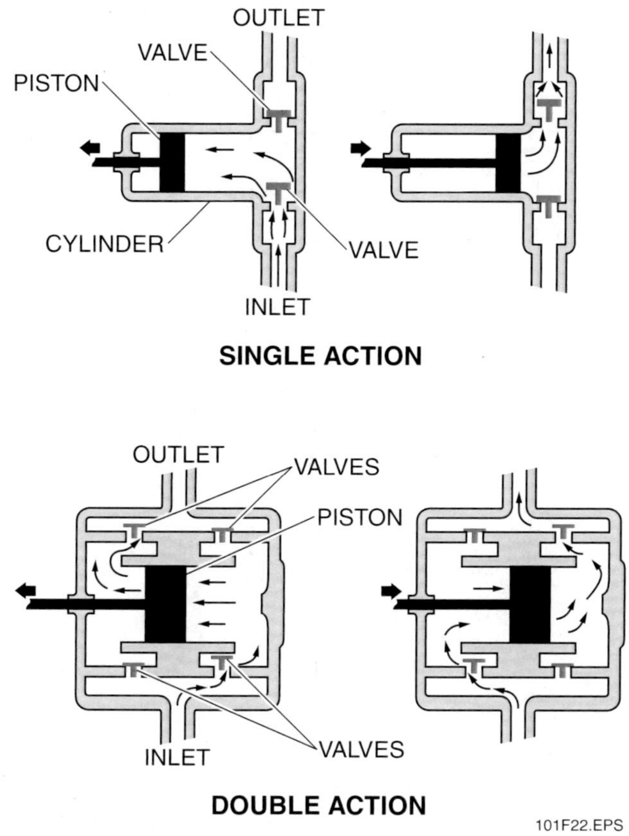

Figure 22 ◆ Two positive-displacement pumps.

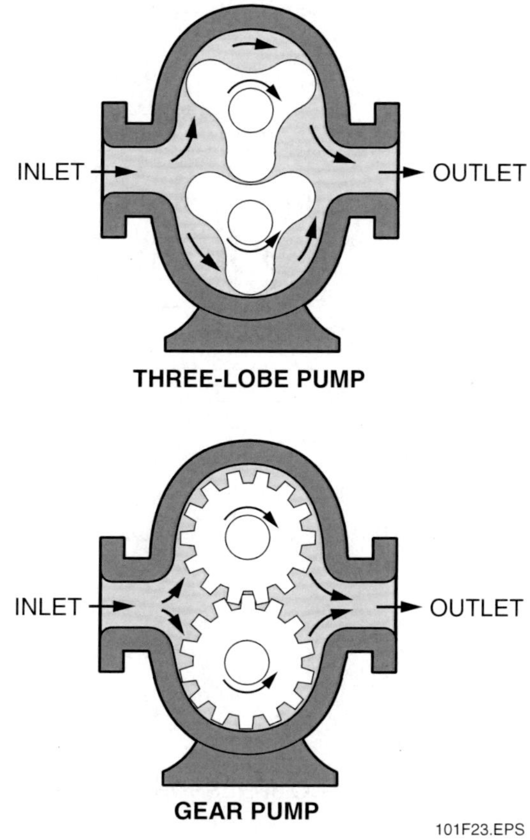

Figure 23 ◆ Two rotary pumps.

7.1.3 Rotary Pumps

Rotary pumps pump air as well as fluids. They have propeller-like blades that rotate on a single shaft. The rotation at the inlet side speeds up the fluid flow in a direction parallel to the shaft, building up pressure between the rows of blades. The pressure moves the fluid to the discharge side of the pump as the blades continue rotating.

A gear pump, which uses gear teeth instead of a piston, is used to pump thick products. The gears rotate in a housing, trapping product between the gear and the casing, moving it to the outlet side. The pressure in the outlet side builds up until it is high enough to discharge the product. Many types of drivers are used to drive these pumps. *Figure 23* shows two rotary pumps.

7.1.4 Drivers

Crude oil, refined liquid products, and gas are pushed through the pipeline grid by powerful pumps and compressors. The size of this machinery depends on the type of pipeline and the products and volumes being transported. Liquids are usually moved through the pipelines by huge pumps, while gases are moved by compressors. These compressors use pressure to push the gas through the pipeline. The pump and compressor drivers may be electric motors with fixed-speed or variable-frequency drive (VFD), diesel or gas engines, or natural-gas-fueled turbines. The choice of driver depends on the power sources available. The most common drive is a fixed-speed electric motor, although variable-frequency drives are becoming more popular. In remote locations, diesel-powered equipment is often used. Diesels are also used as backups during power failures.

7.2.0 Compressors

Compressor stations in a gas line act like pump stations in a fluid line. Compressors boost and maintain the pressure in the line, which keeps the gas flowing. In any compression process, the pressure of gas is increased from one level to another by the addition of energy to the gas being processed.

The centrifugal compressor is classified as dynamic, as opposed to a positive-displacement machine. In a reciprocating compressor, a quantity of gas is drawn into a cylinder, a piston moves, and the trapped gas is compressed by direct volume reduction or positive displacement. By this process, mechanical energy is transformed into pressure energy. *Figure 24* shows a gas compressor.

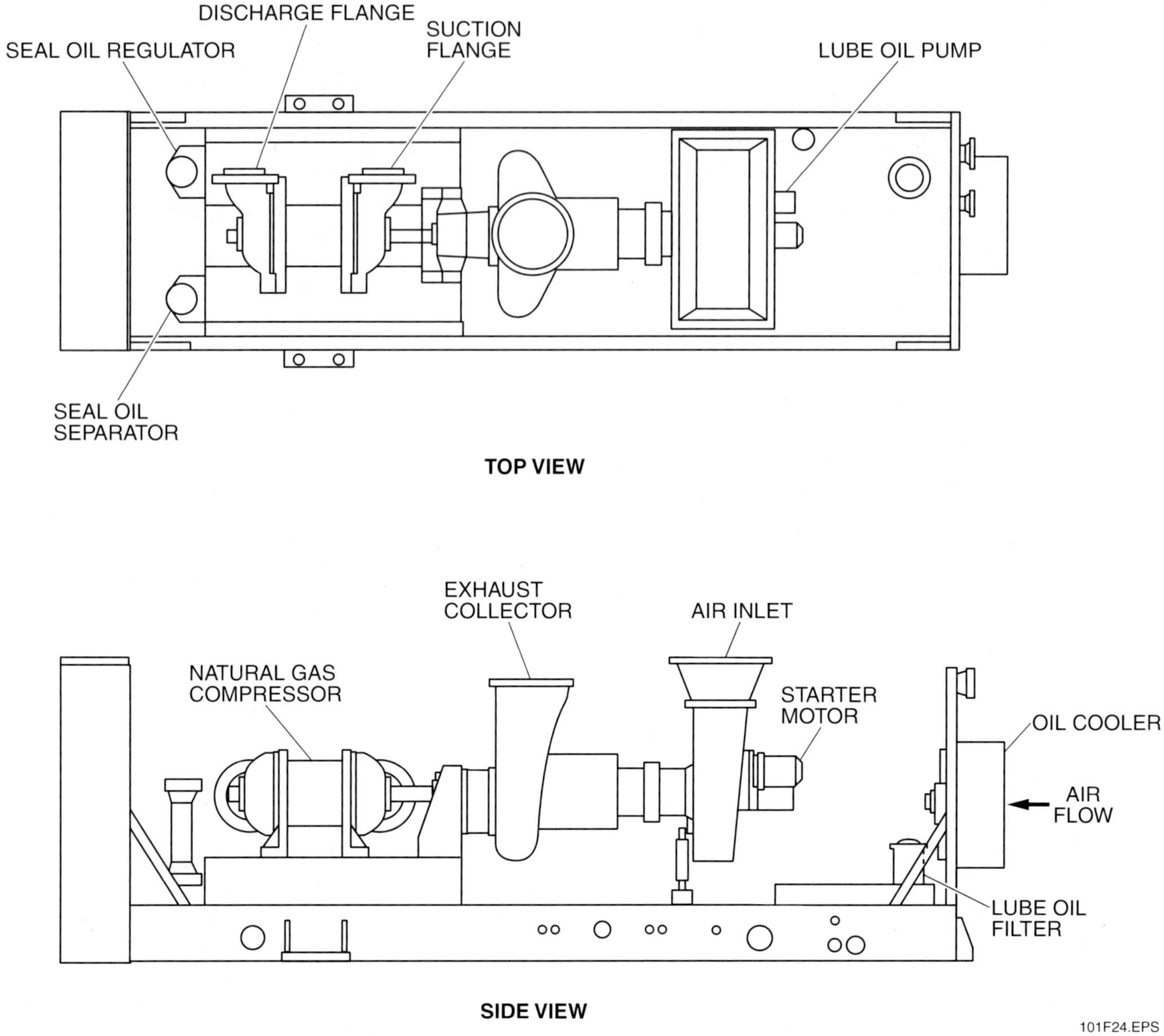

Figure 24 ◆ Gas compressor.

The combination of inlet vane, impeller, and diffuser or diaphragm is referred to as a stage. A single centrifugal stage has a relatively low pressure capability. When larger ratios are desired, additional stages are added in series. The compressor flow can be controlled and varied by changing the pressure differential across the compressor or by changing the power level. Centrifugal compressors apply inertial forces to the gas. The inertial forces are transmitted by an impeller which, by dynamic centrifugal motion, adds velocity energy through acceleration of the gas. The gas flows from the impeller into the diffuser, where the gas decelerates, and velocity energy is transformed to pressure energy.

7.3.0 Valves

Valves are a critical part of pipeline systems. They are used to control the rate of flow in a line, to open or shut off a line completely, or to serve as automatic safety devices. Valves may be identified by type or by function, and can be equipped with locking devices to prevent accidental operation of the valve. Valves may be equipped for manual operation, but most frequently operated valves in the pipeline are equipped with electric-, pneumatic-, or hydraulic-power-driven mechanical operators. Those used extensively include gate, plug, ball, butterfly, check, control, and relief valves.

7.3.1 *Gate Valves*

In a simple gate valve, a wedge-shaped disc or gate is pressed between inclined seats to close the valve. Operating on the same principle but with different configurations of gates and seats, there are other more complex gate valves. Conduit gate valves have an elongated gate and a port in the lower part for a fluid conduit. *Figure 25* shows a simple gate valve.

7.3.2 *Plug Valves*

Plug valves (see *Figure 26*) are similar in design to the stopcocks used to control flow from containers. The plug usually has an elongated port that may be smaller in cross-sectional area than the pipe. The valve is operated by giving the plug a quarter turn. Operation may be by a lever fitted to the stem, by a gear system, or by power-driven mechanical operators. Plug valves are used where reliability and fast operation are important. There is considerable sliding contact area between the plug and the valve body, and as size increases, the operating force required also increases. Many plugs have interior channels and grooves in which lubricants can be forced into the contact areas to improve turning and sealing. Plug valves are generally run in pipe in which scrapers are not run.

7.3.3 *Ball Valves*

Ball valves are an adaptation of a plug valve, using a ball for the plug. The sealing contact area of the ball valve is that of two seal rings. These seals are made practical by use of tough, low-friction, oil-resistant elastomers.

7.3.4 *Butterfly Valves*

Butterfly valves (see *Figure 27*) control product flow by use of a disc mounted on a stem that can be turned to block or allow flow. This simple lightweight valve is used in parts of the pipeline. Butterfly valves have two valuable features. They are quick-operating and take up minimal space.

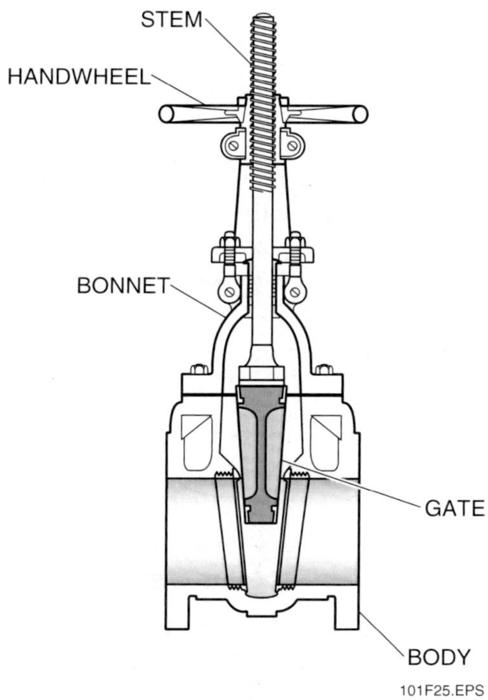

Figure 25 ◆ Simple gate valve.

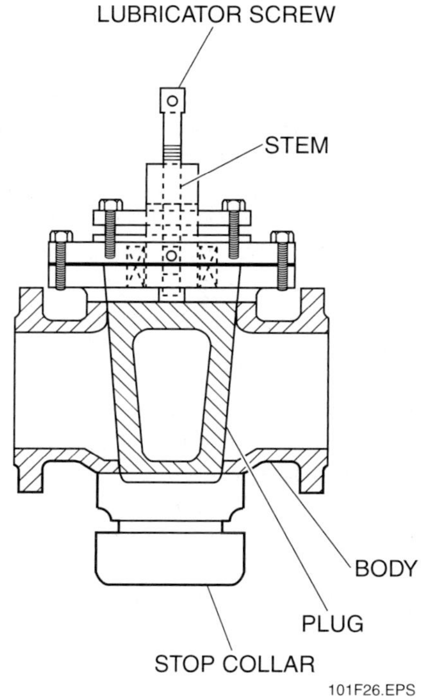

Figure 26 ◆ Plug valve.

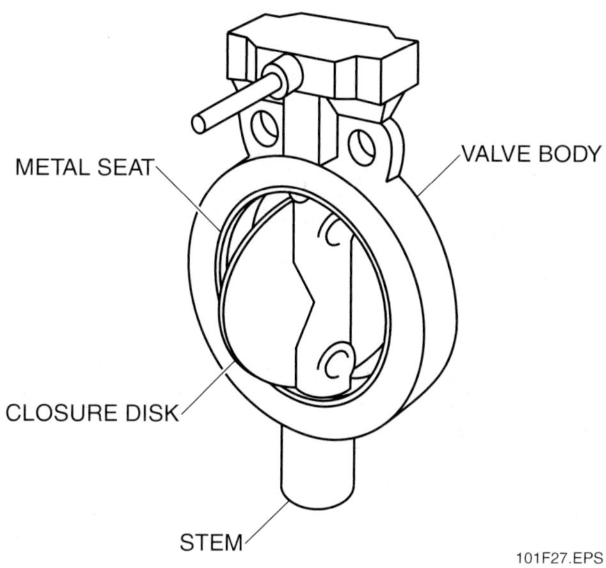

Figure 27 ◆ Butterfly valve.

7.3.5 Check Valves

Check valves (see *Figure 28*) are considered shutoff valves since they permit full flow or prevent flow in one direction. They close automatically when pressure in the flow direction is removed. Check valves can be designed to allow passage of scrapers and are most often used in gathering lines and in station piping. They are often used in conjunction with other shutoff valves.

7.3.6 Control Valves

Automatic control of flow rates and pressures is provided by control valves. They may or may not provide for complete shutoff because flow control is their major function. Port sizes and flow patterns through these valves must be continuously adjusted to provide control by throttling action, increasing their tendency toward damaging vibration, erosion, and corrosion. These valves are used principally at pump stations, although they may be needed at other points in the line.

7.3.7 Relief Valves

Relief valves (see *Figure 29*) may be called safety valves. Their function is to remain closed until pressures rise above design levels. Pressure is released to prevent pressure damage when released products are piped to safe discharge points.

When not opened at set pressure levels, the valves are kept closed by spring pressure, a nitrogen charge, a weight load, or other means. They normally close automatically when the pressure is relieved but may be designed to stay open until reset. They may be operated and monitored remotely.

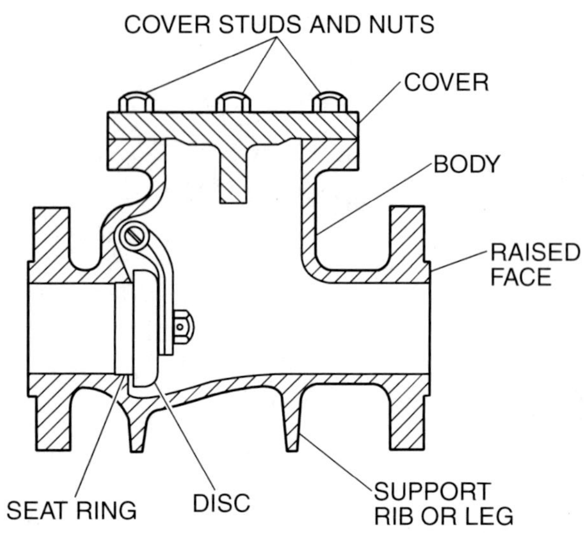

Figure 28 ◆ Check valve.

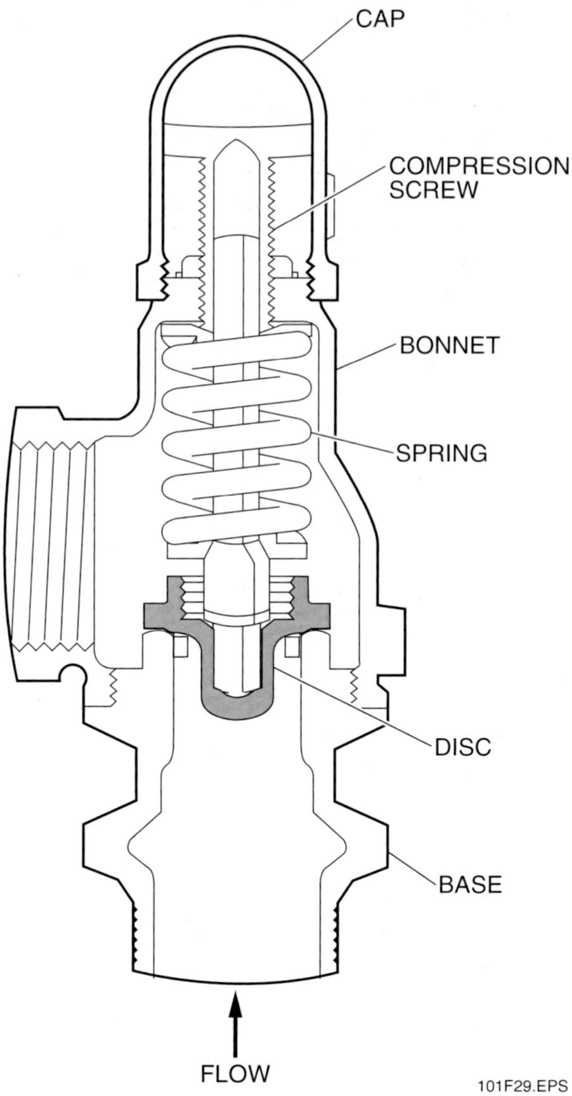

Figure 29 ◆ Relief valve.

7.3.8 Globe Valves

Other valves in the pipeline include globe valves (see *Figure 30*). Globe valves are used for shutoff and throttling, usually in auxiliary systems, such as steam and water systems.

7.3.9 Needle Valves

Needle valves are a type of small, simple, rugged device for throttling and shutting off extremely high pressures compared with other valves. They are used mostly in small auxiliary piping systems.

7.4.0 Actuators/Operators

A valve actuator, commonly called an operator, is a device that opens and closes a valve either manually or automatically. Manual actuators come in the form of handwheels, levers, and wrenches that are attached to the stem. Larger valves that require great effort to turn must be equipped with some type of powered mechanism, including electric, pneumatic, or hydraulic units, mounted on the stem. Some valve actuators use a gearbox with wide reduction ratios. Some manual actuators employ high and low gear ratios to enable personnel to turn the valve.

Valves with automatic actuators are controlled from a console in a control center by remote control, sometimes from many miles away. Such automatic actuators are used when the valve is remote from the main work area, when the frequency of operation would require unreasonable human effort, or if rapid opening and closing of the valve is required. Maintenance crews often carry portable valve actuators to work in outlying installations.

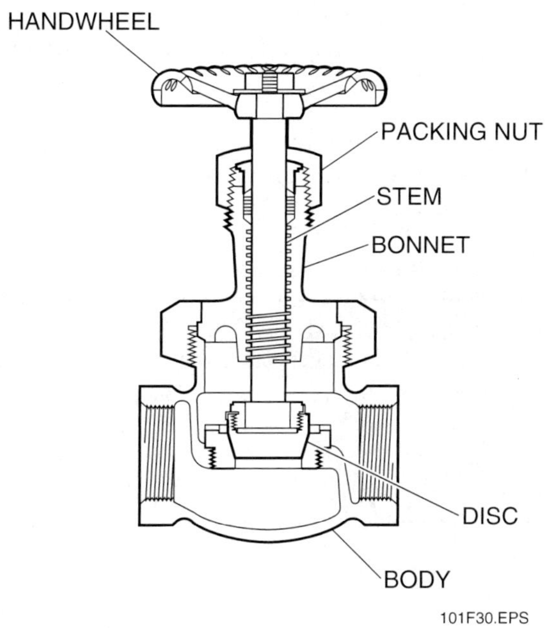

Figure 30 ◆ Globe valve.

7.5.0 Tanks

Tanks are containers used for liquid storage, relief, and breakout. Tanks may exist along the length of the pipeline, at the production site, at pump stations, at terminals, and at delivery points. Although the types and sizes of tanks at each of these sites may vary, they are usually round and made of steel. All atmospheric storage tanks are constructed per *API 650* and repaired per *API 653*.

Tanks must be calibrated for capacity. The dimensions are strapped, or measured, to be used in strapping tables. From these tables, the volume of liquid contained in the tank can be determined. Tank gauges show the tank number and gauge reading from which corresponding volume in barrels can be determined.

Cylindrical tanks may have flat floating roofs, fixed cone-shaped roofs, or both. The first can be used for gasolines and aviation-grade kerosenes, while the second can be used for fuel oils. Some tanks have fixed geodesic domes covering a floating pan. Spheroid tanks are used for volatile products because of their high vapor-pressure capacity.

The floating roof tank (see *Figure 31*) is designed to reduce losses through evaporation of volatile components of crude oil and to reduce the corrosion from sour crude by eliminating the vapor space between the oil and the roof. Seals around the perimeter of the roof prevent vapor loss and rain contamination. The wind girder on the outside rim of the tank adds strength to the shell.

Tanks must also be maintained and repaired per industry standards. Although a mixer helps to keep sediment in suspension in the product, sludge forms in the bottom and must be cleaned out periodically. The tanks are repainted inside and out at specific intervals. Structural inspections and preventive maintenance guard against internal and external damage.

7.6.0 Pressure Vessels

Just as tanks are used for the storage of liquid products, pressure vessels are used along the gas pipeline for the storage of gas products. Liquid vessels are typically cylindrical, but like tanks, can vary in size and shape from site to site, depending on application, capacity, and pressure ratings.

Cylinders and pressure vessels can be fabricated of steel, aluminum, or composite materials. They must be manufactured, inspected, marked, tested, and retested in accordance with United

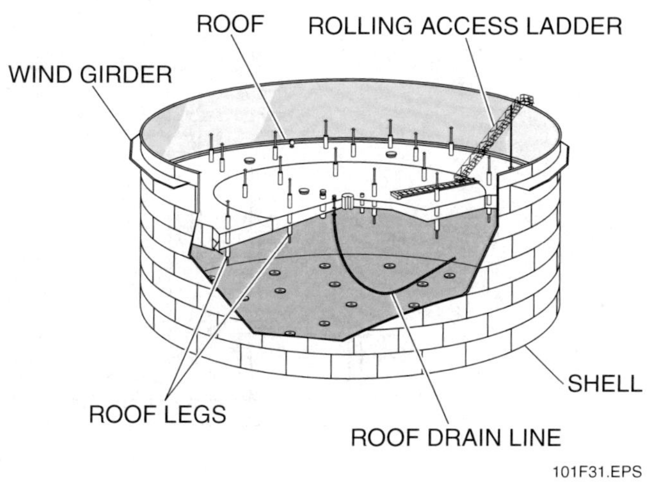

Figure 31 ◆ Open floating roof tank.

States **Department of Transportation (DOT)** regulations and exemptions for the type of service that they are used for. Pressure vessels must also be designed and repaired in compliance with *API Standards 510 and 2510*. These regulations cover the manufacture, marking, maintenance, repair, reconditioning, and testing of seamless pressure vessels, and they prescribe the requirements for transportation of hazardous materials within vessels. *Figure 32* shows a pressure vessel.

7.7.0 Pipe

Line pipe is made of steel through various methods. The main types of pipe used in the oil and gas industry are seamless, electrical resistance welded (ERW), and double submerged arc welded (DSAW). Other piping, less commonly used now, is welded steel pipe.

Seamless pipe is made from a solid length of metal pierced lengthwise by a mandrel with a rounded nose. The manufactured steel tubes and pipes are used in water, gas, and crude oil transmission and distribution lines and as casing pipes in the petroleum industry.

ERW is the most common form of manufacture for pipe in sizes from 2 to 22 inches outside diameter (OD). While seamless pipe is traditionally stronger and more expensive than comparable ERW pipe, ERW technology is improving and being used more and more frequently.

DSAW pipe is pipe that has longitudinal or spiral butt joints. The joints are produced by at least two passes, including at least one on the inside and one on the outside of the pipe. Pressure is not used, and filler metal for the inside and outside welds is obtained from the electrode or electrodes. This pipe is primarily intended for use in the gas and oil transmission industry.

Welded steel pipe is made by bending strips of steel into the form of a tube and welding the longitudinal seam either by electric resistance, by fusion welding, or by heating the tube and pressing the edges together. Two types of older piping that are still found on pipelines are furnace lap-welded and furnace butt-welded piping.

7.8.0 Fittings

Fittings include elbows, tees, reducers, flanges, plugs, and unions. Elbows are connections for directional changes. Tees are used to join three pipe ends. A reducer is used to connect pipes of different sizes. Flanges provide a means of opening and closing piping and attaching pipe ends to mating flanges on valves, tanks, pumps, and other equipment. Plugs and caps close the end of a pipe. A union provides a means of opening and closing piping and permits closing of circuitous piping systems. *Figure 33* shows types of fittings.

Fittings are made in a range of materials based on their service requirements. *Table 3* shows a partial list of materials and their pressure ranges.

Table 3 Materials and Pressures

Material	Pressure
Cast Iron	150 psig
Malleable Iron	250 psig
Cast Steel	150–2,500 psig
Forged Steel	1,000–6,000 psig

101T03.EPS

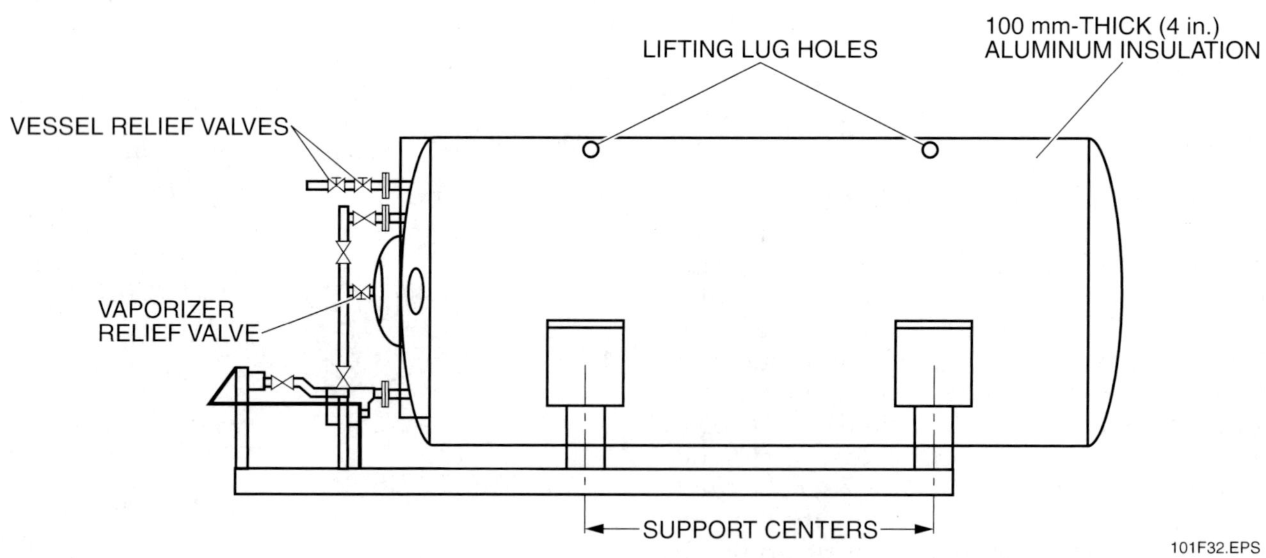

Figure 32 ◆ Pressure vessel.

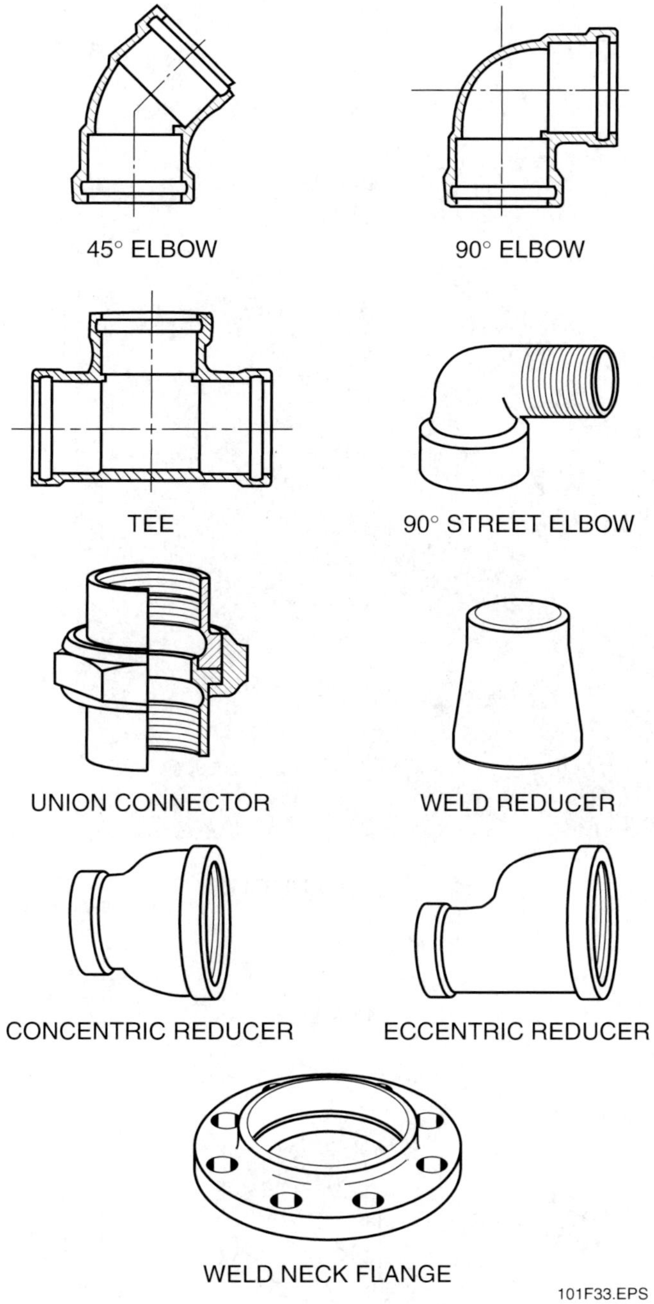

Figure 33 ◆ Types of fittings.

7.9.0 Manifolds

The pipeline uses manifolds to control the flow path of oil, gas, and other petroleum products. Manifolds, with valves and meters, branch off of the line to divert specific volumes of certain products to different locations or other pipelines. *Figure 34* shows a delivery manifold.

The complexity of manifold piping ranges from very simple to very complicated. A booster station manifold may merely allow bypassing of the station. A more extensive manifold at an originating station may connect the main line, field-gathering lines, tanks, and pumps. This manifold may perform any or all of the following operations:

- Pumping product through the main line with tanks inactive
- Receiving product from the field into any tank
- Delivering from the main line into any tank
- Transferring from one tank to another
- Pumping product from a tank into the main line
- Isolating tanks from main-line flow
- Injecting product from any tank into the main line

7.10.0 Traps

Traps in the pipeline are designed to launch or catch scrapers and other pigs that are inserted in the pipeline. Valves and/or manifolds are used to isolate the trap for scraper insertion or removal and then to realign the trap to the pipeline flow. *Figure 35* shows incoming and outgoing scraper/pig traps.

Automation of operation has been extended to the launching of scrapers in some instances. Some traps are arranged so that one or more scrapers may be loaded and then launched at any desired time by remote-controlled valve changes.

7.11.0 Measuring/Sampling Equipment

All volumes entering, leaving, and passing through the pipeline system must be measured and accounted for. Accurate measurement is required for custody transfer, task purposes, leak detection, and delivery schedules.

The primary method of measurement in the pipeline is metering. Meters, whose accuracy is documented with meter provers, measure and record the volume of product passing through. Factors affecting volume are the temperature, pressure, and density of the product being measured. Various instruments are used to measure these factors concurrently with the volume measurement. Product characteristics are also checked periodically with automatic and manual sampling.

7.11.1 Meters

Various meters are used along the pipeline to monitor the volume of fluid or gas. Most product measurements are made by turbine meters or positive-displacement meters, although coriolis-type mass flowmeters, vortex, and ultrasonic meters are also used. Meter readings taken before

Figure 34 ◆ Delivery manifold.

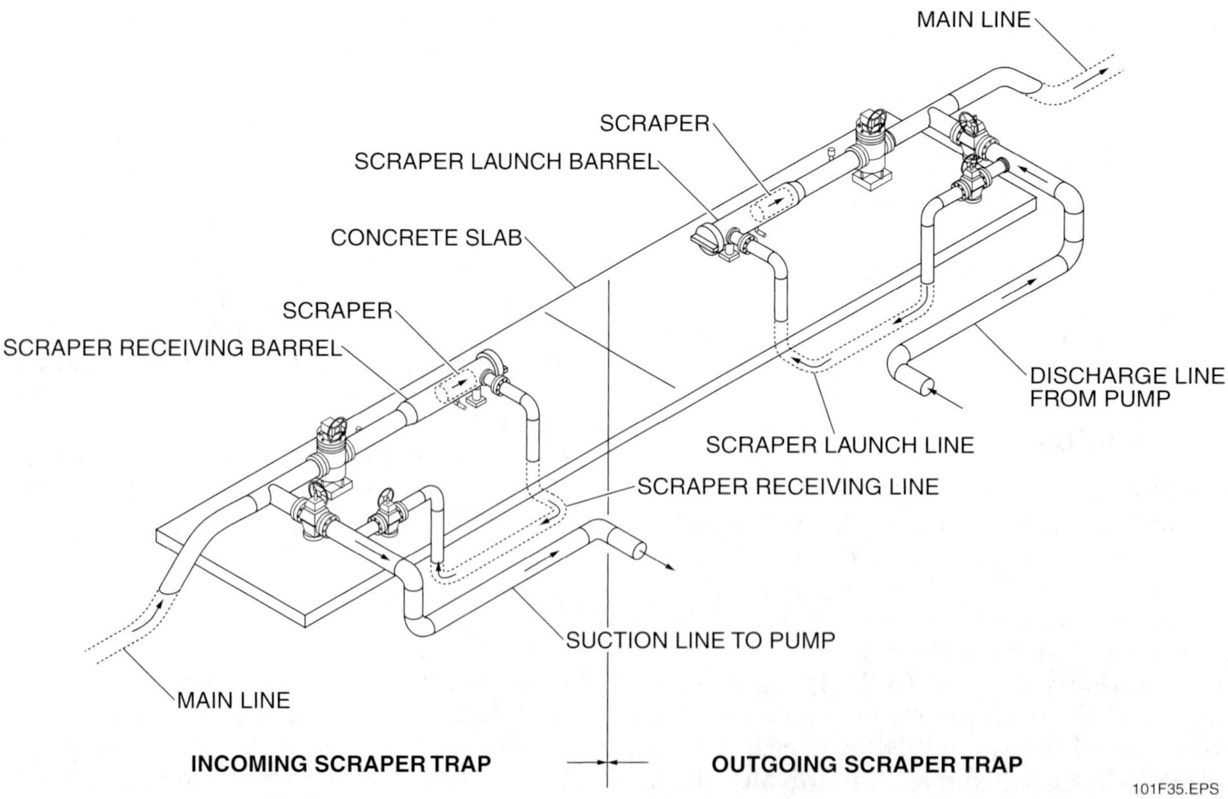

Figure 35 ◆ Incoming and outgoing scraper, or pig, traps.

and after deliveries are used to determine the volume delivered, or **throughput**. Meters are typically equipped with transmitting, remote-reading, and print-out accessories to facilitate the delivery of correct product volumes and for preparation of delivery tickets. Data is often transmitted to remote locations. *Figure 36* shows two types of meters.

Gas is typically measured by orifice, turbine, positive-displacement, ultrasonic, and rotary vane meters. These meters are generally automated to correct gas volumes for moisture content and temperature and to provide remote read-out and printing capability. Strainers are installed upstream of meters to remove any foreign material from the stream before it goes through the meter. The wire mesh in strainers is generally more coarse than filter materials are.

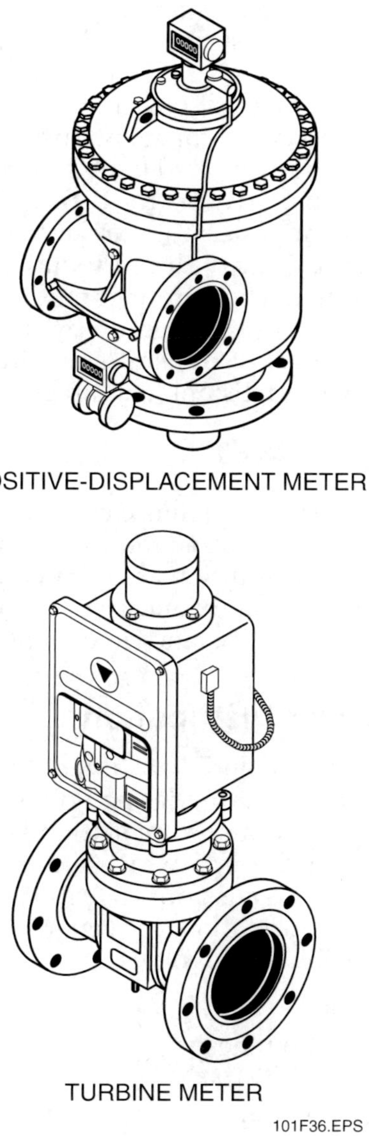

Figure 36 ◆ Two types of meters.

The orifice plate is the vital component of the orifice meter. It is the part that restricts the flow to develop the difference in pressure that is measured to calculate actual flow. The accuracy of orifice meters is dependent on the quality of the meter that is installed. Expert installation and maintenance are necessary to ensure the continued accuracy of the meter.

Orifice plates are secured between halves of flanges of machined carriers or in the body of a special orifice meter housing constructed so that the plate can be removed for inspection. Some meters allow the plate to be removed without stopping the flow. The orifice meter is made up of the carrier housing or flanges, the orifice plate, gaskets, O-rings, devices for removal, and pressure fittings. Pressure taps are drilled at a location exactly 1 inch on center from either side of the orifice plate. *Figure 37* shows some common orifice plate holders designed for easy removal.

Visual inspection of the housing unit involves close examination of the unit looking for any signs of rust, corrosion, cracks, or any buildup of foreign material around the outside housing joints, suggesting the possibility of a leak. Pressure fittings and sensor connections must also be inspected for leaks or damage of any kind. Plates vary from 2.5 inches to 30 inches in diameter.

7.11.2 Provers

A meter prover is used to calibrate meters for specific conditions of flow and product type. Meters are proved by passing a volume of product through the meter at a known rate and comparing this metered volume with the known volume prover. Two types of provers frequently used are the ball prover and the piston prover. A prover loop is a length of pipe or a calibrated U-shaped pipe with a known volume used to check or prove the accuracy of a meter.

8.0.0 ◆ ELECTRICAL POWER SYSTEMS

The electrical power system used by any pipeline company may be quite large or quite small. Power may be self-generated, or partially so, or may be purchased from another source. For the purposes of an introduction, a basic power system will be described from generation to delivery.

An electrical power system is an assembly of equipment and circuits for generating, transmitting, and distributing electrical energy. The elements of a power system form complex networks that require energy control centers to monitor and regulate their operation. The reliability of the system is critical to pipeline operation. *Figure 38* shows an electrical power system. The principle elements of the system can be divided into the following:

- Power generation and transmission
- Power distribution
- Direct current (DC) power distribution
- Utilization equipment

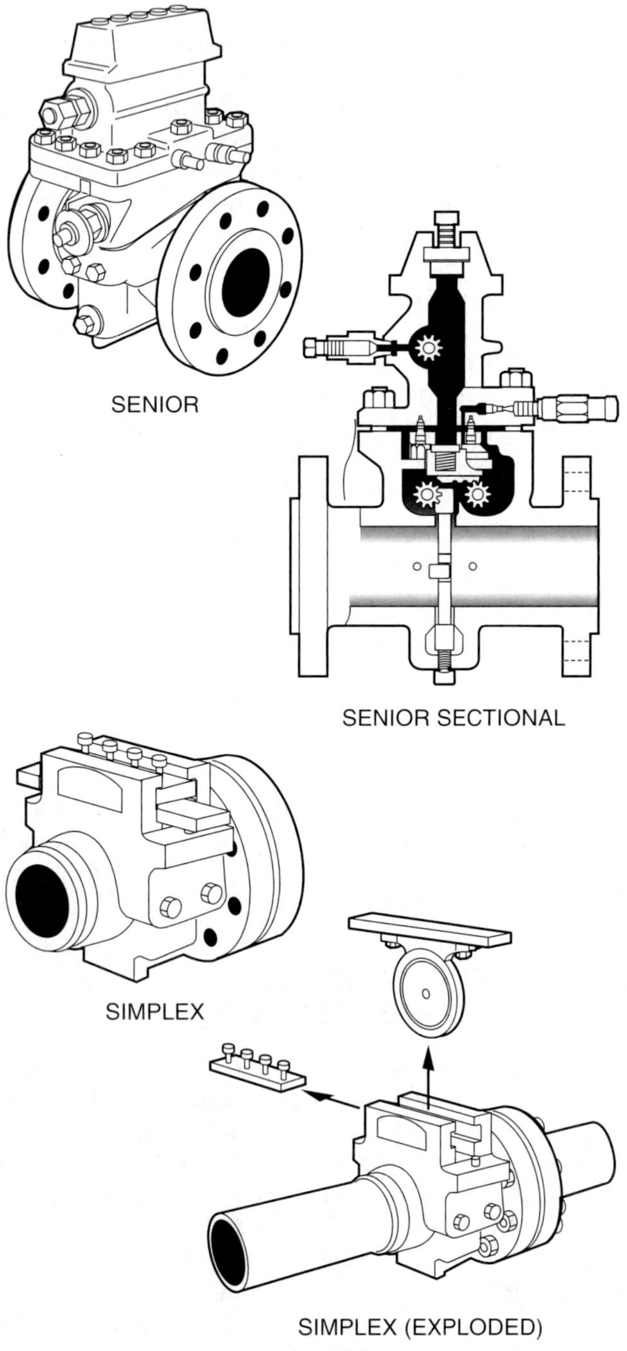

Figure 37 ◆ Common orifice plate holders.

8.1.0 Power Generation and Transmission

Electrical systems require generating stations to produce electric power, and transmission systems to carry it to areas of consumption. Along the way, the voltage of the power delivered must be adapted to the specific needs of consumers. The myriad of uses that consume this energy is the power system load.

8.1.1 Generation

The large quantities of electricity required to supply a system are produced in generating stations, commonly called power plants. These are actually conversion facilities where the chemical or nuclear energy of fuel, such as coal, oil, gas, and uranium, or the kinetic energy of falling water is converted to electricity.

The generators, which supply the electricity, can be driven by steam turbines, hydraulic turbines, gas turbines, or internal combustion engines. Coal, natural gas, and heavy fuel oil are commonly used fuel for steam turbines.

Three-phase 60 hertz (Hz) AC systems are used almost exclusively in the U.S. Consequently, power system generators are wound for three-phase output at a voltage usually limited by design features. The output of modern generating stations is usually stepped up by transformers to the voltage level of transmission circuits used to deliver power to distant load areas.

8.1.2 Transmission

The transmission system carries electric power efficiently and in large amounts from generating stations to consumption areas. Such transmission is also used to interconnect adjacent power systems for mutual assistance in case of emergency.

8.2.0 Power Distribution

Power is reduced and distributed through a series of substations to provide the voltage required in specific applications. A substation is an assembly of equipment through which electrical energy is passed for transmission, distribution, interconnection, transformation, conversion, or switching. Power distribution involves:

- Substations
- Disconnection means
- Transformers
- Cables/conduits
- Switchgear
- Motor control centers

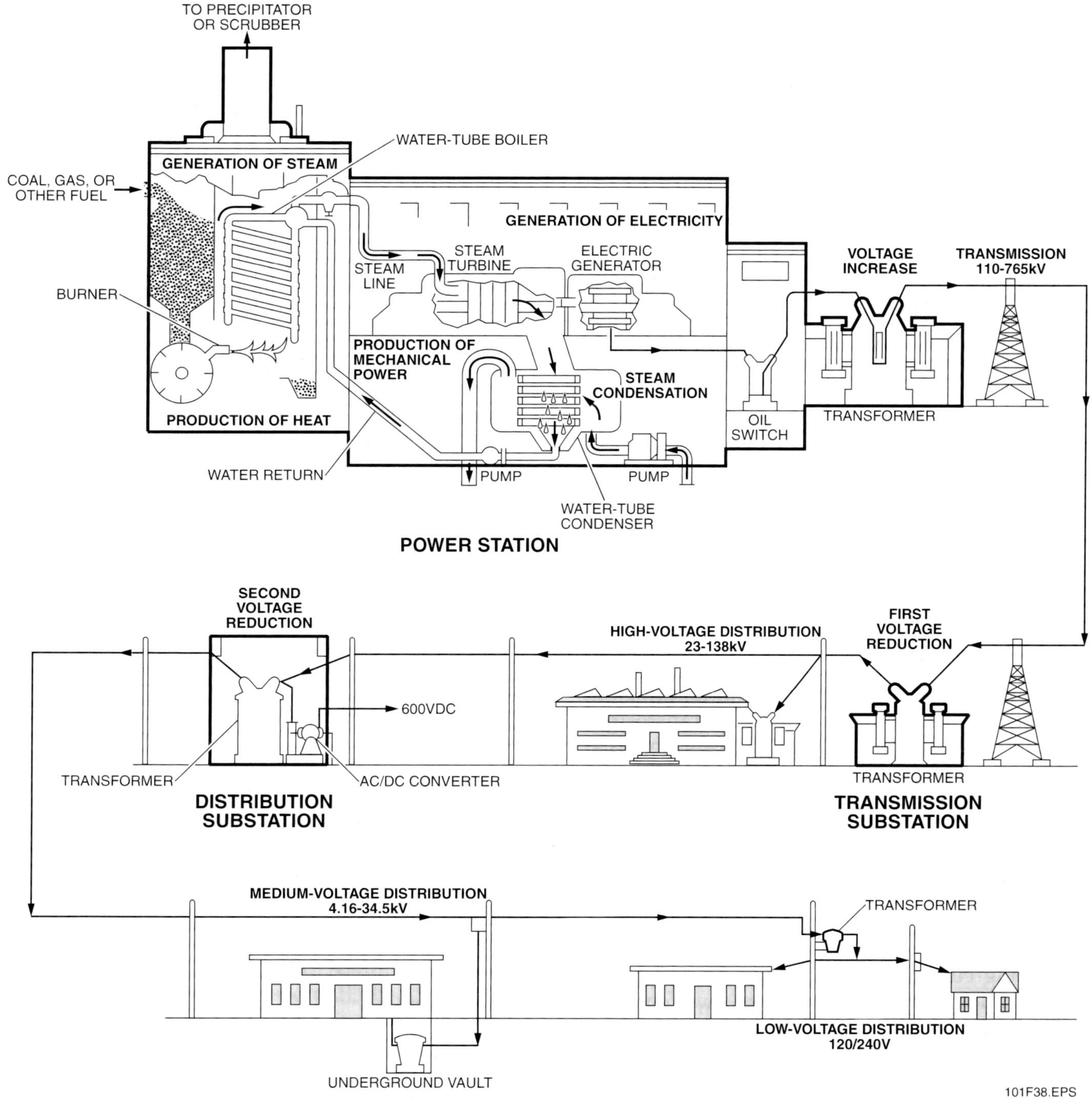

Figure 38 ◆ Electrical power system.

8.2.1 Substations

Substations are used for some or all of the following purposes:

- Connecting generators, transmission or distribution lines, and loads to each other
- Transforming power from one voltage level to another
- Interconnecting alternate sources of power
- Switching for alternate connections and isolating failed or overloaded lines and equipment
- Controlling system voltage and power flow
- Compensating for reactive power
- Suppressing overvoltage
- Detecting faults, monitoring, and recording information, power measurements, and remote communications

9.2.0 Inspection and Testing

Pipeline integrity programs use tools, technologies, and strategies to ensure that pipeline systems are in a condition to perform safely and reliably. Increasingly sophisticated technology is available. Constant monitoring and inspection is essential to any pipeline integrity program.

9.2.1 Microbiologically Influenced Corrosion (MIC)

A variety of methods is available to detect MIC. Sampling and culturing can be performed for identification of the size of the infection and the active strains of organism present. This information is then used to identify preventative or remedial treatments. Chemical analysis and pH of corrosion products can also give important information relative to the nature of the problem. Evidence taken from the metal surface, called sessile surveys, has a better chance of properly assessing the situation versus sampling of the liquids, called planktonic survey.

9.2.2 Sampling

Injection and sampling systems are fundamental to corrosion control and process control programs. They are applicable to a large variety of processes in the petroleum, chemical, and water treatment industries. Injection systems are used to inject a wide range of chemicals into processes. Such chemicals include biocides, demulsifiers, corrosion inhibitors, oxygen scavengers, glycol and monoethylene glycol, dewaxers, methanol, odorizers, and product additives.

Sampling systems are used to take samples of the process product. Such samples are then analyzed in the laboratory for inhibitor concentration levels, the presence of metal ions, oxygen levels, scale forming compounds, and a wide range of other substances.

9.2.3 Corrosion Coupons

Coupons are specimens of materials that are exposed to tests or real environments to assess the effect of degradation on the material. They can be used inside a pipeline or outside if the pipe is underground. Corrosion coupons represent a widely used and relatively simple form of corrosion monitoring. Usually a coupon is exposed to a corrosive environment to represent an actual component/structure. After exposure for a specified amount of time, the coupons can be weighed and examined microscopically to determine and evaluate the type and average rate of corrosion damage. *Figure 39* shows typical coupons.

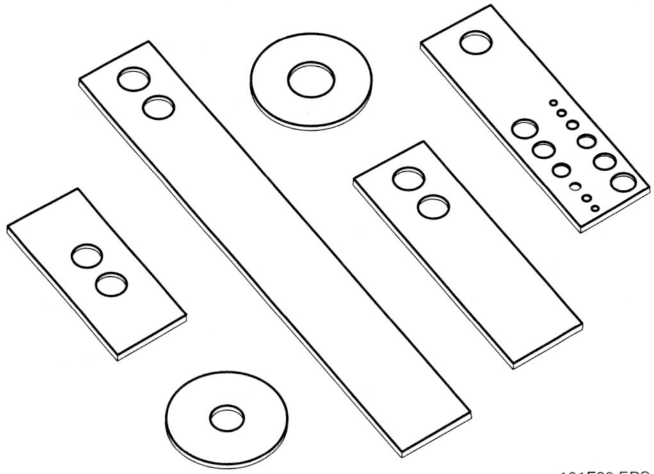

101F39.EPS

Figure 39 ◆ Typical coupons.

9.2.4 In-Line Inspection, or Smart, Pigs

Smart pigs are high-tech inspection tools covered with sensors that gather and record information in onboard computers as they pass through the pipeline. They check for various problems, such as thinning, pitting, and general corrosion. Some smart pigs can detect metal loss in areas around welds. *Figure 40* shows an ultrascan in-line inspection (ILI) pig.

9.3.0 Maintenance and Protection

Monitoring a pipeline for corrosion is essential in preventing failures. If an operator can identify active or growing corrosion defects, then the rate of corrosion can be individually predicted for each defect, allowing an opportunity for corrective measures before the failure occurs.

COURTESY OF PII GROUP LTD

101F40.EPS

Figure 40 ◆ Ultrascan in-line inspection pig.

As soon as production begins in a line, companies begin maintenance programs to reduce the risk of pipeline leaks. Sophisticated electronic control systems monitor changes in pressure, volume, and product flow. A variety of highly technical tools collect information about both the interior and exterior condition of pipelines. In-line inspection (ILI) tools, commonly called smart pigs, assess metal loss, strain, and cracking. This information becomes part of a computer database that is used by experts to identify and plan long-term maintenance programs.

To safeguard the outside surface of the pipe, protective coatings are applied when lines are built. In all cases, as a backup to the protective coating, cathodic protection is installed to help control external corrosion.

9.3.1 Inhibitors

A corrosion inhibitor is a chemical substance that, when added in small concentration to an environment, effectively decreases the corrosion rate. Inhibitors are chemicals that react with a metallic surface or with the environment the surface is exposed to, giving the surface a certain level of protection. Inhibitors often work by adsorbing themselves on the metallic surface, protecting the metallic surface by forming a film. Inhibitors are normally distributed in a solution and injected into a pipeline through a sampling/injection tool. Some are included in a protective coating formulation. Inhibitors slow corrosion processes by one of the following methods:

- Increasing the anodic or cathodic polarization behavior
- Reducing the movement or diffusion of ions to the metallic surface
- Increasing the electrical resistance of the metallic surface

A vapor corrosion inhibitor stops corrosion by releasing protective ions that are attracted to metal. These ions condense on the metal surface, forming a water-repellent layer that contains both positively and negatively charged ions. The layer is maintained and repaired by the constant condensation of new vapor.

Corrosion inhibitors used in deep-water treating of oil and gas wells and deep-water risers differ from the use of land-based inhibitors in several ways, including delivery of the chemical to the point of injection, selection of inhibitor, and monitoring for results.

Each situation must be carefully examined before the proper inhibitor is selected.

9.3.2 Cathodic Protection

Cathodic protection effectively protects underground or submerged metallic structures through the use of a negative potential applied by an external source to the pipeline or structure. Usually, once the pipeline has been made sufficiently negative, it will resist environmental (soil or moisture) corrosion. The method is typically applied to underground pipelines, storage tanks, the interior of water storage tanks, ocean pilings, and electrical transmission towers. Buried steel structures will revert back to their natural state as an iron oxide without proper protection.

Cathodic protection is the most widely applied electrochemical corrosion control technique that reduces the corrosion of a metal surface by making that surface the cathode of an electrochemical cell. Cathodic protection is accomplished by creating and controlling the action of large corrosion cells by connecting the negative terminals of DC power sources to the pipeline and the positive terminals to groundbed anodes (see *Figure 41*).

10.0.0 ◆ REGULATIONS, PROCEDURES, AND DOCUMENTATION

Each pipeline company has published operations and maintenance (O&M) and emergency procedures. These procedures are based on federal, state, and local regulations and on industry standards. Regulations have the effect of laws and are enforced by government agencies. Standards are generally accepted practices gained through experience. All pipeline employees are expected to become familiar with all procedures that are relevant to their job.

10.1.0 Regulations

The U.S. Department of Transportation (DOT) Office of Pipeline Safety also handles pipeline safety and the quality of pipeline design, construction, and operation. This office administers two sets of regulations, *Code of Federal Regulations (CFR) 49 Part 192* and *CFR 49 Part 195*, that address transportation of natural gas and other gases and liquids by pipeline. The pipeline industry is greatly affected by federal regulations. The Federal Energy Regulatory Commission (FERC) controls the natural gas and crude oil and products lines in addition to the regulations of other federal agencies. The Occupational Safety and Health Administration (OSHA) oversees every facet of worker safety and health, from the wearing of personal protective equipment to provision of hazard information to availability of emergency equipment.

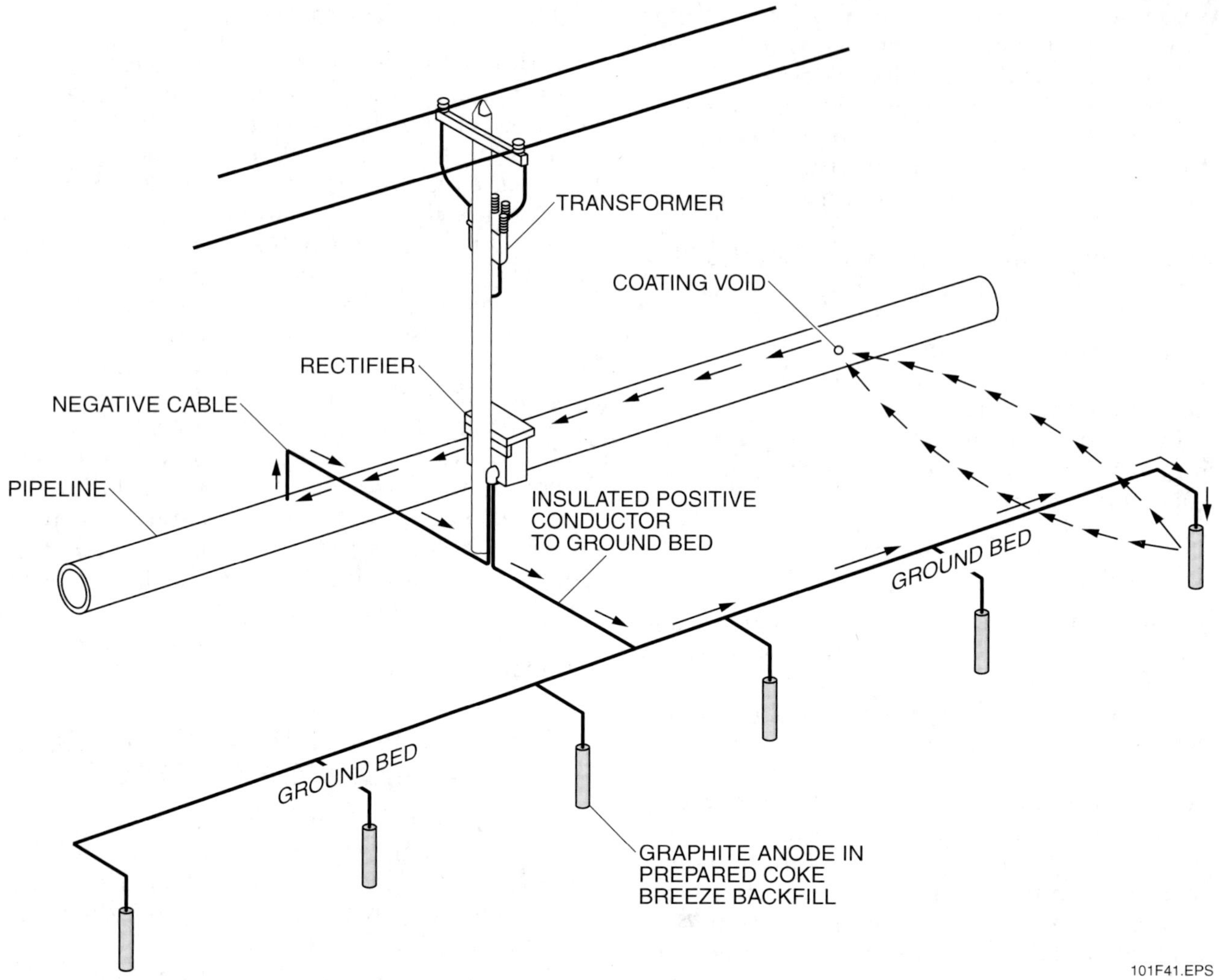

Figure 41 ◆ Cathodic protection arrangement.

State agencies can adopt federal regulations or enforce more stringent regulations than federal because each state has developed their own form of regulation regarding pipelines. Pipelines operate on an interstate or intrastate basis but not both. Interstate pipelines are located through more than one state and fall under federal jurisdiction and classification. Intrastate pipelines are located in a single state, and different states may classify the pipeline differently. Procedures should be relevant to the states involved.

10.2.0 Abnormal Operating Conditions

Abnormal operating conditions (AOCs) are specified by federal law in *49 CFR Part 195.503 and CFR 192.803*. The Operator Qualification Rule requires that an individual recognize and respond to any AOC that could happen while performing a task. An AOC refers to any condition identified by the individual that may indicate failure of a component, or deviation from normal operations that may indicate a condition exceeding design limits or that may result in a hazard to persons, property, or the environment. The following list identifies the AOCs that can occur when operating a pipeline system:

- Abnormal facility condition
- Activation of a safety device
- Communications failure
- Control system failure
- Power interruption
- Fire, explosion, or natural disaster
- Pipeline system damage
- Unexpected hazardous liquid or carbon dioxide encountered
- Unexplained flow rate deviation
- Unexplained pressure deviations
- Unexplained status change
- Other company specific circumstances

10.3.0 Company Policies and Procedures

Company policy and procedure manuals typically include:

- Operating procedures manual
- Maintenance procedures manual
- Safety procedures manual
- Emergency procedures manual

Each procedure will typically include procedure details, purpose, responsibilities, and other supporting information. There are many different requirements depending on the company and the location of the pipeline.

10.4.0 Qualifications

With the introduction of the new Operator Qualification Rule in *49 CFR 195.503*, pipeline operators are now required to have a written qualification program. Required elements of the qualification program include:

- Identification of covered tasks
- Qualification in the performance of covered tasks through operator evaluations
- Evaluation measures for individuals following a DOT-reportable accident or for other reasonable causes
- Communication processes
- Identification of intervals at which reevaluations will occur

10.5.0 Documentation

Documentation is an important part of the pipeline operation from origin to end point. The DOT requires that many operations and maintenance procedures be documented to record their correct and timely performance. Problems must be documented to track down the cause and to prevent recurrence.

Some documents are forms that must be completed by operations personnel and technicians. Others are forms, reports, and summaries that are printed out periodically through the control system. Some of the documentation required for operations includes:

- Accident/incident reports
- Operations logs
- Work orders
- Daily events/event logs
- Personnel qualifications
- Inspection and repairs

10.5.1 Accident/Incident Reports

Accident/incident reports are required, as specified in the Office Pipeline System (OPS) regulations, whenever there is a reportable incident along the pipeline. An incident is defined as an unexpected situation that causes, or could be expected to cause, an operational problem, harm to personnel, or damage to the environment. Reportable situations are defined in the regulations.

10.5.2 Operations Logs

Operations logs record the daily events of operation. These are usually summarized on reports printed from the control system; they include all activities communicated to and from the control room as associated with the pipeline.

10.5.3 Work Orders

Work orders and/or work permits are issued when specific maintenance or other tasks need to be performed. Each work order or permit has its own number and is recorded on a work order or permit log along with a description of the job and other details of the assignment.

These forms are required to ensure systematic maintenance of the pipeline, to assist with scheduling, and to record work done. Work orders and schedules for tasks done at specific time intervals can be triggered by the computer system. Not all pipeline companies use a work order system.

10.5.4 Daily Events and Event Logs

Daily events are recorded in the event log as specific tasks or activities that are performed during the day's operations. These might include repairs, inspections, AOCs, investigations, maintenance, or many other events.

10.5.5 Personnel Qualifications

The OPS regulation provides for formal documentation of the qualifications of personnel responsible for pipeline operations and maintenance in each company. The regulation was designed to reduce the risk of accidents on pipeline facilities caused by human error. Each company must have a written program specific to its own system for the qualification of individuals, including employees, contractors, and subcontractors who work on that system. These qualifications often cover about fifty different safety-related tasks, usually by position, such as pipeline controllers, and corrosion, maintenance, and operations personnel. The rule requires that a number of specific elements be present in all operator plans, including:

- Personnel must demonstrate the ability to recognize and respond to abnormal operating conditions.
- Companies must keep records that demonstrate compliance to the OPS.

10.5.6 Inspection and Repairs

When repairs are made, the pipe specifications, defect description, stress levels, and action taken must be documented and maintained in case of an audit. Other information, in accordance with the requirements of regulatory authorities, must also be prepared. Some documents are forms that must be filled in by operations and maintenance personnel. Others are forms, reports, and summaries that are printed out periodically through the control system.

Federal guidelines require that all inspections, treatments, and repairs be documented. Some of the documentation required for corrosion control includes:

- Cathodic protection system inspections
- In-line inspection results
- Pipeline coating inspections
- Incident reports
- Maintenance reports

Record-keeping is a huge and important task. Tremendous amounts of inspection and monitoring data are generated over the life of a pipeline, including:

- Development of written pipeline integrity procedures and coordination with operation, maintenance, contingency, and emergency response procedures
- Identification and implementation of repairs and modifications
- Program evaluation

Many software programs are now available to make record-keeping easier and to help companies stay in compliance with regulations. Failure to maintain good documentation can result in significant penalties for your company.

Summary

The term *pipeline* has many meanings. In the simplest terms, it applies to the pipe and fittings used to construct facilities for transporting commodities. In a broader sense, it applies to all of the facilities in addition to the pipe and fittings that are needed to transport the commodities from the point of origin to the point of delivery. Pipeline facilities include the means of measuring, controlling, and accounting for quantities and qualities of product delivered from the custody of the original transporter to the shipper.

In order to efficiently operate and maintain the pipeline facilities, personnel must understand the basic principles of pipeline hydraulics when working with liquid pipelines, and pipeline pneumatics when working with gas pipelines. Personnel must also familiarize themselves with basic pipeline equipment and its proper use.

Electrical power systems supply power to the pipeline equipment and facilities. Electrical and instrumentation technicians maintain and repair this equipment to ensure that the pipeline remains in service. Another major task of pipeline personnel is to monitor corrosion and practice measures to prevent or minimize the presence of corrosion on the pipeline and equipment.

Each pipeline company has published operations and maintenance and emergency procedures. These procedures are based on federal, state, and local regulations and on industry standards. All pipeline personnel must familiarize themselves with these standards in order to operate the pipeline in a manner that will ensure the safety of the employees and the general public and protect the environment.

Review Questions

1. Oil and gas are formed from _____.
   a. decaying organic matter
   b. buried rocks
   c. salt domes
   d. groundwater

2. Different types of pumps or valves are used to cap the well, depending on the _____ of the fluid being withdrawn.
   a. temperature
   b. destination
   c. properties
   d. amount

3. Lines from the oil fields are called _____ lines.
   a. field
   b. main
   c. distribution
   d. gathering

4. The pipe must be _____ tested to verify the pipeline integrity.
   a. strength-
   b. X-ray-
   c. hydrostatically
   d. sonar-

5. Different petroleum products can be shipped consecutively in the same pipeline. This is called _____.
   a. bunching
   b. mixing
   c. batching
   d. blending

6. Large volumes of natural gas are typically stored in _____.
   a. underground storage fields
   b. surge tanks
   c. relief tanks
   d. open roof tanks

7. Oil containing a large amount of sulfur compounds is called a _____.
   a. sweet crude
   b. sour crude
   c. paraffin
   d. light crude

8. There is a set of _____ for each hazardous substance, listing its properties and hazards.
   a. chemical descriptions
   b. OSHA labels
   c. product papers
   d. MSDSs

9. The marine environment is particularly sensitive to _____.
   a. high pressure
   b. overproduction
   c. oil spills
   d. vapor accumulation

10. A(n) _____ depicts the functional layout of a piping system and its components as clearly and accurately as possible.
    a. P&ID
    b. alignment sheet
    c. strip map
    d. blueprint

11. _____ is the software used to monitor and control the flow of millions of barrels of oil and products through the pipeline.
    a. RTUS
    b. SCADA
    c. SEACA
    d. PLCS

12. The science of the behavior of fluids at rest or in motion is referred to as _____.
    a. pneumatics
    b. hydraulics
    c. compressibility
    d. viscosity

13. Boyle's law states that when temperature is kept constant and volume is decreased, pressure will _____.
    a. decrease
    b. increase
    c. stay the same
    d. not be affected

14. In a typical positive-displacement pump, a _____ slides back and forth in a cylinder.
    a. rotor
    b. vane
    c. piston
    d. blade

15. _____ are used to control the rate of flow in a line, to open or shut off a line, or to serve as automatic safety devices.

a. Motors
b. Valves
c. Pumps
d. Manifolds

16. Tanks must be calibrated for _____.

a. strength
b. sealing ability
c. capacity
d. wall thickness

17. _____ are used to join three pipe ends.

a. Flanges
b. Unions
c. Elbows
d. Tees

18. The pipeline uses _____ to control the flow path of oil and petroleum products.

a. manifolds
b. pumps
c. tanks
d. vents

19. Various _____ are used along the pipeline to monitor the volume of fluid.

a. gauges
b. tests
c. meters
d. provers

20. Communications failure is one example of an _____.

a. emergency
b. AOC
c. accident
d. interruption

Trade Terms Introduced in This Module

Anomaly: A defect in the pipeline or pipeline coating caused by corrosion, wear, or other external means.

Barrel: The common measure of petroleum products: one barrel equals 42 U.S. gallons or 0.15899 cubic meters or 9,702 cubic inches.

Cathodic protection: A means of preventing the destructive electrochemical process of corrosion of metal by using the metal as the cathode of a cell with a sacrificial anode.

Corrosion: The gradual destruction, dissolving, or wearing away of a substance, usually a metal or an alloy, caused by a chemical or electrochemical reaction with its environment. Corrosion can be a problem in different locations along the pipeline, both externally and internally.

Crude oil: Unrefined liquid petroleum. Crude oil ranges in color from yellow to black, and may have a paraffin, asphalt, or mixed base. Crude oils may be referred to as heavy or light according to API gravity, with the lighter oils having the higher gravities.

Cubicle: An enclosure for high-voltage equipment.

Department of Transportation (DOT): The federal agency that sets standards for, and regulates, transportation.

Hydrostatic testing: The most common final quality check on a complete pipeline. Water is pumped into the line and pressured up to a specified level. The pipe is checked for leaks or ruptures.

Main line: The primary pipe between distant points; it is also called the trunk line.

Manifold: An accessory system of piping to a main piping system that serves to divide the flow into parts, to combine several flows into one, or to reroute a flow to several possible locations.

Material safety data sheet (MSDS): A form containing information on a potentially hazardous material, including names, contents, test results, characteristics, potential hazards, and handling procedures.

Natural gas: A highly compressible, highly expandable mixture of hydrocarbons with a low specific gravity. It occurs naturally in a gaseous form. Besides hydrocarbon gases, it may contain nitrogen, helium, carbon dioxide, hydrogen sulfide, and water vapor. The gases are variable in form and may be found as gases or liquids under various conditions of pressure and temperature. Natural gas liquids include butane, propane, and natural gasoline.

Natural gas liquids (NGLs): The portions of reservoir gas that are liquefied at the surface in lease separators, field facilities, or gas plants.

Petroleum: Unrefined liquid hydrocarbons, including *natural gas liquids (NGLs)*.

Pig: A scraping tool with blades or brushes that is forced through a pipeline to clean out accumulations of wax, scale, and debris from the pipe walls; it is also called a line scraper or go-devil.

Product: Any fluid, refined or not, that is transported through a pipeline.

Module 62104-02

Tools of the Trade

Introduction to the Pipeline Operator Qualification Rule and NCCER

It could be said that petroleum products keep America running. Petroleum products like oil and natural gas are the principal sources of energy in the United States and are used in many manufacturing and chemical processes. Petroleum products are primarily transported by pipeline. There are more than a million miles of liquid and natural gas pipeline in this country, ranging in diameter from two inches to four feet. These pipelines bring crude oil and natural gas from production fields to processing plants and refineries—and ultimately to the consumer.

Pipeline Operator Qualification

Petroleum products are considered hazardous, so pipeline personnel must take extra precautions and follow specific procedures when working with them. To help ensure the safe handling and transportation of petroleum products, the United States Department of Transportation (DOT) mandates safe practices as federal regulations.

One such regulation, the Operator Qualification (OQ) rule, requires pipeline operators to develop a written qualification plan to evaluate their personnel and their contract personnel (see *49 CFR Part 192 Subpart N—Gas,* and *49 CFR Part 195 Subpart G—Liquid*). The evaluation hinges on the operator's ability to perform certain covered tasks and to be able to recognize and respond to abnormal operating conditions that may be encountered while performing these covered tasks.

Covered Tasks

A pipeline performance task is deemed covered if the activity passes the following four-part test:

1. Is it performed on a pipeline facility?
2. Is it an operations or maintenance task?
3. Is it performed as a requirement of *49 CFR Part 195 (Liquid)* or *49 CFR Part 192 (Gas)*?
4. Does it affect the operation or integrity of the pipeline facility?

If the answer to all four questions is yes, then the activity is considered covered and must figure into the evaluation process of pipeline personnel.

Consortium on Operator Qualification

The Consortium on Operator Qualification (COOQ) became the guiding force for the identification and development of covered tasks for the liquid pipeline industry. Their work product, API publication 1161, *Guidance Document for the Qualification of Liquid Pipeline Personnel*, formed the basis for the resulting Pipeliner Training and Assessment Program.

The covered task list created in API publication 1161 has now evolved to include gas, as well as construction covered tasks. NCCER maintains a version of the covered task list, which can be accessed at **www.nccer.org**.

NCCER

With the guidance and support of the Consortium on Operator Qualification, NCCER was able to design and implement the Pipeliner Training and Assessment Program.

NCCER is a not-for-profit 501(c)(3) education foundation established in 1995 by the world's largest and most progressive construction and maintenance contractors and national contractor associations. NCCER was founded to address the severe workforce shortage facing the construction and maintenance industries and to develop a standardized training process and curricula. Today, NCCER is supported by hundreds of leading construction and maintenance companies, manufacturers, and national associations and developed more than 70 construction and maintenance titles.

NCCER also maintains a Registry, which contains the records of trainees and instructors who have received NCCER training and certification. All completion information for every trainee is sent to NCCER and filed within the Registry. The Registry can then confirm training and skills for workers as they move from company to company, state to state, or even within their own company.

NCCER Standardized Curricula

NCCER standardized curricula are developed using the expertise of industry subject matter experts (SMEs). The use of SMEs as an integral part of every new development or revision ensures that the curriculum is current, industry-driven,

and indicative of industry practices. The development and/or revision process is facilitated from start to finish by NCCER project managers, who work closely with the SMEs and a select team of technical writer(s) on the development and presentation of the curriculum.

Pipeline: Five Training Paths

The pipeline curriculum focuses on essential skills and competencies within five skill areas:

- Corrosion Control
- Electrical and Instrumentation
- Maintenance
- Mechanical
- Operations (Field and Control Center)

SMEs from the pipeline industry have determined the skills and competencies needed based on API's covered tasks and the consensus of the group/industry. Specific company information is generally not incorporated into the curriculum unless the SMEs determine that the information is the industry standard.

Training Modules and Covered Tasks

The smallest unit of instruction in the curriculum is a *module*. Each curriculum is composed of modules (the equivalent of stand-alone chapters) that are logically sequenced into complete *levels* of instruction. Every module is task-oriented and requires the successful completion of a written (knowledge-based) examination and a performance (hands-on) test. Modules may be individually selected and/or combined for efficient task training. Modules in the pipeline curricula, to the extent possible, align with industry-developed covered tasks. While not all modules relate to covered tasks, all covered tasks relate to at least one module.

Benefits of Training to the Pipeline Marketplace

Applying NCCER's standardized training process allows pipeline owners and contractors in the United States to establish the benchmark for acceptable skill and competency levels within the pipeline industry. This process will assist training organizations in developing and/or improving standardized pipeline training programs that effectively match their resources and operating conditions while responding to regulatory demands and to the needs of the industry. The process also recognizes training organizations that meet the minimum standards of competence, warranting regulatory, public, and professional confidence. This, in turn, fosters national unity within the pipeline industry for the mutual benefit of individual technicians, owners, contractors, and ultimately pipeline-users and society.

Additional Resources

To review the Pipeliner Training and Assessment Program Implementation and Resource Guide and the latest version of pipeline covered tasks, visit NCCER's website at **www.nccer.org,** or contact:

NCCER
13614 Progress Boulevard
Alachua, FL 32615
(888) 622-3720
Website: **www.nccer.org**

Additional sources of information on the pipeline operator qualification regulation include:

DOT – Office of Pipeline Safety
www.ops.dot.gov

Transportation Safety Institute
www.tsi.dot.gov

COURSE MAP

This course map shows all of the modules in the first level of the Pipeline Maintenance curriculum. The suggested training order begins at the bottom and proceeds up. Skill levels increase as you advance on the course map. The local Training Program Sponsor may adjust the training order.

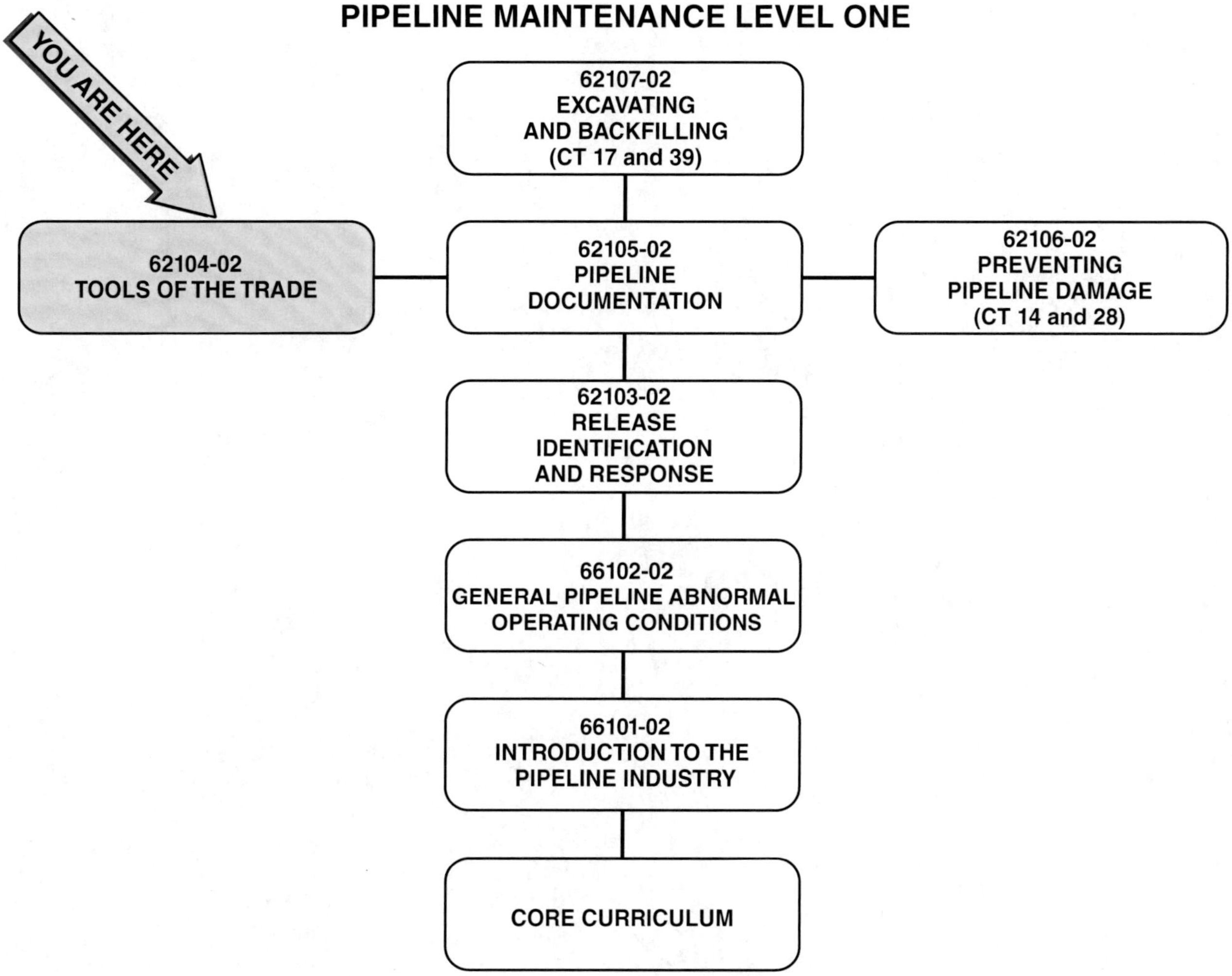

104CMAP.EPS

MODULE 62104-02 CONTENTS

Figures

Table

Tools of the Trade

Objectives

When you have completed this module, you will be able to do the following:

1. Use and maintain hand and power tools.
2. Identify and explain the use of welding equipment.
3. Explain the use of meters and testers.
4. Explain the use of nondestructive testing tools.
5. Identify and explain the use of lifting and excavation equipment.

Prerequisites

Before you begin this module, it is recommended that you successfully complete the following: Core Curriculum; Pipeline Core, Modules 66101 and 66102; Pipeline Maintenance Level One, Module 62103.

Required Trainee Materials

1. Pencil and paper
2. Appropriate personal protective equipment

1.0.0 ◆ INTRODUCTION

This module focuses on the use, care, and maintenance of hand and power tools, welding equipment, meters and testers, nondestructive testing tools, and heavy excavating equipment. Use and care of basic hand tools and power tools are explained in the *Core* curriculum, and this module describes the more specialized tools used by pipeline maintenance technicians.

Using the correct type and size of tool that has been properly maintained can prevent accidents and damage to your equipment, system components, and tools. The money spent on the tools required to maintain a pipeline represents a significant cost to your employer. Performing the required maintenance and using the tools properly will protect you and your co-workers, your employer's investment, and will also make the job run more smoothly.

2.0.0 ◆ USING AND CARING FOR HAND TOOLS

Occupational Safety and Health Administration (OSHA) *CFR 1926.301-Hand Tools* provides four very simple, common-sense rules that must be followed when using hand tools. It is the pipeline maintenance technician's responsibility to ensure that these rules are strictly followed:

- Employers shall not issue or permit the use of unsafe hand tools.
- Wrenches, including adjustable, pipe, end, and socket wrenches, shall not be used when the jaws are sprung to the point at which slippage occurs.
- Impact tools, such as drift pins, wedges, and chisels, shall be kept free of mushroomed heads.
- The wooden handles of tools shall be kept free of splinters or cracks and shall be kept tight in the tool.

> **NOTE**
> OSHA writes these rules directed at the employer. However, it is the employee's responsibility to not use defective tools and to notify the employer of any defective tools on the job site.

The following hand tools are described in this section:

- Probe rods
- Banding tools
- Wrenches
- Flange alignment pins
- Rotary pipe cutters
- Flange spreaders
- Drilling and **tapping** machines
- Sandblasting
- Air bags
- Bonding clamps
- Pit gauges
- Pipeline **skids**

2.1.0 Probe Rods

Probe rods (*Figure 1*) made of metal or fiberglass are used to locate existing utilities. The probe rod is pushed into the soil and used to probe for the utility. When the utility is located, its exact placement can be determined and marked by measuring the amount of buried probe rod in the **excavation**. Follow these guidelines when using probe rods:

- Always use a fiberglass or insulated rod in the vicinity of electric, phone, or television cables.
- Never use a probe rod to locate electric utilities.
- The probe rod should always be stored when not in use to prevent damage.
- Do not use excessive force when using the probe rod.
- Always use proper lifting procedures.
- The probe should have blunt end.

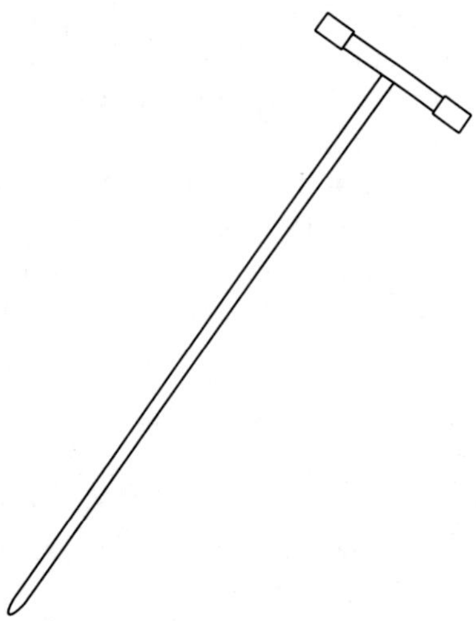

104F01.EPS

Figure 1 ◆ Probe rod.

2.2.0 Banding Tool

The banding tool (*Figure 2*) is used to perform a number of tasks on the job site. It is commonly used for repairing suction and discharge hoses and for banding together stacks of pipe or bundles of wood. The banding tool is also used to install diapers, to band skids to carrier pipe for installation in casings, and occasionally for strengthening concrete forms. The banding tool operates by using a ratchet action to tighten the banding material. Once the band is tightened, it is held in place by striking the anvil. A small protrusion on the back side of the anvil places a dimple on the band clip, preventing the band from loosening.

Follow these steps to use a banding tool:

Step 1 Place the band around the material to be banded.

Step 2 Load the band into the receiver.

Step 3 Depress the tension handle repeatedly until the band is tight.

Step 4 Strike the tool anvil sharply with a hammer to dimple the band clip.

Step 5 Release the tool.

Step 6 Clean and store the banding tool.

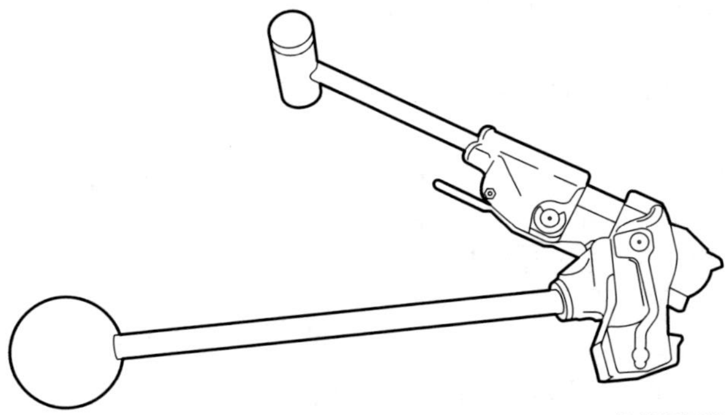

104F02.EPS

Figure 2 ◆ Banding tool.

2.3.0 Wrenches

Several specialty wrenches are used in pipeline maintenance. These tools are designed to make a specific task easier. Examples of wrenches include the valve wrench, T-handle torque wrench, bolt-thru ratchets, hydrant wrenches, and hammer/slug wrenches.

2.3.1 Bolt-Thru Ratchets

Bolt-thru ratchets (see *Figure 3*) are commonly used to fasten the nuts on a variety of pipe joints. The unique aspect of these wrenches is the opening in the head. This opening allows the bolt or rod to extend through the wrench so that the nut can be threaded any distance required. The wrench is available with a variety of interchangeable sockets and is reversible.

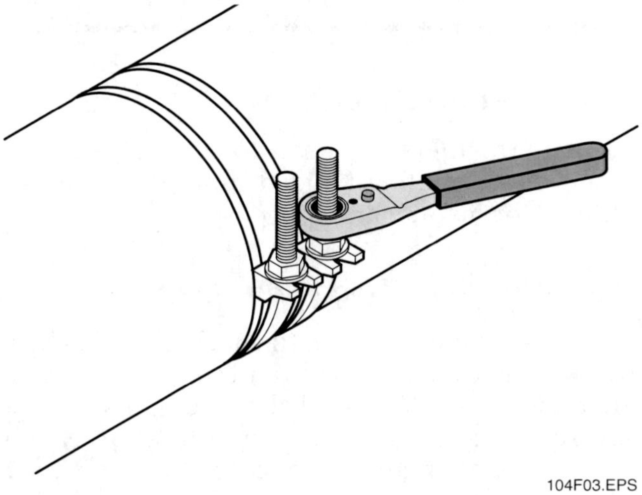

Figure 3 ◆ Bolt-thru ratchet.

2.3.2 Hammer/Slug Wrench

A hammer/slug wrench is designed with a heavy box end on one end that completely surrounds the bolt head or nut and a large striking area on the other end for striking with a hammer or mallet (*Figure 4*). The slug wrench is the only wrench that is designed to be hit with a hammer.

WARNING!
Never hammer or strike any other type of wrench.

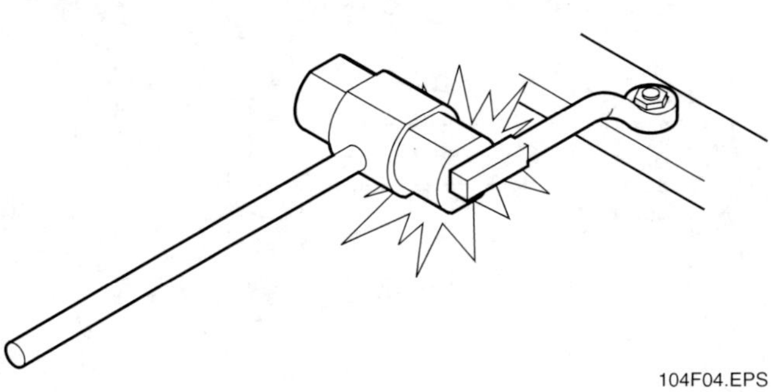

Figure 4 ◆ Hammer/slug wrench.

2.4.0 Flange Alignment Pins

A variety of flange alignment pins are manufactured for the pipeline industry. A drift pin, also commonly referred to as a line-up pin, is a round steel piece that is tapered on one end. A barrel pin (*Figure 5*) is a drift pin that is tapered on both ends. Drift pins and barrel pins are used to align bolt holes when connecting flanges, structural steel, and other holes in mating surfaces. Drift pins and barrel pins are available in various sizes to fit different size bolt holes. The pin selected for a specific job should fit all the way through both holes being aligned and still be large enough and strong enough to pull the mating pieces together.

Two-hole pins (see *Figure 6*) are used to level or plumb two adjacent bolt holes on flanges. The pins line up with the centers of each of the holes when tightened, and then a level is placed across the top of the pins to assist in aligning the flange.

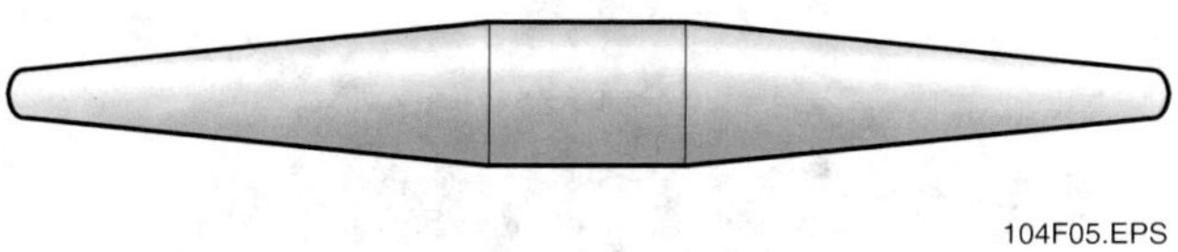

Figure 5 ◆ Common barrel pin.

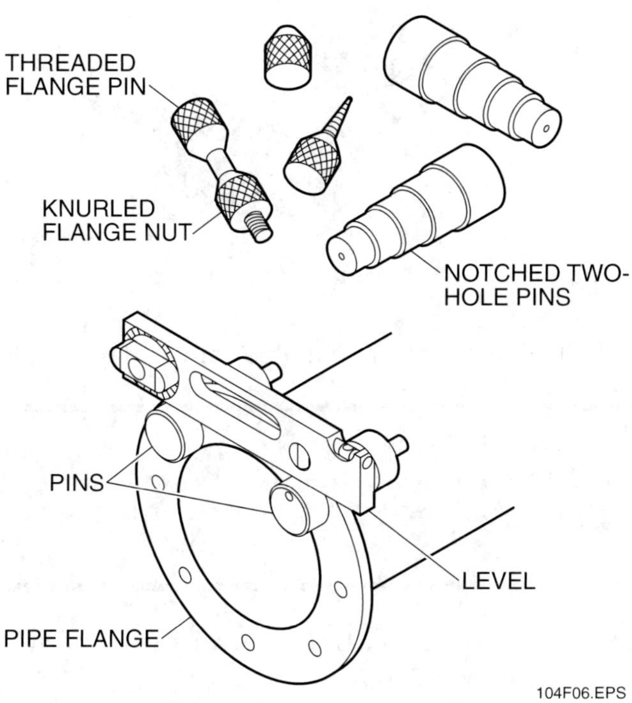

Figure 6 ◆ Two-hole pins.

2.5.0 Rotary Pipe Cutters

Rotary pipe cutters come in a wide range of sizes (see *Figure 7*) and can be used to cut steel, stainless steel, ductile iron, and cast-iron pipe in tight spaces. The tool usually needs only about 4 to 8 inches of clearance to operate, and will cut pipe up to 34 inches in diameter with wall thicknesses to about ½ inch. The cutting blades are interchangeable so that blades for a specific material can easily be installed.

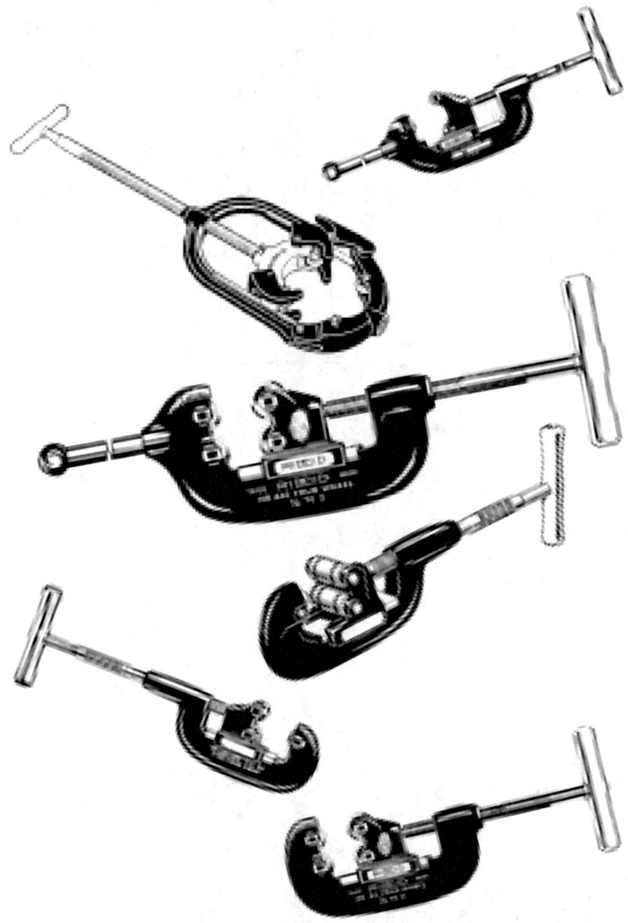

104F07.EPS

Figure 7 ◆ Rotary pipe cutters.

Follow these steps to cut pipe with a rotary cutter.

> **WARNING!**
> Always make sure that all pressure on the pipeline has been released before working on the line.

Step 1 Place the cutter around the pipe and tighten it in place.

Step 2 Using a ratchet action, move the pipe cutter back and forth around the pipe.

Step 3 Feed the blade further into the cut by tightening the tool around the pipe.

Step 4 Repeat Steps 2 and 3 until the pipe is cut.

> **CAUTION**
> A pipe may snap off before the cut is all the way through the pipe. To reduce the risk of injury, be sure that you are standing clear of the pipe should it unexpectedly snap off.

2.6.0 Flange Spreaders

Flange spreaders are used to separate flanges when disassembling flanged piping systems for removal or maintenance purposes.

2.6.1 Hex Turn Flange Spreaders

Flange spreaders are small mechanical devices used to part flanges quickly and easily. These tools are typically made of steel alloys to withstand the severe forces encountered during use. Spreaders are also heat treated to add to their ability to undergo heavy use without distortion or breakage.

Hex turn flange spreaders (see *Figure 8*) are easy to set up and use on flanges that do not easily separate because of corrosion or coating that might be binding them together. The user removes one bolt from the flange face and inserts the flange spreader support. Next, the remaining flange bolts are removed. Finally the hex nut is turned with the appropriate tool to drive the spreader between the flange faces.

> **NOTE**
> A spreader is required on each side of the flange for even distribution of pressure.

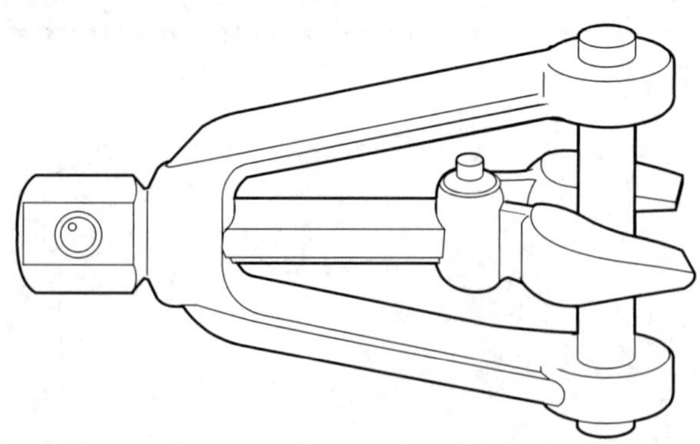

104F08.EPS

Figure 8 ◆ Hex turn flange spreader.

2.6.2 T-Handle Flange Spreaders

T-handle flange spreaders (see *Figure 9*) are another variation on the flange spreader line of tools. Used in pairs to keep the pressure equal on both sides of a flange, they allow the user to part flanges quickly for small jobs like gasket replacement. T-handle spreaders do not require additional tools to exert pressure on tight flanges.

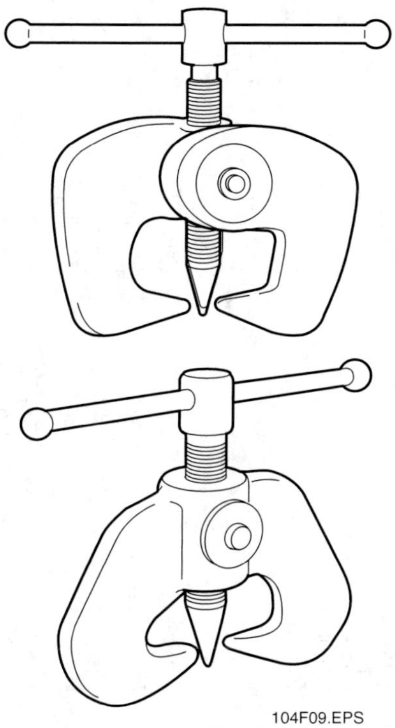

Figure 9 ◆ T-handle flange spreaders.

2.6.3 Ratchet Action Flange Spreaders

Ratchet action flange spreaders (*Figure 10*) are also small and lightweight but offer the user a way of applying greater force to separate flange faces through the use of a ratcheting tool that advances the spreader point into the gap between the flange faces. These flange spreaders are typically used to separate pipe connections to permit insertion of blinds or gaskets.

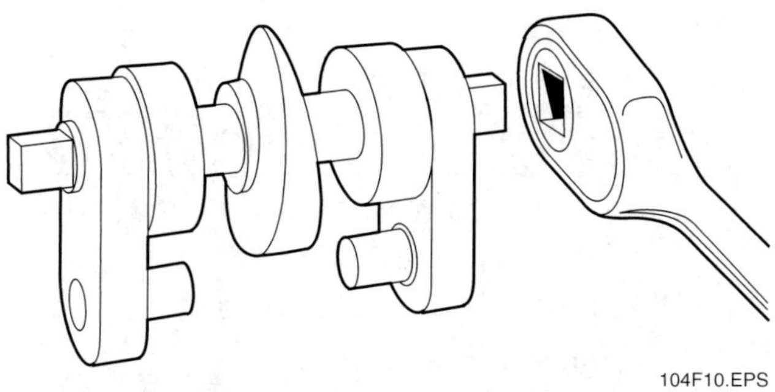

Figure 10 ◆ Ratchet action flange spreader.

2.6.4 Hydraulic Flange Spreader

Hydraulic flange spreaders (see *Figure 11*) are very effective tools for separating flange faces where abnormally high spreading loads are required, such as on heat exchangers, pressure vessels, and other applications in the marine, oil, and gas industries. Different models of flange spreaders can exert pressures as high as 10 tons and can separate flanges that are up to 6⅝ inches in thickness. These tools require minimal clearance between flanges for insertion.

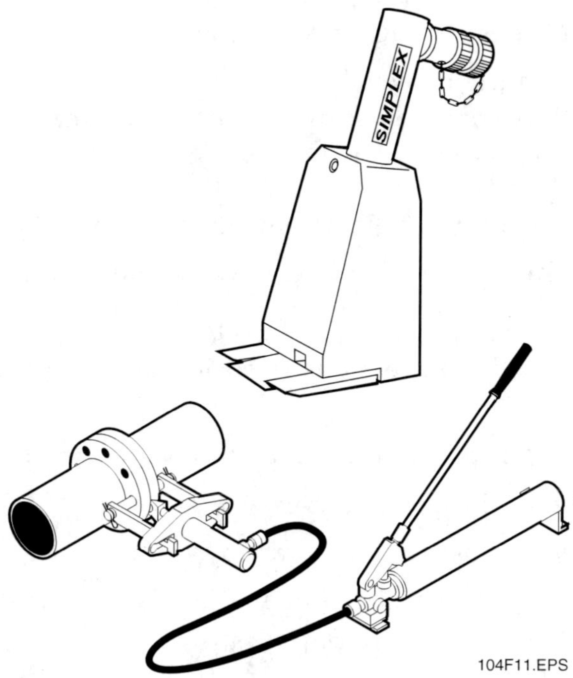

Figure 11 ◆ Hydraulic flange spreaders.

2.7.0 Drilling and Tapping Machines

Drilling machines are used to make connections to pipelines, tanks, and plant piping without shutdown and are used to make hot **taps** in preparation for plugging machine applications. With the proper attachments, large drilling machines can also be used to install line stops and valves into main lines. Most taps and connections can be made with the main line pressurized and in service.

The machines are widely available for cuts or taps ranging from ¼ inch to 12 inches. Much larger equipment is available, but these are not common tools and are usually operated by crews who specialize in that type of work. Drilling and tapping machines can cut most pipe types including cast iron, cement-lined cast iron, ductile and cement-lined ductile iron pipe, steel, concrete, reinforced concrete, asbestos cement, and plastic.

2.7.1 Line Plugging/Stopping

When maintenance must be done on a section or part of the pipeline without shutting down the operation, equipment can be installed to provide a temporary bypass line for the product. This allows the section being worked on to be drained and removed. This is accomplished by using valves, line plugging or stopping machines, and an appropriate bypass line. Once repairs are made, the plugs and bypass are removed, and the product flow restored to its original path.

Stoppling is the term applied to the use of TD Williamson (TDW) equipment and fittings to make either permanent or temporary changes while performing the following:

- Connecting new lines
- Removing, replacing, or repairing damaged or corroded lines
- Relocating lines
- Completing other maintenance or repair tasks

This general procedure represents a typical stoppling operation to clear an obstruction or make a modification to a section of line. *Figure 12* shows the process described in the step-by-step procedure.

Step 1 Install the required fittings on the line.

Step 2 Install temporary valves.

Step 3 Install a temporary bypass.

Step 4 Make taps through the valves.

Step 5 Install STOPPLE® plugs.

Step 6 Close the plugs to direct product through the bypass.

Step 7 Complete the task on the isolated section of pipe.

Step 8 Retract the plug heads.

Step 9 Close valves and remove the temporary bypass.

Step 10 Install LOCK-O-RINGS® in the fittings.

Step 11 Remove all equipment and install blind flanges to complete the job.

Step 12 Notify the control center that the line is returned to full service.

Step 13 Complete all documentation as required by your company.

Always follow your company's policies and procedures when performing this type of task.

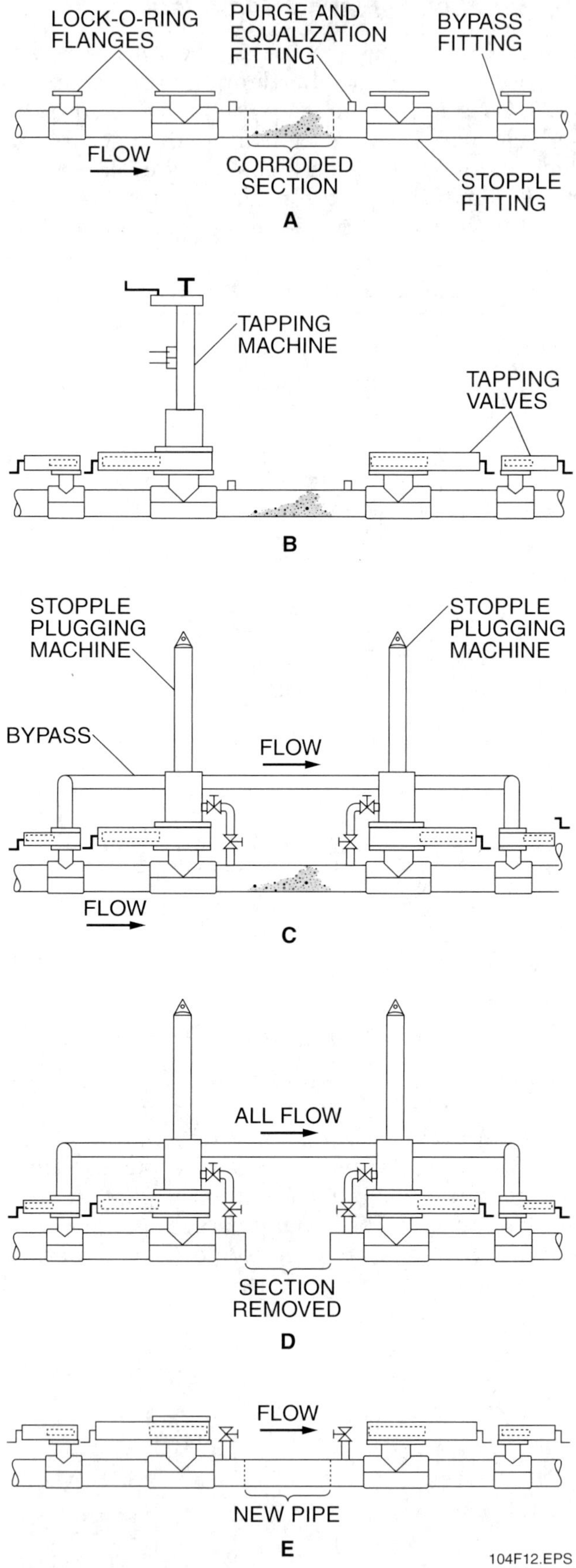

Figure 12 ◆ Stoppling process.

2.7.2 Hand Drilling and Tapping Machines

Hand drilling and tapping machines are available as manual or power-driven tools that allow the pipeliner to tap into pipes for repairs or adding lines while the pipe is under operating pressure and in service. These tools come in a wide range of models and sizes for making cuts or taps from ½ inch to 4 inches, and larger, in diameter. Hand-operated cranks are replaced by handheld air motors for taps of 6 inches or larger. These smaller tapping machines are also used to install completion plugs once tapping work is complete.

One widely used small machine, the TD Williamson Model T-101, is available in two versions: the standard T-101B with 18" of travel and the XL Model with 28" of travel. These and other commonly used drilling machines meet NACE standards for oil field equipment. *Figure 13* shows TDW T-101 series drilling machines.

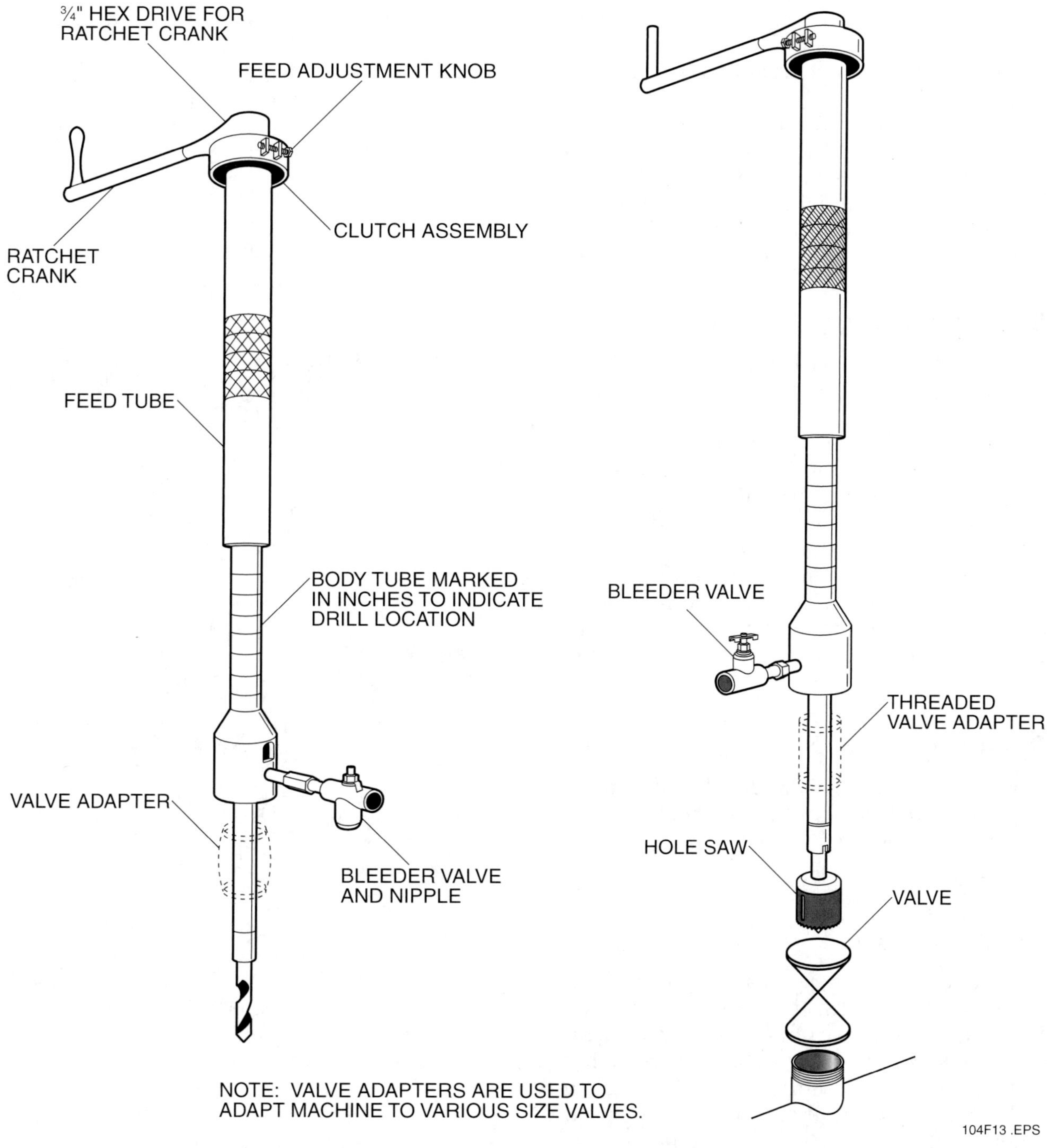

Figure 13 ◆ T-101 series drilling machines.

2.7.3 Hydraulic Powered Tapping Machines

Larger tapping machines are more suited for making hot taps into main lines, storage tanks and plant piping. These taps can be made without shutdown of the system.

The TD Williamson Model 1200H tapping machine is widely used throughout the petroleum industry where larger taps are necessary. It is hydraulically operated and can make taps ranging from 12 inches to 36 inches in size. It is used to prepare for plugging operations, in support of maintenance requirements, or to allow the addition of new lines. Using this equipment, taps may be made in a vertical or horizontal position or at any other angle where there is adequate space for equipment installation and safe work practices.

This model, like other large machines, is equipped with an access ladder and work platform for operator use. The Model 1200H meets NACE standards for oil field equipment. See *Figure 14* to view a typical setup for tapping into a pipeline with a large tapping machine.

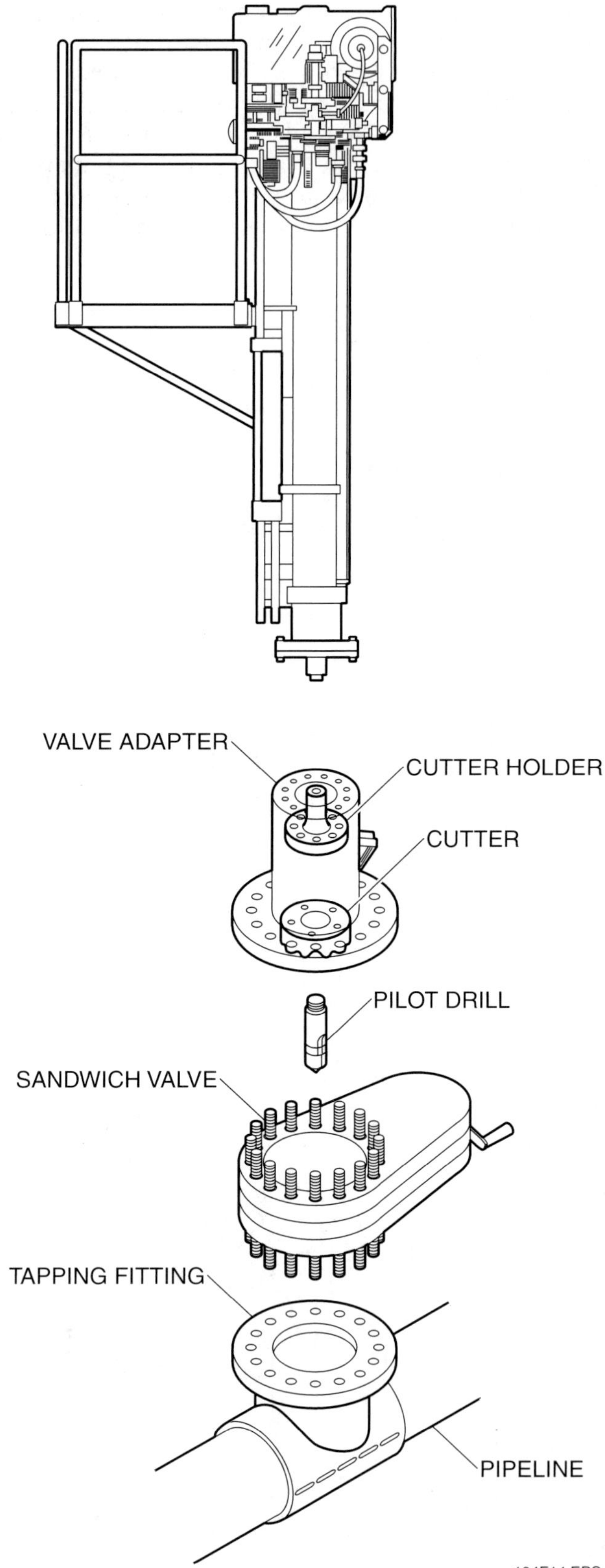

Figure 14 ◆ Typical setup for tapping.

2.8.0 Sandblasting

Sandblasting, also referred to as abrasive blast cleaning, is used to prepare pipe surfaces for coating applications and/or repairs. Sand, plastic pellets, and walnut hulls are widely used for treating different surface materials. The abrasive material used impacts the surface at a high velocity and cuts or knocks the old paint, rust, and dirt from the pipe while creating a good surface profile for the coating material on which to adhere.

Follow your company's procedure for use of personal protective equipment and review applicable sandblasting guidelines prior to starting the process.

2.9.0 Air Bags

Air bags, or airbag jacks, were developed to meet the need for a tool that could meet three critical requirements. Those are:

- Low height
- Fast inflation
- Large lifting capacity

Some models of airbag jacks are capable of lifting more than 70 tons and even larger loads when used in sets of two or more. The gas and oil pipeline industry is but one of the industries that find airbag jacks a very useful tool for lifting heavy loads. *Figure 15* shows multiple airbags being used to lift a piece of heavy equipment for maintenance.

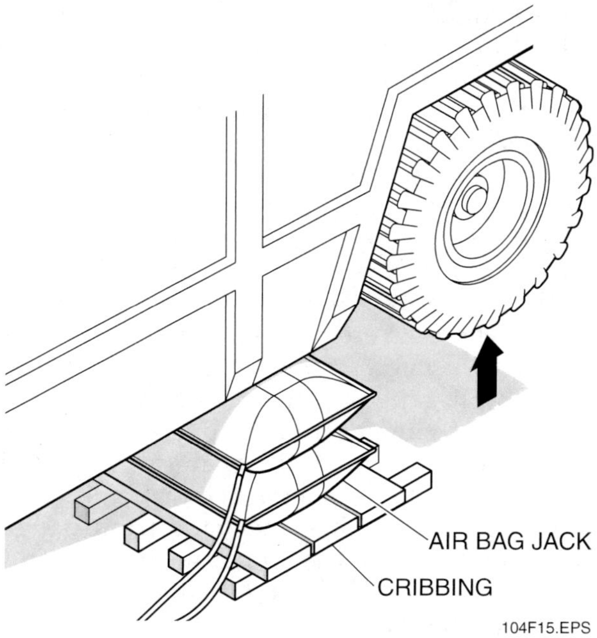

Figure 15 ◆ Multiple airbag jacks under a load.

The one disadvantage when using airbag jacks is the loss of lifting power as the bags inflate. To offset this, the user places cribbing between the airbag and the load to shorten the distance the bag must inflate before contacting the load. Refer to *Figure 15* to see how cribbing was used to shorten the distance the bag must be raised before coming into contact with the load.

Because of variables in load size and contact area, it is a good practice to place a steel plate (or other stabilizer material) on top of the bag. This increases the area of surface contact between the bag and the load (plate) and offers a more stable support for oddly shaped loads while spreading the load over a greater part of the bag surface. For pipe, cradles can be used in conjunction with the plates to prevent the pipe from rolling off to the side while being elevated. *Figure 16* shows this concept.

These illustrations are very general in nature. Always refer to your company's policy on airbag jack use before attempting to raise a load with these tools.

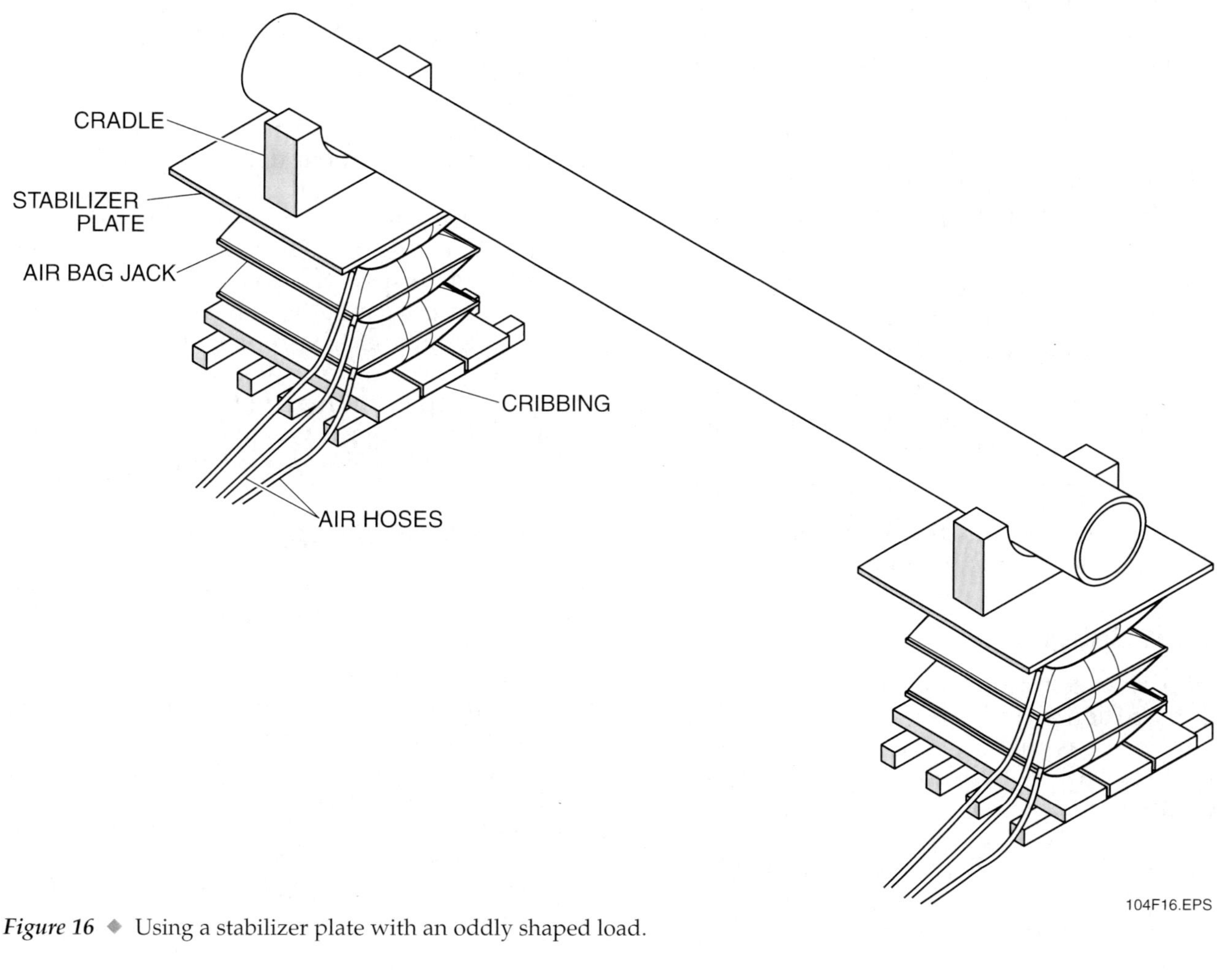

Figure 16 ◆ Using a stabilizer plate with an oddly shaped load.

> **NOTE**
> Airbags are effective lifting tools because each pound of air pressure (psi) pumped into the bag creates a lifting force that is multiplied over the entire lifting surface of the bag, resulting in a much greater lifting force that can be used to lift the load.

2.10.0 Bonding Clamps

Bonding and grounding are acceptable controls for static electricity that can build up in a variety of ways. Static electricity can be produced when any of the following occurs:

- Liquid flows through a pipeline or hose
- Tanks or drums are filled
- Dry material passes through a chute or a pneumatic conveyor

Bonding is the term used to indicate the connection of two or more objects with a conductor. This conductor, such as a copper wire, equalizes but does not eliminate the potential electrical charge (static electricity) between the connected objects, reducing the chance that sparks will jump the gap between objects. Grounding, on the other hand, connects one or more of the objects to the earth using ground rods or steel building components, allowing the static to drain away.

Clamps used for bonding and grounding must provide a good conductive path in order to perform their protective function. To ensure this:

- Remove all dirt, paint or corrosion from the area where the clamp is placed.
- Use strong connectors that will maintain a firm grip, keeping the flow path intact.
- Use the appropriate size and type of connector based on the location and conditions.
- Always connect the metal clamp to other metal.

Figure 17 shows various types of clamps and connectors used for bonding and grounding.

2.11.0 Pit Gauges

One of the most common tools used to check the extent of corrosion to a pipe is the pit gauge. Most pit gauges are easy to use and to carry. These gauges, if properly maintained, can provide accurate measurements.

When corrosion or damage is discovered, the corrosion technician or maintenance team member needs to determine the remaining wall thickness of the pipe in order to set the priority for repair or replacement of the corroded section of pipe. The mechanical pit gauge is a small, sturdy tool and easy to use in making a quick determination of the extent of corrosion damage present. These gauges are usually made of polished stainless steel and can easily be carried in a tool kit. Care must be taken to use these gauges properly and to recognize and understand their limitations.

Inaccurate readings can result if the gauge is not aligned properly or is resting on an uneven surface. Where severely pitted or partially coated pipe and associated structures are involved, these handheld pit gauges will provide the user with only an approximation of pit depth. The depth measurement scale on the mechanical pit gauge is marked in sixteenths of an inch and in mils (thousands of an inch). Most mechanical pit gauges include formulas, tables, and data giving outside diameter (OD), wall thickness, and weight per foot for various sizes of pipe.

When pit depth measurements are critical and accuracy is a priority, dial pit gauges, which are designed for a more accurate measurement of corrosion pit depth, may be used. Most dial pit gauges are capable of measuring depth to extremely close tolerances. On the down side, these indicators are more fragile than mechanical gauges and should be handled with care to ensure their repeated accuracy. *Figure 18* shows a dial pit gauge positioned on a pipe.

The accuracy of any pit measurement is only as reliable as the instrument that is used to take the measurement. Before a pit gauge is used for measurement, it must be inspected for any damage and any worn points. During this inspection, always closely inspect the point of the instrument for wear. If a pit gauge does have a worn point, the pit gauge should be replaced. Only use an instrument that is in excellent condition; otherwise, your inspection may not be accurate. Prior to using the pit gauge, clean the pit in accordance with your company's approved policy.

2.12.0 Pipeline Skids

Pipeline skids are timbers used to support the pipe off the ground. These timbers, which typically measure 4" × 6" × 36", are usually cut from oak or other hardwood for strength and durability. They are built up one level at a time until the structure is sufficient for the task at hand.

For larger jobs, pipeline skids can be used in multiples, in a cross-buck style, to support larger sections of pipe for construction or maintenance and repair jobs.

Figure 19 shows examples of both single-skid and multiple-skid, or cross-buck, applications.

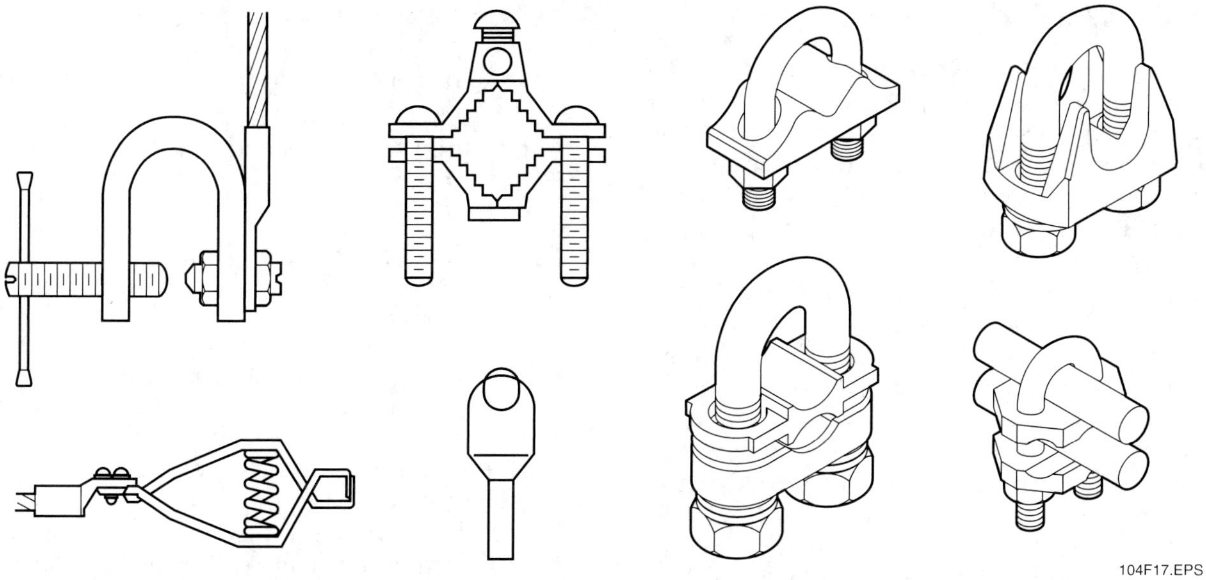

Figure 17 ◆ Various types of clamps and connectors used for bonding and grounding.

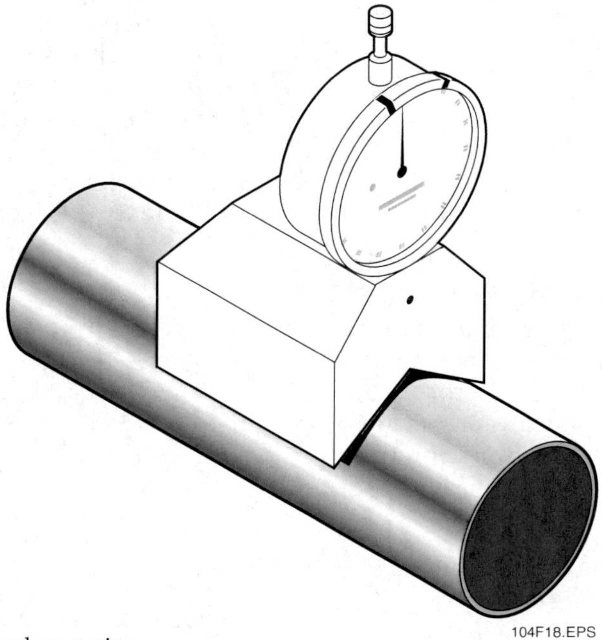

Figure 18 ◆ Dial pit gauge positioned on a pipe.

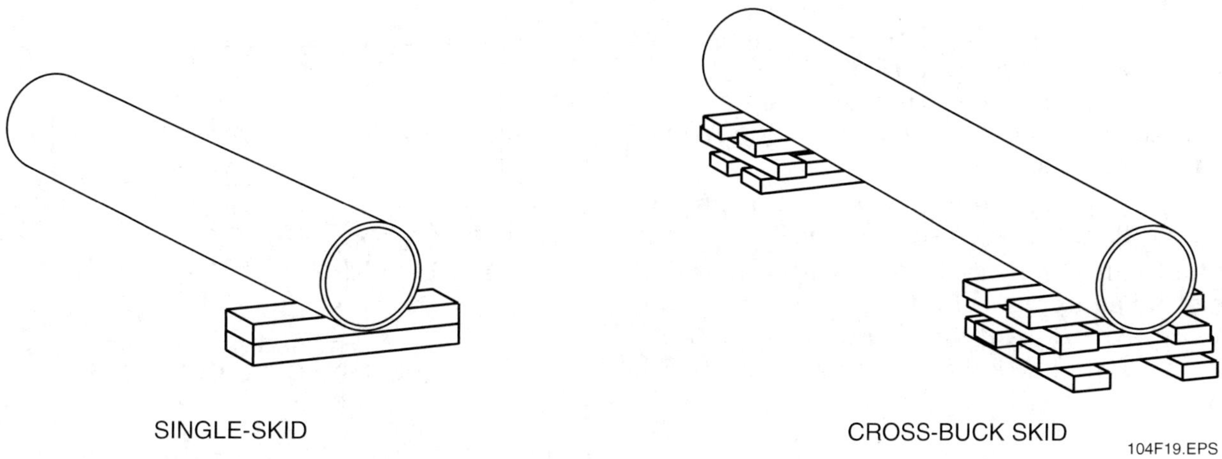

Figure 19 ◆ Single-skid and cross-buck skid arrangements.

2.13.0 De-Watering

Before any excavation, **trenching**, or grading can begin on a construction project, such as laying or relining pipelines, groundwater must be removed. Material that contains too much moisture will slow or prevent construction and leave the project site without the necessary preparation for building or compaction. Optimum construction project conditions can be maintained with the proper use of de-watering pumps. Once a trench has been dug, or grading has been done and open bodies of groundwater have collected, de-watering is needed to evacuate the moisture and return the material to a drier state for further work.

De-watering pumps are self-contained units typically mounted on a frame or trailer. The power source turns a shaft connected to the pump end. Inside the pump end is an impeller, a fan-like piece that rotates at a very high speed and forces the air out of the pump end. This creates a vacuum inside the pump end and draws water into the pump. The water is then discharged through piping or hose and emptied at another location.

3.0.0 ◆ USE AND CARE OF POWER TOOLS

Pipeline maintenance technicians work with and around various types of powered hand tools throughout their careers. Care and maintenance of these tools is a critical factor in completing job tasks in a safe and timely manner. Before you operate any power tool, make sure you understand and follow the safety precautions for that tool. The following checklist should be reviewed before operating any power tool:

- Are grinders, saws, and similar equipment provided with appropriate safety guards?
- Are power tools used with the correct shield, guard, or attachment that is recommended by the manufacturer?
- Are portable circular saws equipped with guards above and below the base shoe? Are circular saw guards checked to assure they are not wedged up, leaving the lower portion of the blade unguarded?
- Are rotating or moving parts of equipment guarded to prevent physical contact?
- Are all cord-connected, electrically operated tools and equipment effectively grounded or of the approved double-insulated type and plugged into a ground fault interrupter (GFI) outlet?
- Are effective guards in place over belts, pulleys, chains, and sprockets on equipment such as concrete mixers, generators, and air compressors?
- Are pneumatic and hydraulic hoses on power-operated tools checked regularly for deterioration or damage?

3.1.0 Air Compressors

Compressors are machines that are designed to compress air from an initial intake pressure and release it at a higher discharge pressure. Portable air compressors consist of a compressor and drive motor that can be readily moved as a unit. Portable air compressors are driven by either gas, diesel, or electric motors. A portable air compressor can be a small engine compressor mounted on two air tanks that can be handled by one person or a larger engine compressor mounted to a trailer that can be towed by a truck or a tractor. The larger types are more widely used on job sites.

The principle behind air compressors is to force air into an area smaller than it normally occupies so that the pressure increases significantly. This compressed air is then released and allowed to expand to its normal free state, thus producing work. *Figure 20* shows portable air compressors.

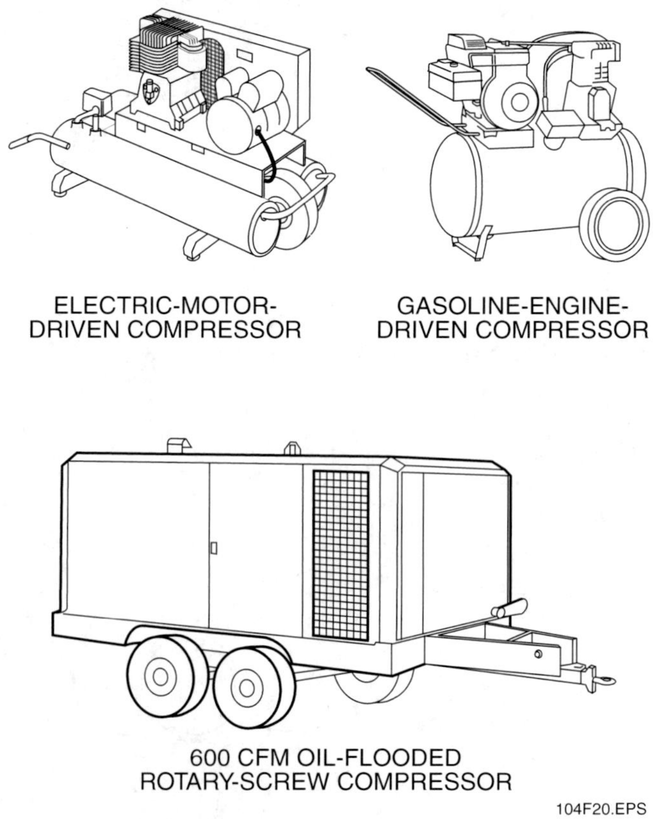

Figure 20 ◆ Portable air compressors.

The rated operating pressure for most portable air compressors is 100 psig (pounds per square inch gauge) with a maximum allowable pressure of 120 or 125 psig. Portable air compressors have a safety valve setting, usually 10 psig above the maximum allowable pressure. The safety valve automatically opens to release the pressure if the pressure reaches the safety valve setting. The actual free air delivery is measured in cubic feet per minute (cfm). The smaller air compressors deliver air between 2.5 cfm and 6 cfm, while the larger, trailer-mounted compressors usually deliver air between 100 and 200 cfm.

Portable air compressors are classified as positive-displacement compressors, and positive-displacement compressors are classified as either reciprocating or rotary.

3.2.0 Pipe Preparation Machines

Line travel pipe preparation machines include cleaning/priming machines, cleaning/priming/tape wrapping machines, and coating/wrapping machines.

3.2.1 Powered Pipe Wheel Cutters

Pipe wheel cutters were developed for fast and easy pipe cutting. Different models of wheel cutters come in a variety of sizes capable of cutting pipe ranging from 2 to 36 inches in diameter. The Wachs wheel cutter (see *Figure 21*) is used for special pipe cutting situations where it is important that there is no metal chip contamination. The Wachs tool uses three cutting wheels that are fed equally into the cut using Wachs' starwheel feed system. As the cutting wheels are fed into the cut line, the pipe wall is pierced and displaced by the cutting wheels.

The combination of a parting blade and a beveling blade coupled to the Wachs LCSF/3 split frame is a more widely used tool in the oil and gas pipeline industry. Using this tool set to bevel the pipe after cutting may still produce some chips, but much of the time the metal is removed in ribbons, making it easier to control chip contamination.

3.2.2 Cleaning/Priming Machines

Cleaning/priming machine (CPL) units clean the full pipe circumference with wire cup brushes and/or scraper knives and, simultaneously, apply the primer coating. Three- or four-speed forward travel control is available. These are available for pipe sizes 3 to 60 inches.

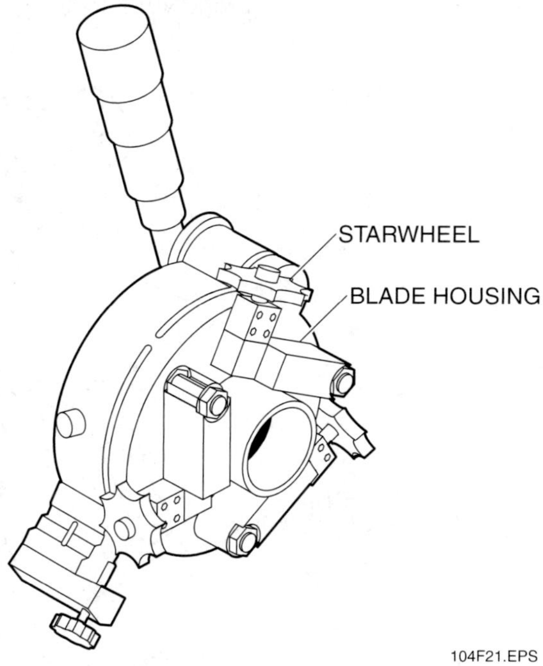

Figure 21 ◆ Wachs powered pipe wheel cutter.

3.2.3 Cleaning/Priming/Tape Wrapping Machines

Line-traveling cleaning/priming/tape wrapping (CPT) machines perform pipe cleaning, primer application, and tape wrapping in a single pass. These units are equipped with powered pressure rollers, three- or four-speed forward travel, tape tension adjustment, tape overlap adjustment, and exhaust fan(s). These are available for pipe sizes 3 to 60 inches.

3.2.4 Coating/Wrapping Machines

Line-traveling coating/wrapping (CWL) machines apply coal tar enamel to 3⁄32" thickness, a wrap of reinforcing material, and an outer wrap in a simultaneous operation. Independent clutches govern travel speed, coating, and wrapping. Double coating and wrapping machines are also available. These are available for pipe sizes 4 to 60 inches. Hand coating/wrapping equipment is also used. These types of wrappers are common in remote areas of the pipeline. *Figure 22* shows a hand coating/wrapping machine.

3.3.0 Air Movers

Air movers and fans are used in pipeline facilities for ventilation and air removal. Many areas require explosion-proof air movers and fans due to volatile fumes in the working environment. Two main types of air movers are air horns and axial fans.

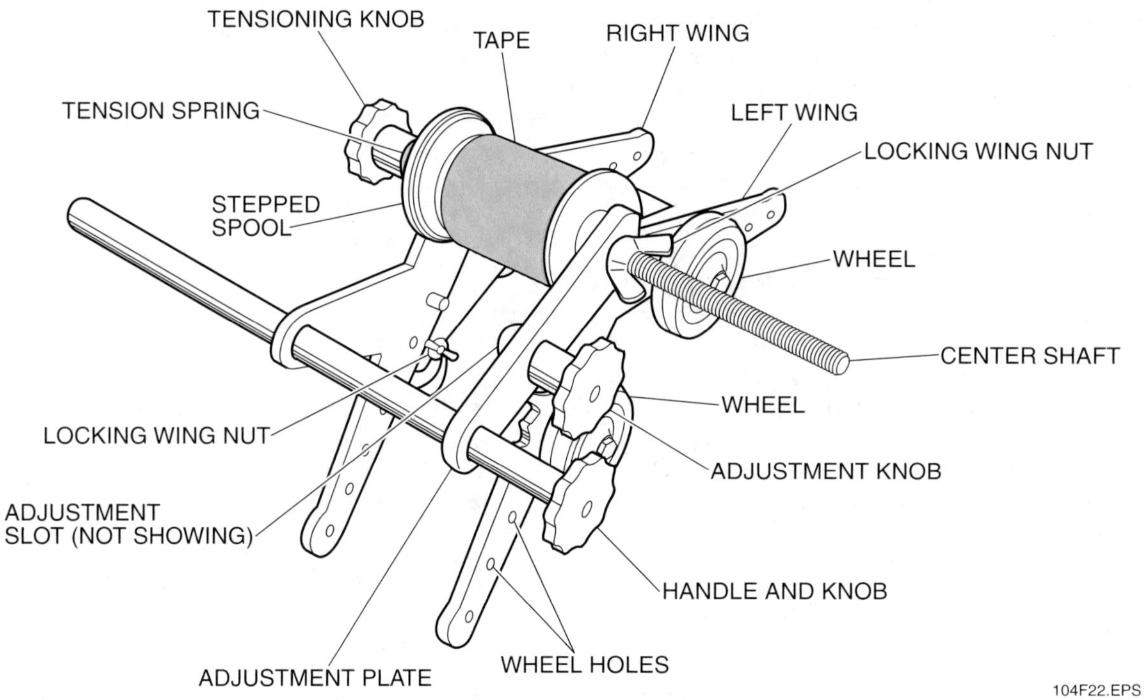

Figure 22 ◆ Hand coating/wrapping machine.

3.3.1 Pneumatic Air Horns

Pneumatic, explosion-proof air horns (see *Figure 23*) are compact, quiet, and highly efficient for moving large volumes of air. Operating on compressed air, they are ideal for ventilation in hazardous and/or volatile areas. Having no moving parts, they are safe and easy to maintain.

Figure 23 ◆ Pneumatic air horn.

3.3.2 Axial Fans

Pneumatic axial fans (see *Figure 24*) that are spark- and explosion-proof are available. The pneumatic fans are driven by a rugged, low-maintenance turbine directly from compressed air. These fans are especially suitable for ventilation, exhaust, and moving large air volumes through ductwork. They can be used both in normal atmospheres and where an explosion hazard is present in pump stations, refineries, and tank farms. Features available include protective grills on the intake side and provisions to mount flanges for installation.

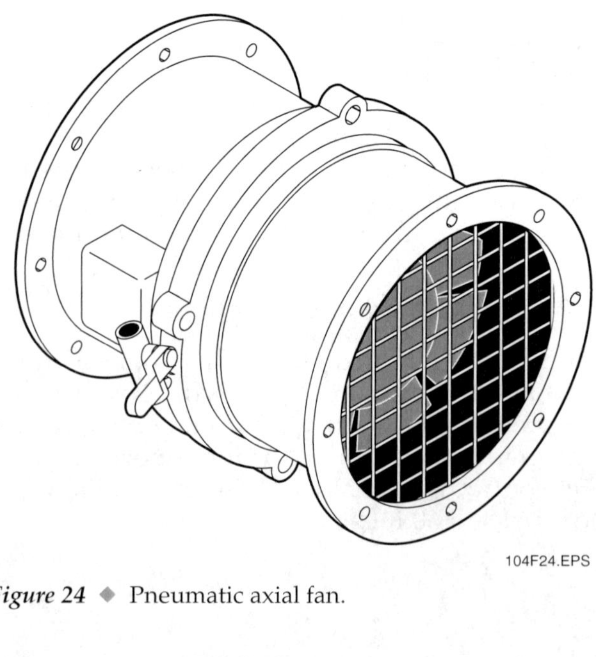

Figure 24 ◆ Pneumatic axial fan.

4.0.0 ◆ WELDING EQUIPMENT

Several different types of welding equipment and accessories are on the market to help pipeline maintenance technicians and fabricators weld pipe to the high quality standards required in the pipeline industry. You must follow company policies and procedures, which dictate the certifications required to make welds on pipe. *Figure 25* shows an engine-driven welding machine.

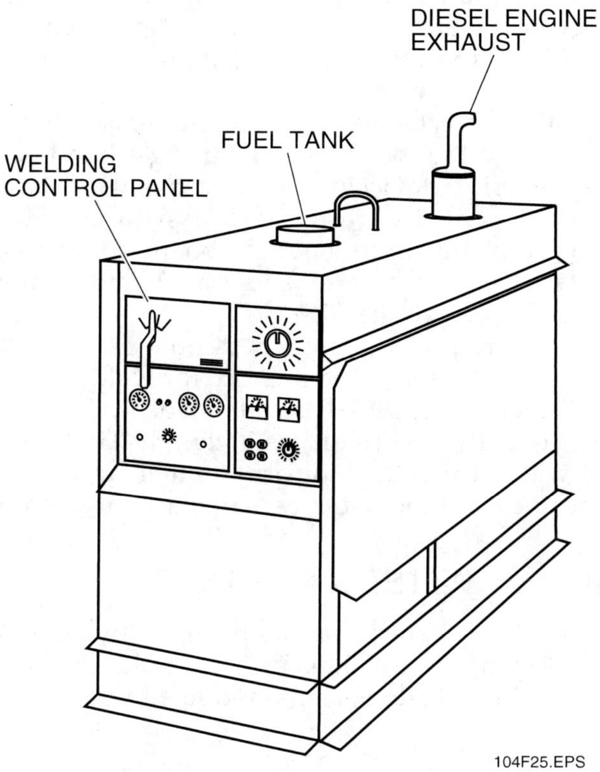

Figure 25 ◆ Engine-driven welding machine.

4.1.0 Beveling Machines

Various types of beveling machines are used to prepare pipe ends before a joint can be fitted up for welding. Pipe ends can be beveled using grinders, small-diameter pipe bevellers, nibblers and cutters, thermal beveling machines, pipe facing machines, and band-type beveling machines.

4.1.1 Beveling Using Grinders

Portable hand grinders can be used to prep bevel the pipe, but great care must be taken to make the bevel at the specified angle and keep the bevel square prior to welding.

4.1.2 Thermal Beveling (Beveling Machine)

Thermal joint preparation using an oxyacetylene torch is a quick and easy way to bevel pipe ends; however, the cutting torch leaves an oxide coating on the pipe that must be ground away before welding. The oxyacetylene pipe-beveling machine is fitted over the pipe end and tightened to the pipe (*Figure 26*). The torch is then adjusted to cut the correct angle. An electric drive motor moves the cutting torch around the pipe end to cut the preset bevel.

There are many different types of oxyacetylene pipe-beveling machines on the market. The best way to learn how to operate the pipe-beveling machine is through on-the-job training with qualified personnel or the equipment manufacturer.

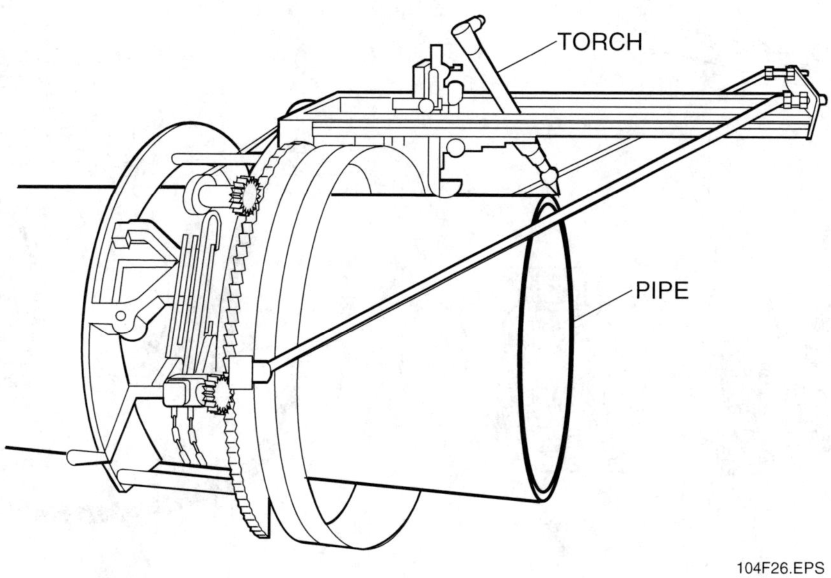

Figure 26 ◆ Cutting/beveling machine.

4.1.3 Pipe Facing Machines

Pipe facing machines (PFMs) (see *Figure 27*) ensure a precise and consistent narrow-gap bevel for high quality welds. The machine mounts to the pipe end, square with the pipe axis, by means of hydraulically operated shoes. A rotating faceplate, fitted with as many as six cutters, machines the desired bevel on the pipe end.

Major projects normally require the use of two PFMs. The units are mounted on sideboom tractors or A-frames. A facing operation is completed in two to five minutes, depending on pipe wall thickness and skill of the operator. Hydraulic power for each unit can be supplied by diesel or electric drivers. They can be tractor-mounted on the ground in close proximity to the workpiece.

Pipe facing machines are available for pipe diameters up to 60 inches.

4.1.4 Band-Type Machines

Band-type machines (see *Figure 28*) feature a rugged design for accurate cuts and long life. The transmission/torch unit, with an appropriate band, cuts any size pipe. A 9-inch clearance requirement allows for minimum space use, such as small bell holes.

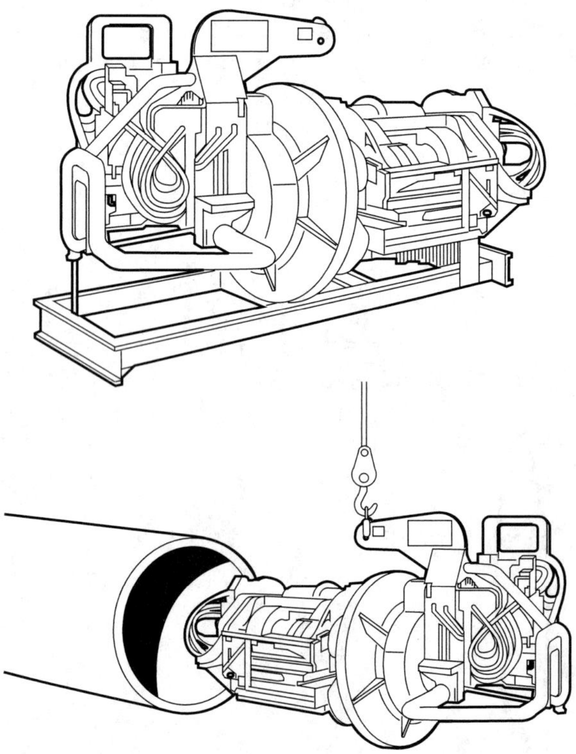

Figure 27 ◆ Pipe facing machine.

4.1.5 Line-Up Clamps

Prefabricated pipe welding clamps are available for both external and internal use to position pipe ends to be joined by welding. Internal alignment clamps are fitted with various types of wheels, depending on whether the pipe is coated internally or not, to allow the unit to easily be rotated and positioned inside the pipe. External clamps are constructed to overlap the two pipe faces to be joined by fitting around the exterior surface. Both types of clamps hold the pipe faces firmly in place to permit welding around the complete circumference of the pipe without having to adjust the clamp position.

Different types of external clamps match up and secure the two pipe ends using a hand-actuated levering device or, in the case of some larger models, with hydraulic jack assistance. Some internal clamp units have on-board compressed air tanks to retract or extend wheels and to move the unit through the pipe.

When required, line-up clamps can be site-fabricated to hold pipe in place for welding. Site-fabricated clamps are typically made up from iron bars, usually 6 to 16 inches in length. The clamps are welded together to form a circle around the pipe and hold the two pieces in place for welding.

5.0.0 ◆ METERS AND TESTERS

Meters and testers frequently used by pipeline maintenance technicians include multi-gas detectors, **holiday** testers, jeeps, and line locators.

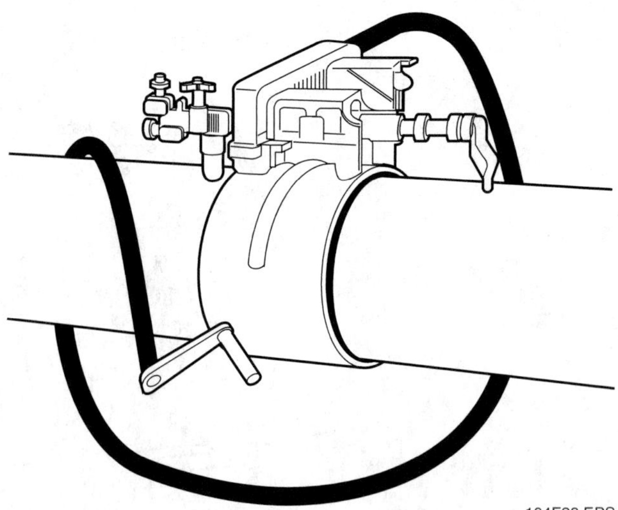

Figure 28 ◆ Band-type machines.

5.1.0 Multi-Gas Monitors

Multi-gas monitors (see *Figure 29*) are used to sample the air for **hazardous atmosphere**. The most common gas detector, the Draeger tube, consists of a disposable direct reading detector tube and a piston-type volumetric pump.

To use this type of gas detector, simply break off the ends of the detector tube and place the tube in the pump. When the pump handle is squeezed, a predetermined 50 cc or 100 cc sample of air is collected. The absorbing media immediately reacts with the gas or vapor being analyzed, and a constant color stain is produced. The stain will vary in length to show the concentration being measured.

It is imperative when using this type of gas detector that you use the proper type of detection tube for the vapor you are analyzing. After taking the reading from the detection tube, the tube is disposed.

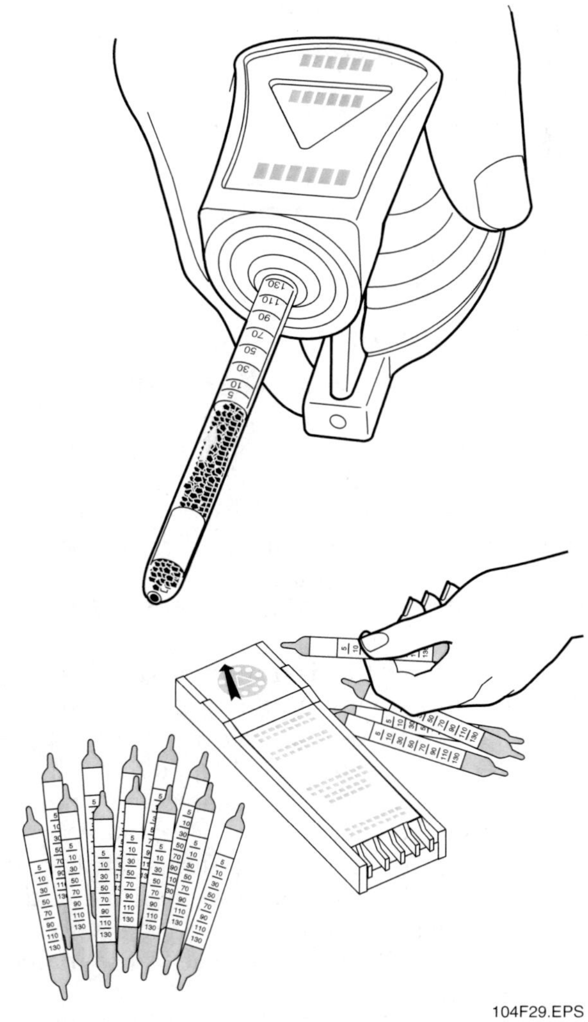

104F29.EPS

Figure 29 ◆ Multi-gas monitor.

Other models of multi-gas monitors include those designed to detect combustible gases, carbon monoxide, hydrogen sulfide, and low levels of oxygen. Some of these detectors also are equipped with dataloggers and modems that allow them to communicate directly with a base unit that can receive real-time readings, a function that can be very valuable in emergency situations. Selected models include lithium ion battery packs that enable the monitor to be used for up to 20 hours of continuous operation without requiring a re-charge.

5.2.0 Portable Gas Detectors

Many models of portable gas detectors can monitor as many as five atmospheric conditions simultaneously, including oxygen, combustible gas, and selected toxic gases. Toxic gas monitoring capability typically covers both carbon monoxide and hydrogen sulfide sensors. Most portable gas detectors are powered by interchangeable nickel-cadmium or alkaline battery packs that provide up to 12 hours of continuous operation. *Figure 30* shows a widely used portable gas detector, the Biosystems PhD5.

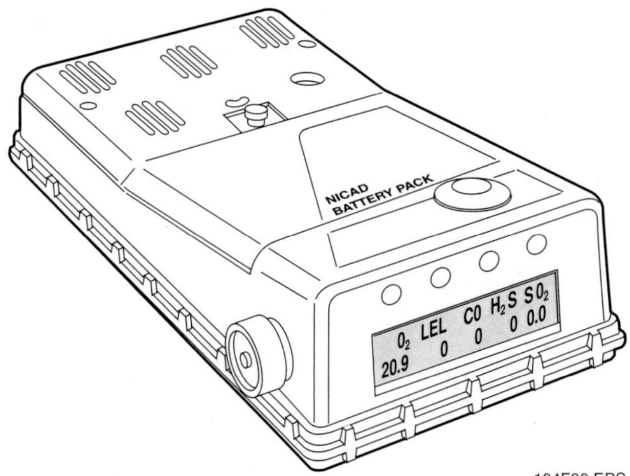

104F30.EPS

Figure 30 ◆ Portable gas detector.

5.3.0 Holiday Detectors

Pipeline maintenance technicians are sometimes called upon to perform conductivity tests, or holiday tests, on piping systems. Conductivity testing ensures that there is no unprotected piping that is not covered with coating. Conductivity tests use amperage to test the conductivity of a material. The holiday testers have variable amperage ratings, and the amperage rating used must comply with the specifications of the piping system.

The basic principle behind this technique is based on establishing an electric potential between the pipe and an external electrode with the help of the holiday detector. The holiday detector (see *Figure 31*)consists of a metallic ring that encircles the pipe; a handheld voltage source, known as a wand; and an alarm unit. The amount of voltage applied to the ring can be adjusted according to the type of coating.

Thicker coatings, such as coal tars, generally require a higher setting. Too much voltage, however, can damage the thinner coatings, such as tapes. The pipeline maintenance technician pushes or rolls the holiday detector along the pipe. When the detector passes over a gap, or holiday, in the coating, an electric spark completes the circuit between the electrode and the pipe. An alarm on the unit is then activated, and the exact location of the flaw is marked.

5.4.0 Jeeps

The output voltage of the holiday detector is subject to influence by weather conditions, moisture on the pipe, improper grounding, and soil conditions. To ensure the absolute highest accuracy and performance from the holiday detectors, special jeeps are used to **calibrate** detector voltage output. These jeeps are designed to accurately measure the output voltage of any holiday detector.

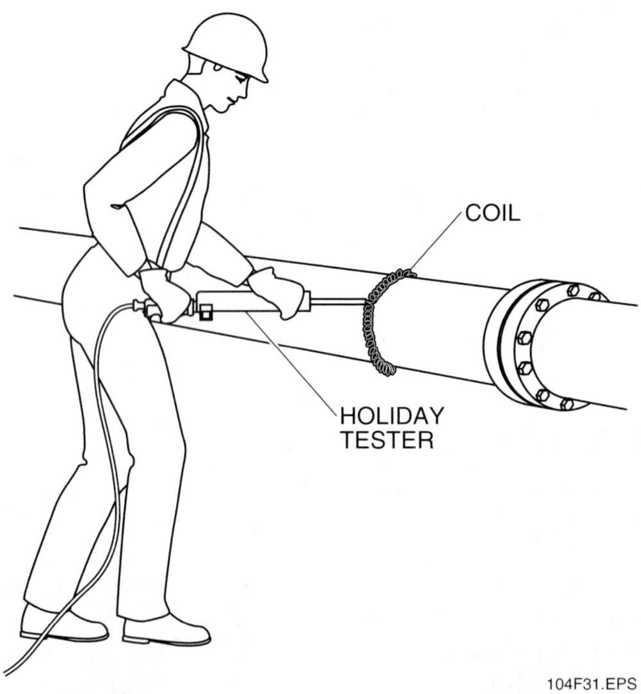

Figure 31 ◆ Holiday detector.

The digital jeep provides high accuracy and versatility. It features a digital display of measured voltages. The precision measurement display minimizes interoperation errors and is good for daily tests. Frequent calibration is necessary to ensure that holiday detectors are actually giving the output voltage they were set to give.

Other types of jeeps measure both pulse and DC voltage outputs. Pulse-type holiday detectors that are used to inspect shrink sleeves, tapes, and coal tars will measure up to 40 kilovolts (kV). DC holiday detectors that are used to inspect thin film epoxy coatings have a DC range up to 5 kV. *Figure 32* shows a digital jeep.

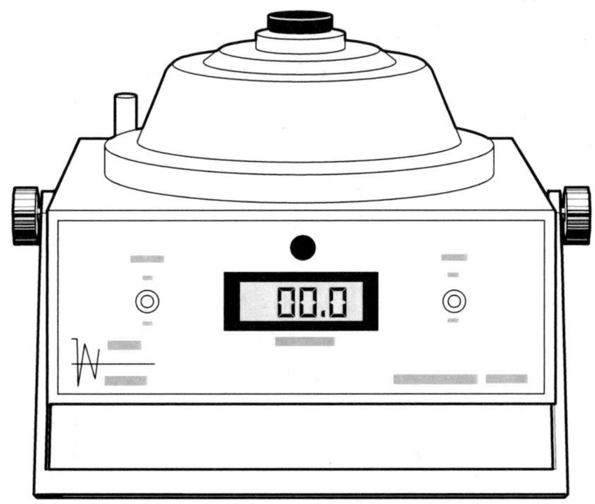

Figure 32 ◆ Digital jeep.

5.5.0 Line Locators

It is sometimes necessary to determine the location of buried items such as the pipeline to be worked on and the location of foreign utilities, ground bed cables, or other concealed metallic structures. Pipe and cable locators are used when trying to locate existing pipeline utilities. Pipe and cable locators impress an electrical signal onto the target line, and this permits a handheld receiver to detect their location.

The type of signal that is transmitted to the pipe or cable can vary from manufacturer to manufacturer and can even be variable within a single locator. Alteration of the signal frequency allows the operator to set the most appropriate signal for the application. For example, poorly coated pipes, well-coated pipes, shorted pipelines, and insulated cables might each be best located using different frequencies. It is imperative to remember to set both the transmitter and the receiver to matching frequencies.

Other locator features can be useful in cathodic protection work. Some transmitters have a damp accessory that can be dipped around a conductor to induce the signal, such as when tracing a live electrical cable. This technique saves time as well as enhances safety. The damp can be placed around an electrical conduit leading to a rectifier, for instance, so that unlocking and opening the rectifier enclosure becomes unnecessary.

Advances in locator technology have also enabled a variety of attributes to be incorporated into pipe and cable locators Depth measurement capability has greatly improved and allows the operator to measure, with increasing confidence, the depth of the target. Other pipe locators have the ability to measure and trace the amount of signal current being conducted by the pipeline or cable. These locators can trace the pipeline as well as improve the operator's ability to analyze and evaluate the pipe's current-carrying characteristics.

6.0.0 ◆ NONDESTRUCTIVE EXAMINATION METHODS AND TOOLS

Nondestructive examinations (NDE) are any tests that are used to test equipment and components without damaging them. The approach of nondestructive examination is a system of one or more types of tests that quantify defects or detect future failures.

With NDE, it is possible to detect defects or changes that will lead to premature failure. The results can be used to find the real cause of the change or abnormality and correct the problem, not just the symptom. *Table 1* lists the types of NDE.

Table 1 Types of NDE

Fluid	Type of Reading	Advantages	Limitations	Examples of Uses
Liquid penetrant	Detects surface openings due to cracks, porosity, seams, or folds.	Is inexpensive, easy to use, readily portable, and sensitive to small surface flaws.	Flaw must be open to the surface. Is not useful for porous materials.	Turbine blades for surface cracks or porosity.
Magnetic particles	Detects leakage of magnetic flux caused by surface or near-surface cracks, voids, inclusions, and material or geometry changes.	Is inexpensive and sensitive to both surface and near-surface flaws.	Is limited to ferromagnetic material; surface preparation and postinspection demagnetization may be required.	Railroad wheels for cracks. Detection of weld defects.
Ultrasonics	Detects changes in acoustic impedance caused by cracks, nonbonds, inclusions, or interfaces.	Can penetrate thick materials; excellent for crack detection; can be automated.	Normally requires coupling either by contact to surface or immersion in a fluid. Orientation, detection, or interpretation of the defect can present problems.	Adhesive assemblies for bond integrity. Detection of cracks.
Radiography	Detects changes in material density from voids, inclusions, material variations, and placement of internal parts.	Can be used to inspect a wide range of materials and thicknesses; is versatile; film provides a record of inspection.	Radiation safety requires precautions, is expensive, and detection of cracks can be difficult.	Pipeline welds for penetration, inclusions, and voids. Verification of parts in assemblies.
Visual-optical	Detects surface characteristics, such as finish, scratches, cracks, or color, and strain in transparent material.	Is often convenient and can be automated.	Can be applied only to surfaces, through surface openings, or to transparent material.	Paper, wood, or metal for surface finish and uniformity.

104T01.EPS

6.1.0 Dye Penetrant Test

Liquid penetrant inspection is a method of highlighting cracks and gaps that are not visible to the naked eye in the surface of smooth or nonporous samples. The liquid penetrant seeps into surface cracks, porosity, delaminations, or shrinkage areas by capillary action. It is also used to highlight wear patterns and measure stress cracking.

Liquid penetrant inspection involves wetting the surface of a test object so that the liquid flows over that surface to form a continuous and uniform coating and migrates into cracks and surface defects. The liquid is then washed off the surface, and a developer is placed on the surface. As the developer draws the penetrant out of the defects, it is stained by the penetrant and highlights the cracks and surface irregularities to form a visual image of the flaw.

With liquid dye penetrants, two types of indications are revealed: true indications and false indications. True indications are caused by the penetrant bleeding out from actual discontinuities in the metal. The standard defects indicated are fatigue cracks, underbead cracks, pits, and porosities. Large fatigue cracks are indicated by wide lines that become apparent quickly after the developer is applied. Underbead cracks are undersurface cracks that bleed through the surface. Underbead cracks are indicated by a line of dots that appear a few minutes after the developer is applied. Porosity is indicated by dots that come to the surface almost immediately.

The major reasons for false indications are failure to apply the penetrant correctly or rough, irregular surfaces of the test metal. The procedure for liquid dye penetrant testing must be carefully performed in order to achieve the proper results.

6.2.0 Magnetic Particle Test

When a magnetic field is generated in and around a part made of a ferromagnetic material, such as steel or iron alloys, and the lines of magnetic flux are intersected by a defect such as a crack, magnetic poles are generated on either side of the defect. Magnetic particle inspection works only for materials that can be magnetized, such as steel and nickel alloys.

In a magnetic particle test, the sample is magnetized by an electrical current, and magnetic particles are applied over the surface. These particles can be applied either dry or in a liquid carrier such as oil or water. The flaws on or just under the surface of the sample modify the magnetic field so that the particles outline the defect in a pattern and highlight the shape and location of the flaw.

This method is very sensitive to minor surface cracks and flaws that are near the surface. The disadvantages are the limitation to magnetic samples and the inability to read deeper flaws in the sample.

Maximum sensitivity is obtained when the part is oriented perpendicular to the magnetic field and the strength of the field saturates the entire part. The magnetic particle equipment uses either AC or DC and can be applied to the part with coils, probes, or flexible conductors that can wrap around a test piece.

Due to the variety of shapes of samples, the inductor for the test must come in a variety of sizes and shapes. Some sample parts, such as gears, balls, and ring-shaped parts, can be placed in the magnetic field without contact from a probe. Other samples use coils, clamps, central conductors, or other forms of direct contact with the sample. A welded joint would use a yoke to induce a magnetic field around the weld seam and show any discontinuities as extra particles scattered outside the weld seam. *Figure 33* shows the electromagnetic yokes of a magnetic particle tester.

6.3.0 Handheld Thickness Gauge

Handheld thickness gauges use high-frequency sound waves to detect the noise generated by plant equipment. Handheld thickness gauges measure thickness in materials or listen for turbulence in pipes to detect internal leaks and malfunctions. Other ultrasonic test instruments send focused, amplified sound waves into solid materials and compare them with the return wave to find flaws below the surface, such as corrosion, bad welds, or casting flaws. The frequency range for ultrasonic monitoring is from 20,000 Hz to 100 kiloHz.

A beam of ultrasound is similar to a beam of light. Both are waves and travel in a constant velocity in a consistent medium. The velocity depends on the medium, not on the properties of the wave form. Like a beam of light, ultrasonic waves are bent, reflected from surfaces, refracted when they cross boundaries of a different surface, and diffracted at edges or around obstacles. This beam, when reflected back, creates a signature that is interpreted as an image by the ultrasonic monitor.

Ultrasonic monitors can penetrate deep and with high sensitivity and accuracy. The output is readily digitized, and operation of the monitor is entirely nonhazardous. Some of the different types of ultrasonic techniques are pulse-echo, pulse-transmission, ultrasonic attenuation, continuous wave resonance, and ultrasonic spectroscopy. Ultrasonic monitors are very portable but require experienced technicians to monitor equipment. Reading an ultrasonic flaw detector (see *Figure 34*) requires certification because the screen image shows the peaks of back reflection.

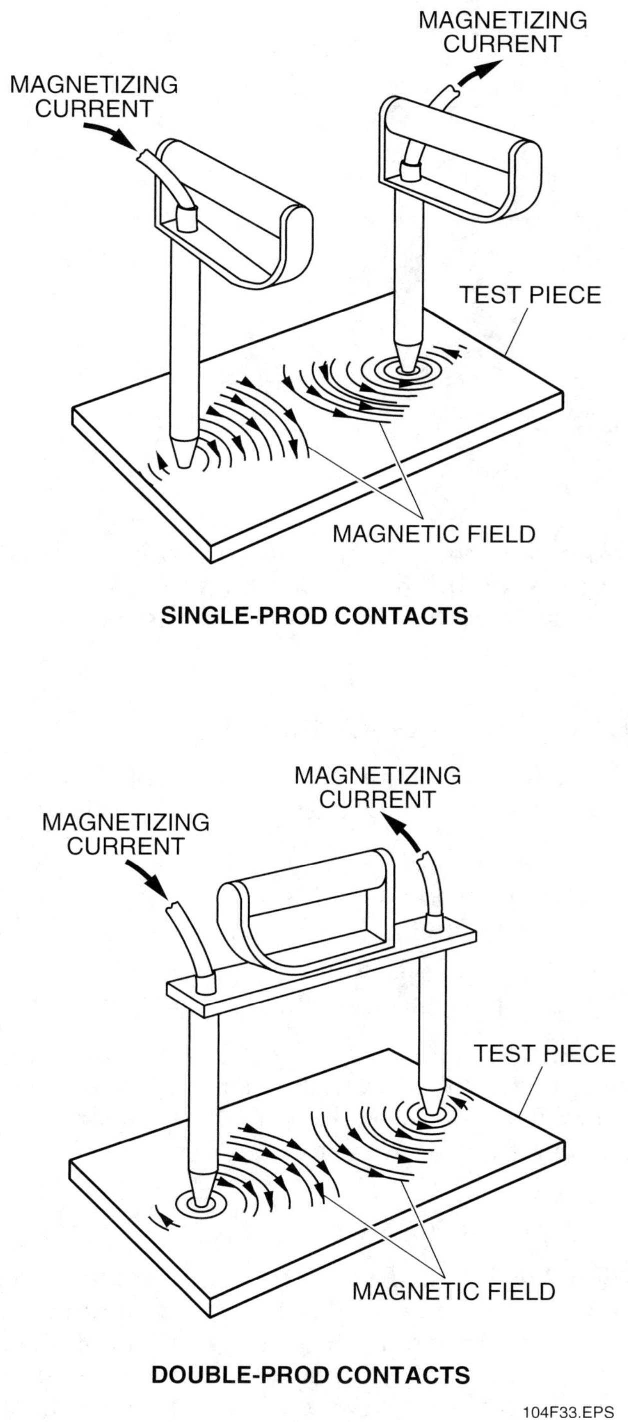

Figure 33 ◆ Electromagnetic yokes of a magnetic particle tester.

Figure 34 ◆ Ultrasonic flaw detector.

6.4.0 Radiography

Radiography is a nondestructive examination method that uses a beam of X-ray radiation that penetrates the weldment. X-rays are very sensitive and can be used to inspect the density of almost any ferrous, nonferrous, organic, or inorganic material. Some of the radiation energy is absorbed, and the intensity of the beam is reduced. The variations in the beam intensity are recorded on film. These variations are seen as differences in shading that are typical of the types and sizes of any discontinuities present.

X-rays can detect both surface and subsurface discontinuities. Because of the extremely powerful nature of gamma rays, much radiographic work is now done with radium or Cobalt 60 sources. The principal advantages of gamma ray inspection are the low cost and portability of the source. Safety is an essential part of radiography, and special precautions must be taken when working with X-ray materials to protect yourself from overexposure to radiation.

The X-ray method is used to detect internal defects and to check for proper alignment of assembled parts. X-rays are also used extensively to examine welds, check wall thickness, and examine the quality of casting and forging. *Figure 35* shows how an X-ray is made using gamma rays.

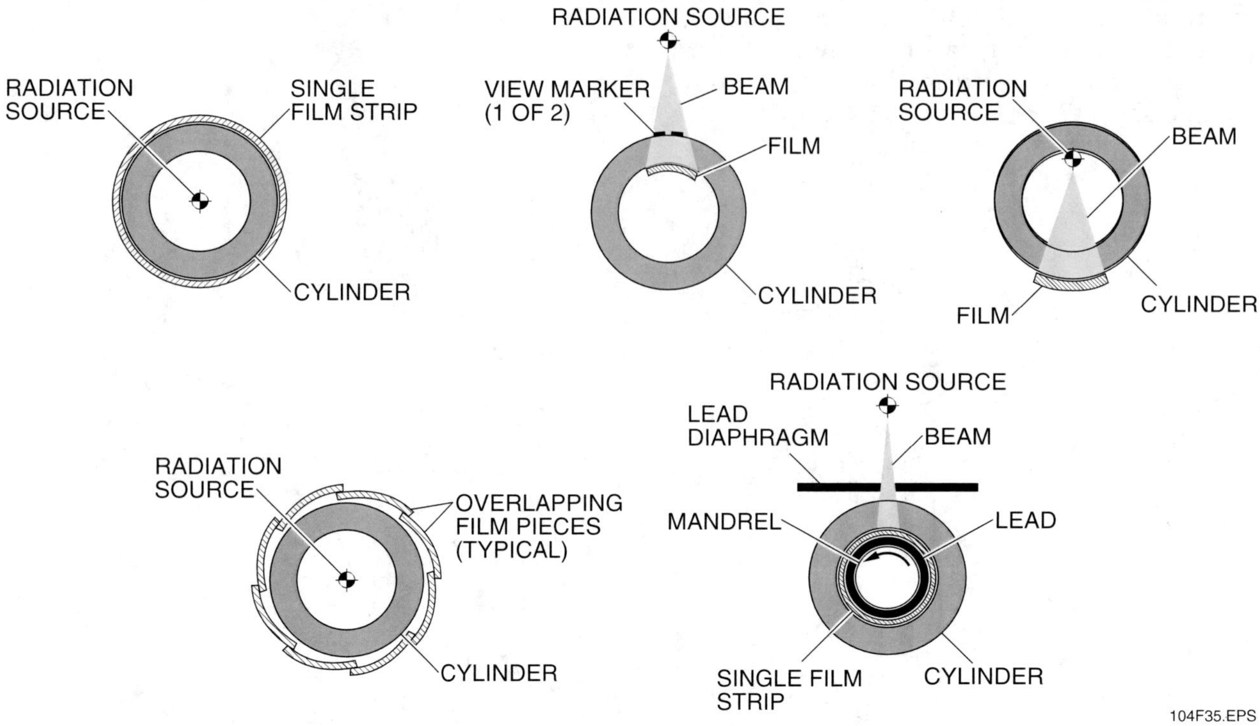

Figure 35 ◆ X-ray made using gamma rays.

6.5.0 Visual and Optical Inspection

Visual and optical inspection of equipment is the easiest to perform but is the most likely to be overlooked as an inspection technique. There should be a systematic approach to visual inspection so that no details or events are missed. Make a detailed observation while the equipment is running and follow the path of the product, carefully watching every step and noting any abnormalities. Then, with the equipment off or in a safe position, look for any external detail that could provide a clue to potential trouble. Excessive corrosion, bent shafts, leaking seals, and any sign of abnormality should be documented in detail.

Hard-to-reach, internal, or hot parts of the equipment can be inspected with optical borescopes and fiberscopes. The optical borescope (see *Figure 36*) is designed to reach deep inside a machine, magnifies like a microscope, and has its own light source. The fiberscope is flexible because of its fiber-optic cable and can provide a close look in areas that can normally be seen only after the equipment is disassembled.

7.0.0 ◆ HYDRAULIC CRANES AND HEAVY EXCAVATING EQUIPMENT

Pipeline maintenance personnel must understand the types, applications, and safety hazards of cranes and heavy excavating equipment, although they are not required to operate this equipment because operators must be licensed. Pipeline maintenance personnel must also be familiar with the special types of rigging harness used to carry pipe.

7.1.0 Hydraulic Cranes

The basic operational characteristics of all mobile cranes are the same. These include the following:

- Adjustable boom lengths
- Adjustable boom angles
- Ability to lift and lower loads
- Ability to swing loads
- Ability to travel about the job site

The most common types of mobile cranes that a pipeline maintenance technician is exposed to are boom trucks, cherry pickers, carry decks, and pipeline side boom tractors.

7.1.1 Boom Trucks

Boom trucks (see *Figure 37*) are unlike any of the other mobile cranes in that they are mounted on carriers that are not solely designed for crane service. These cranes are mounted onto a commercial truck chassis that has been strengthened to carry the crane. The boom truck can also be used to haul loads and be driven on the highway system when the boom is in the transport position. Boom trucks typically range in capacity from 6 to 22 tons.

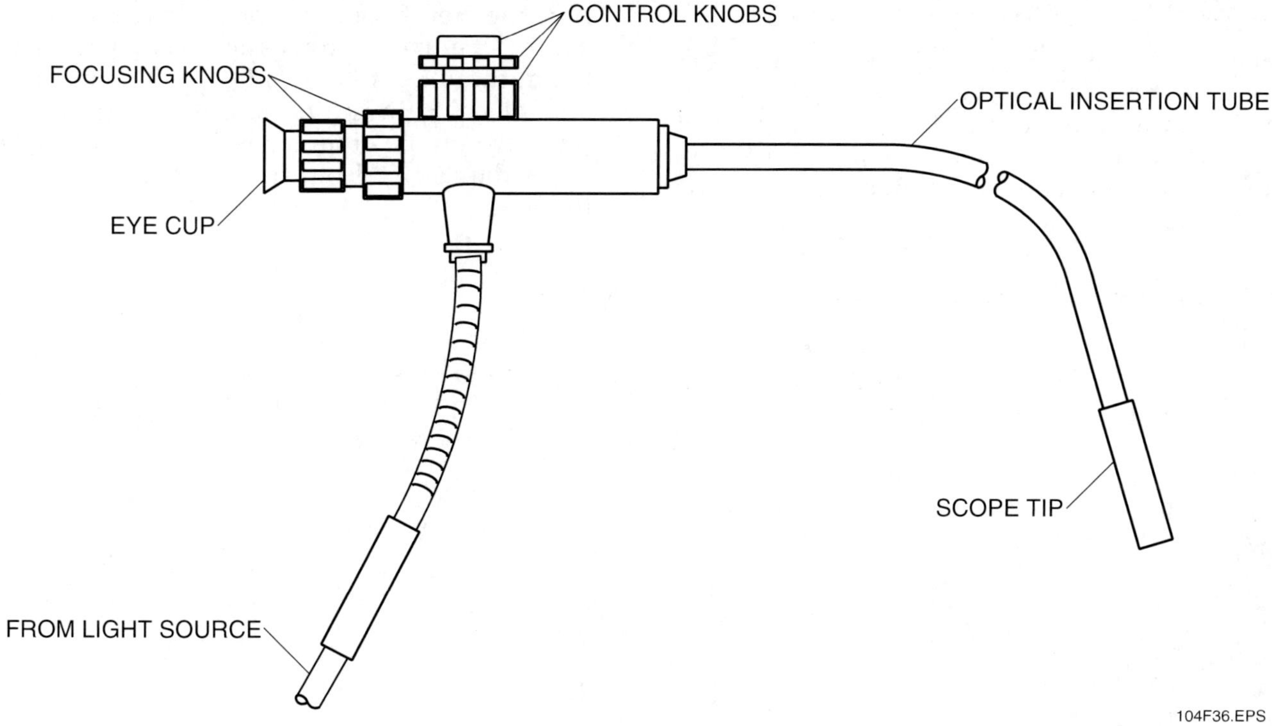

Figure 36 ◆ Optical borescope.

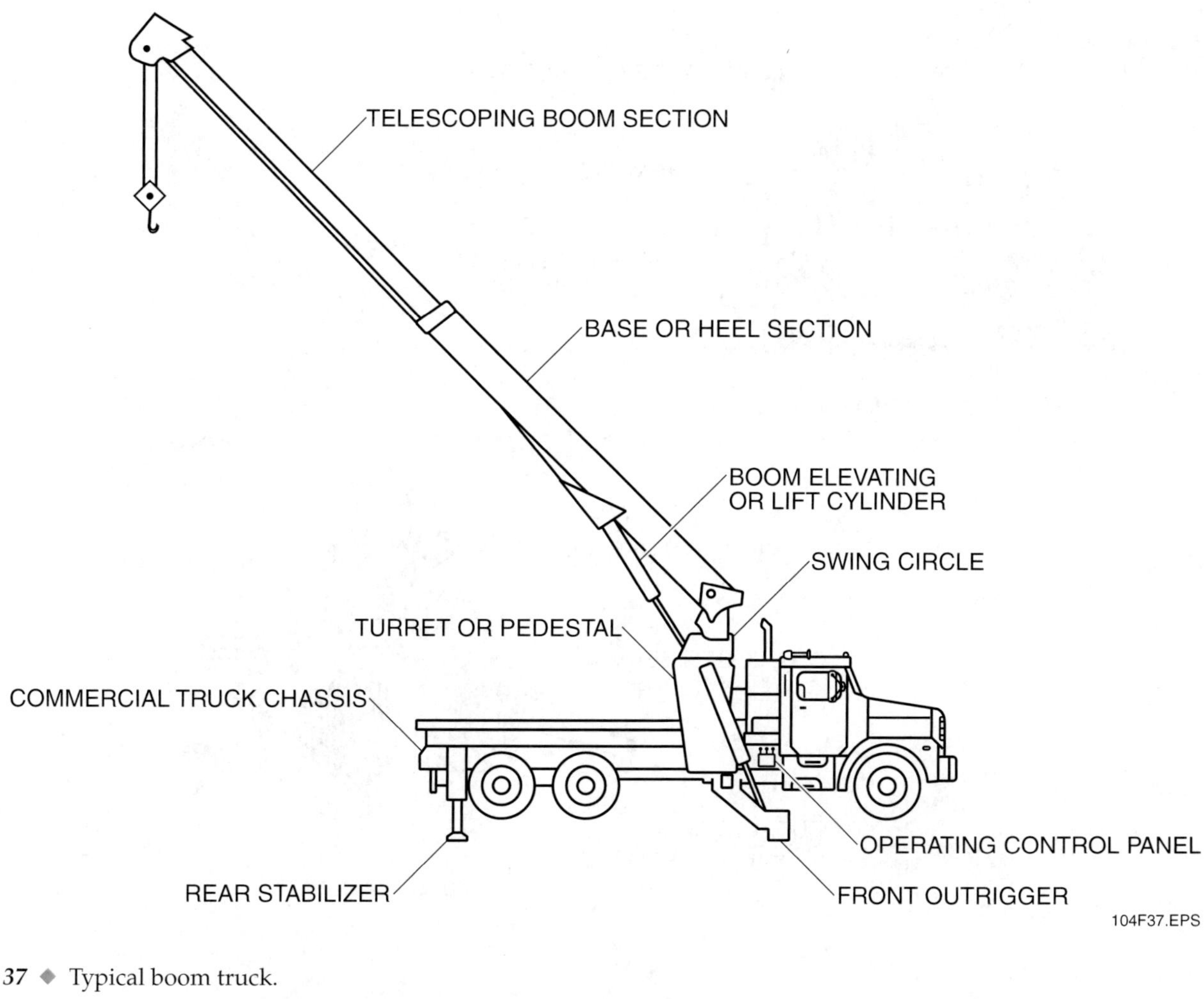

Figure 37 ◆ Typical boom truck.

7.1.2 Cherry Pickers

Cherry pickers (see *Figure 38*) are rough-terrain mobile cranes. They have oversized tires that allow them to move across the rough terrain of construction sites and other broken ground. Two types of cherry pickers are the fixed cab and the rotating cab.

7.1.3 Pipeline Side Boom Tractors

A pipeline side boom tractor (see *Figure 39*), also known as a pipe layer, lays pipe down a trench while moving along the side of the trench. The pipeline side boom tractor is driven by a diesel engine. It moves on tracks and has a winch-driven boom mounted to one side. The boom lifts pipe and places it in the trench. A counterweight is mounted on the other side of the tractor and prevents the tractor from tipping over when lifting the pipe.

7.1.4 Trackhoes

Trackhoes (see *Figure 40*) come equipped with a hydraulically powered bucket with a hydraulically operated closing lower "jaw" attached to help with grasping.

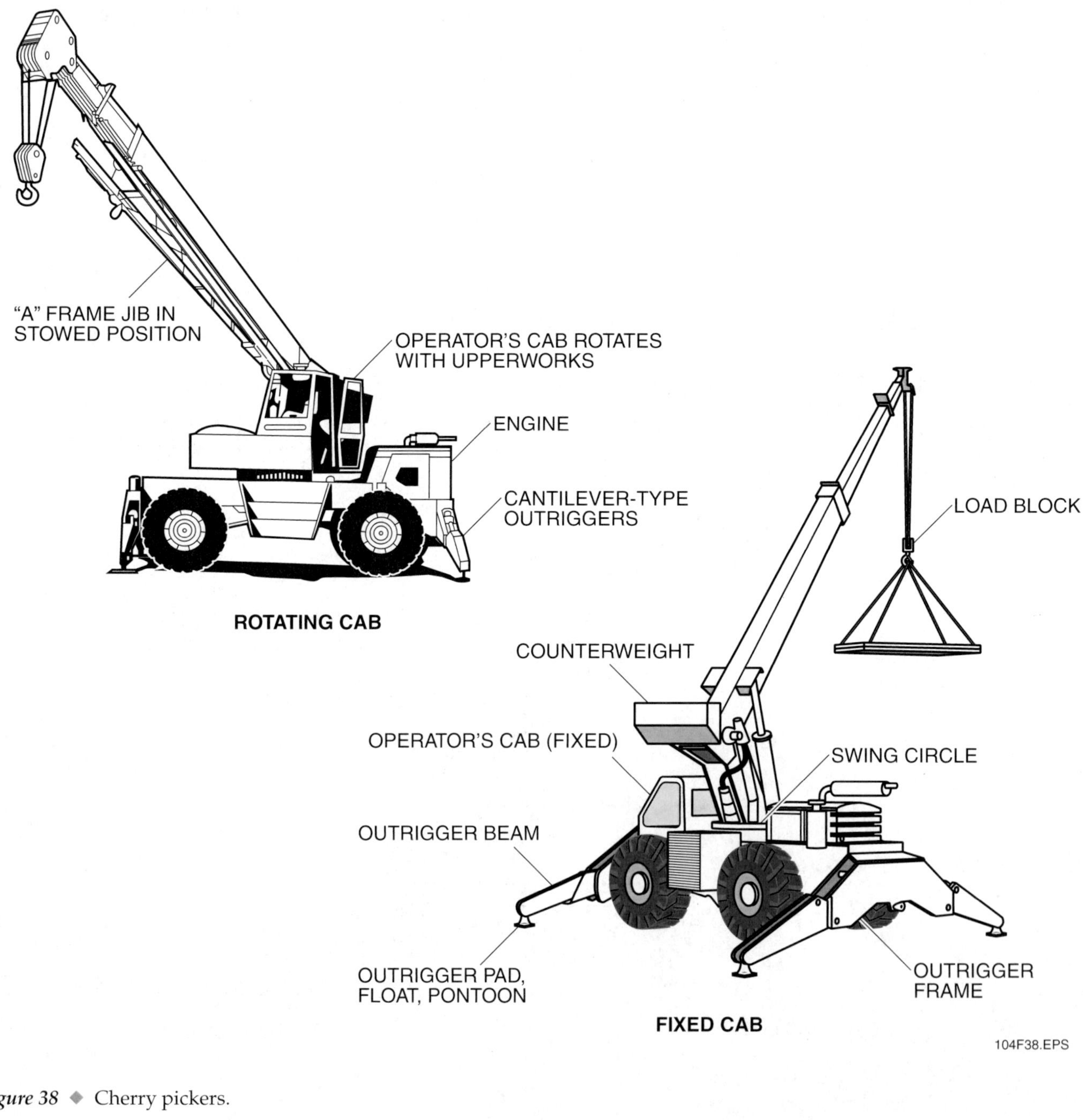

Figure 38 ◆ Cherry pickers.

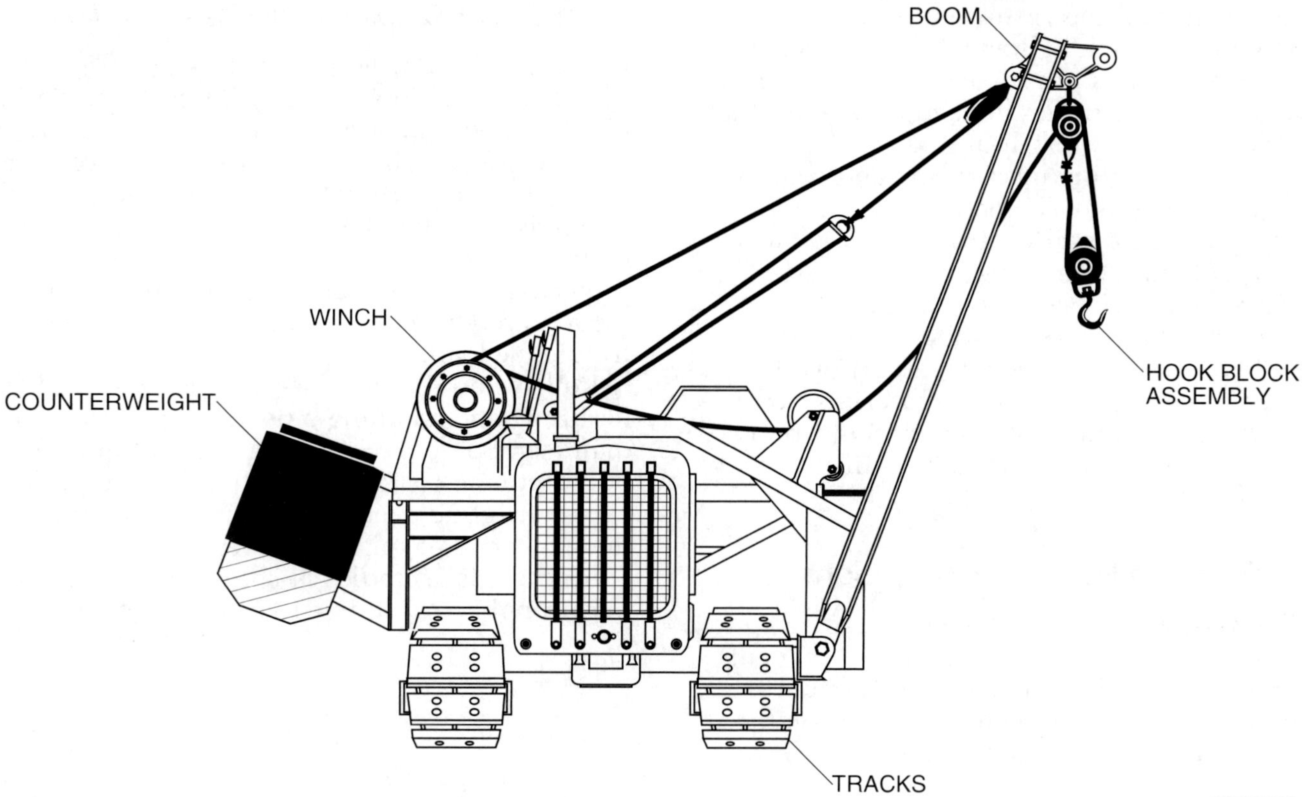

Figure 39 ◆ Pipeline side boom tractor.

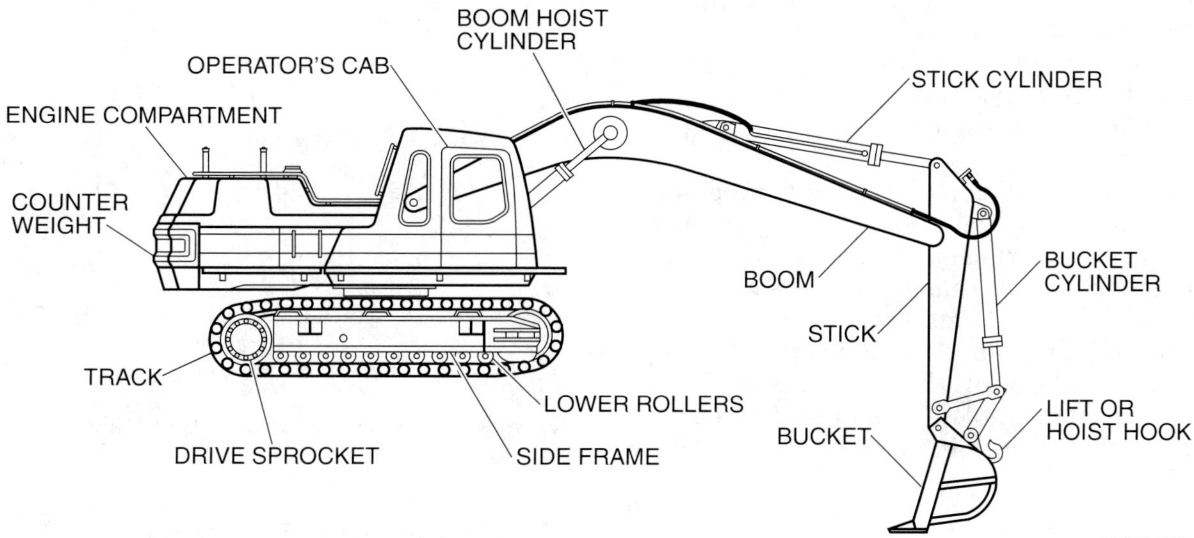

Figure 40 ◆ Trackhoe.

Trackhoes are used for various applications such as:

- Digging flowline
- Digging dugouts
- **Backfilling** line leaks
- Loading trucks with clay
- Pulling subsoil/topsoil material back on leases to allow other equipment to be able to get behind material on fenced locations

7.2.0 Overhead Cranes

Electrically powered overhead cranes, with a wide range of lifting capacities, have become a mainstay throughout industry. Overhead cranes are available in both free-standing (self-supporting frame) and building-supported configurations. Overhead cranes employed in industrial settings range from small, 1-ton capacity light cranes serving a shop or work center to heavy-duty, rail-supported models used on production lines, in maintenance shops, and in warehouses capable of lifting 50 tons or more.

Overhead cranes are made up of the rail or frame system, the trolley that moves along the rails, electric trolley drive unit, electric brakes, hoist motor and lifting mechanism, and controls. The largest cranes have an on-board cabin for the operator while others are controlled from the work floor by pendant-mounted controls or through the use of a remote radio-signal control unit.

Crane operators are certified in accordance with OSHA guidelines. Never use an overhead crane to move equipment or materials before reviewing your company's policies and procedures covering the training and certification of crane operators.

7.3.0 Pipeline Rigging Equipment

Pipeline rigging equipment ensures secure control in placement of the pipe and also prevents damage to the **pipe coating** and structural integrity of the pipe. Some of the more common pipeline rigging equipment is explained below.

7.3.1 Lowering-In Belts

Lowering-in belts are made up of a steel lifting iron and a fabric **sling** formed of several bands, or layers, of high-grade rip-resistant nylon. Each band should be manufactured to the same length, width, and stretching capacity as other bands in the sling.

The slings are flexible and easy to handle and conform to the pipe being lifted, even at very low temperatures. Lowering-in belts (see *Figure 41*) are available to handle a range of pipe sizes from 12 to 66 inches in diameter. Larger sizes may be special ordered. They are designed to permit easy hooking and unhooking by one person. Each model can be disassembled for storage and transport to and from the job site.

104F41.EPS

Figure 41 ◆ Lowering-in belt.

7.3.2 Roller Cradles and Wheel Cradles

Roller cradles and wheel cradles (see *Figure 42*) are designed to lower-in and lift-out pipe during pipeline construction and maintenance activity. These cradles easily meet a broad range of lifting and other operational requirements. Cradle design is such that the a range of pipe sizes can be handled with the same size cradle. The pipe is protected from damage if loaded properly using the required number of cradle units to evenly distribute the load.

Roller cradles and wheel cradles are available for lifting and handling pipe from 4 to 60 inches in diameter. Some models of cradles are fitted with polyurethane-coated rollers that have high load capacity, good wear resistance, and elasticity under load that minimize breakage during use. Some models of cradles can be used with a matching strongback to couple two cradles together.

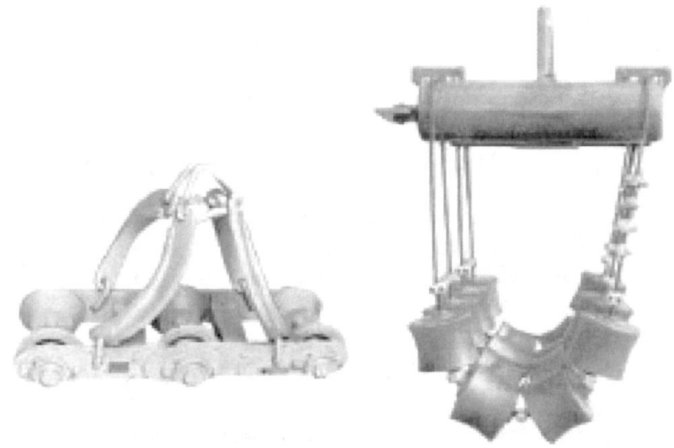

104F42.EPS

Figure 42 ◆ Roller cradles and wheel cradles.

7.3.3 Lined Steel Choker Belts

Lined steel choker belts (see *Figure 43*) can be used to align pipe precisely for tasks like bending or welding. A distinct advantage to using choker belts is that these belts allow you to roll the pipe, pull the pipe horizontally, and secure extreme bends in the pipe without damaging the pipe or any applied coating.

Choker belts are adjustable to accommodate differences in coating material thickness. Choker belts are available in sizes that will handle pipe diameters up to 60 inches and lift loads up to 27,000 pounds.

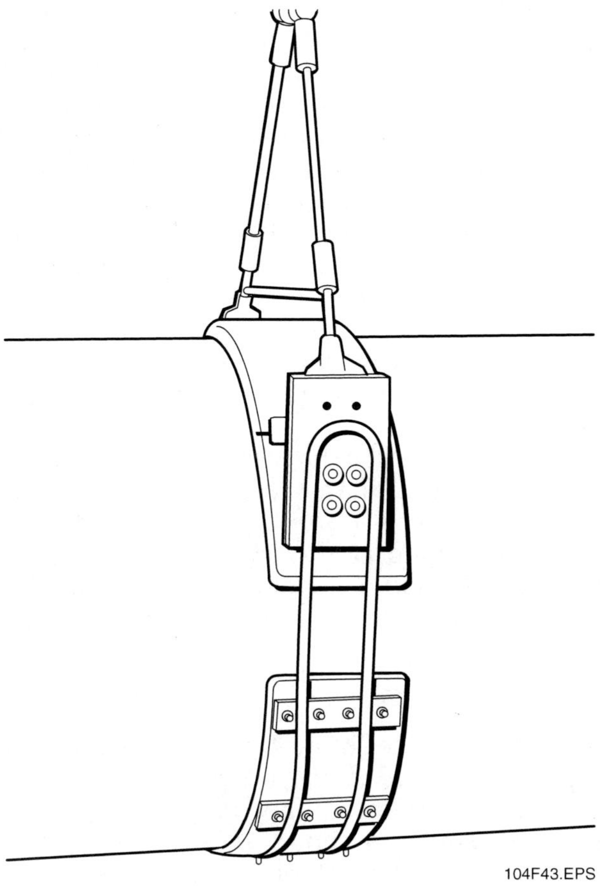

Figure 43 ◆ Lined steel choker belts.

7.3.4 Pipe Handling Calipers

Pipe handling calipers, sometimes referred to as pipe handling tongs, are used along with lowering-in belts, cradles, and slings to lift and position pipe for storage, construction, and maintenance tasks. Calipers are usually available to handle pipe up to 38 inches in diameter. Larger calipers can be special ordered when required.

Some calipers are designed so that a single individual can position a section of pipe and then release the calipers, eliminating the need for a second person to be stationed in a position to release the pipe.

7.4.0 Excavation Equipment

This section provides a general overview of heavy excavating equipment used to install, replace, and maintain pipelines and pipeline facilities. Types of excavating equipment explained include backhoes, loaders, and bulldozers.

7.4.1 Backhoes

A backhoe (see *Figure 44*) is a hydraulically operated unit mounted on the back of a wheel-type tractor (or sometimes on a track-type tractor). The most common activities that backhoes are used for include:

- Digging and cleaning trenches
- Excavating hard to reach areas
- Hoisting light loads

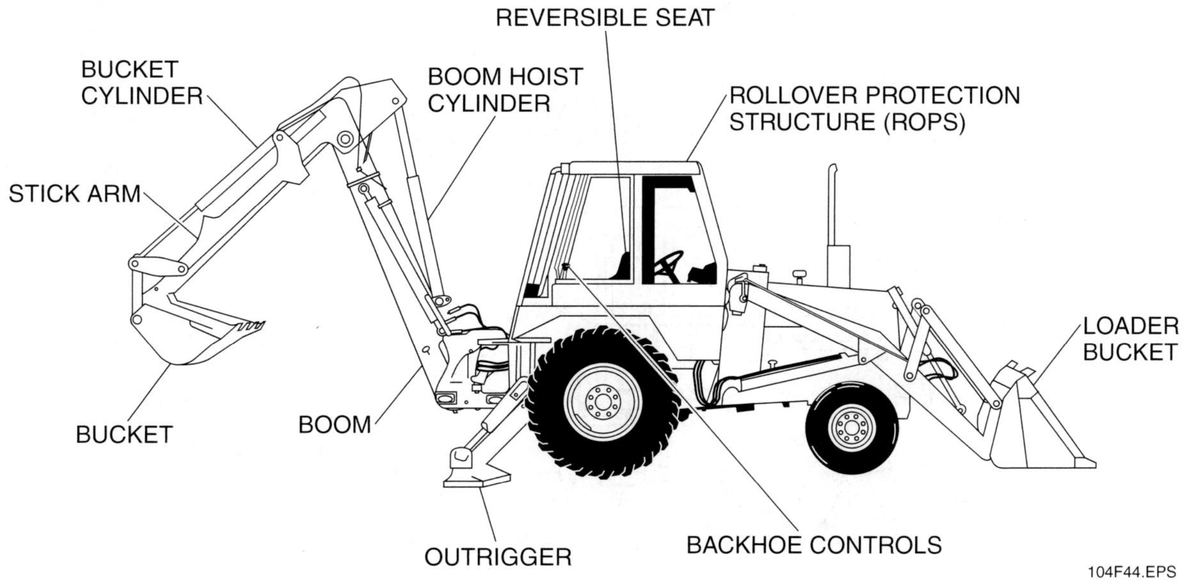

Figure 44 ◆ Backhoe.

There are numerous type of backhoes such as:

- Rubber-tired tractor-type, hydraulically operated
- Truck-mounted, cable-operated
- Track-type, hydraulically operated
- Trenching machine with backhoe accessory

The tractor-mounted backhoe is smaller than most other hydraulic excavating equipment. It has a limited swing compared to the swing of the tracked excavator or telescoping excavator and is less powerful. The tractor-mounted backhoe is a useful piece of equipment because it is very mobile and can be maneuvered easily.

The backhoe attachment is usually mounted on a rubber-tired two- or four-wheel drive tractor. It may also be fitted onto a crawler tractor or a truck. The tractor usually has permanent mountings, such as a three-point hitch, to which the backhoe can be attached with pins. Most models are backhoe front-end loader combinations with permanent attachments.

7.4.2 Loaders

Backhoe loaders are used for a wide range of jobs. They include:

- Loading trucks
- Stockpiling
- Rough grading
- Digging
- Hauling materials and equipment
- Removing snow
- Backfilling

Understanding the operation of a backhoe loader and use of the bucket and other accessories is very important. The operator not only needs to know how to operate the machine and its accessories effectively, but also must understand the material being handled. This includes knowing the general properties of such materials as clay, sand, mud, and different types of aggregate.

The two categories of backhoe loaders are the wheel-type and the track-type. The wheel-type loader has rubber tires that are driven by either a gasoline or diesel engine. Early models of wheel-type backhoe loaders were on a rigid frame with only the rear wheels being driven. *Figure 45* shows the basic parts of a wheel loader.

Newer wheel-type backhoe loaders are articulated and allow for tighter turning and better maneuvering.

The track-type backhoe loader (see *Figure 46*) is mounted on tracks and looks very much like a bulldozer with a loading bucket instead of a blade. The engine for this loader is usually diesel. This type of backhoe loader comes in many sizes with bucket capacities up to four cubic yards.

7.4.3 Bulldozers

The bulldozer (see *Figure 47*) is designed to move large amounts of earth and provide maximum power and stability at speeds of up to approximately 8-10 miles per hour. Although it is not very fast, its good traction, stability, and lower center of gravity make it especially suited to work on irregular terrain and steep slopes.

Bulldozers may also be called crawler tractors, track-type tractors, dozers, or crawlers. They come in many different configurations and sizes and are made by a number of manufacturers. However, they all have the same basic features and perform the same operations.

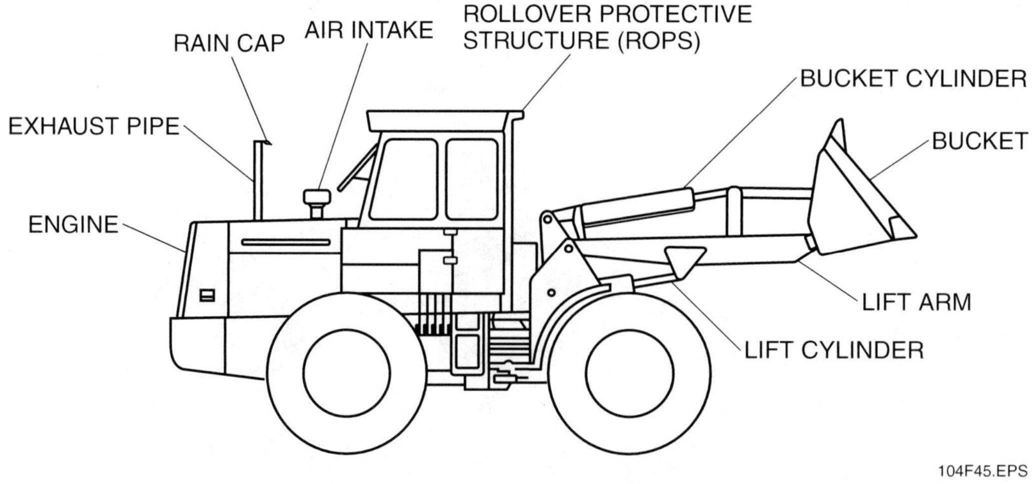

Figure 45 ◆ Basic parts of a wheel loader.

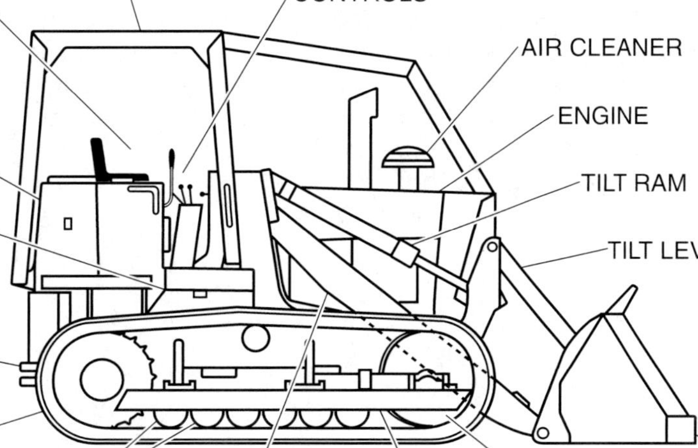

Figure 46 ◆ Track-type backhoe loader.

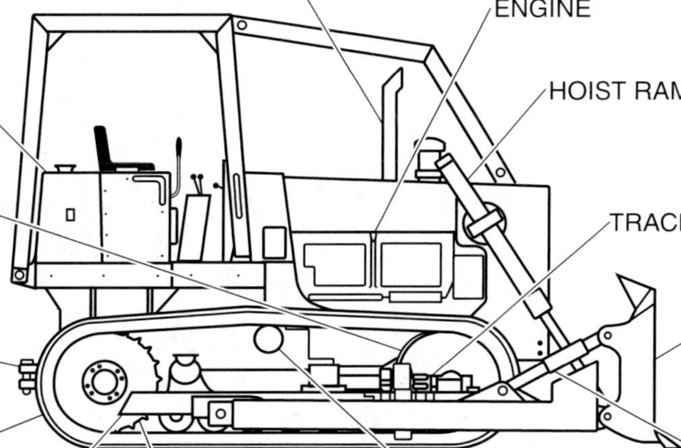

Figure 47 ◆ Bulldozer.

Understanding how a bulldozer operates and how to set up the blade and other accessories for proper operation is very important. The operator must know how to operate the machine and its accessories effectively, aso well as the characteristics of the material being moved. This includes understanding the general properties of materials such as clay, sand, mud, different types of rock, and different types of manufactured materials like asphalt and concrete.

Because the bulldozer is equipped with tracks instead of wheels, it provides the maximum amount of traction. The primary uses of the bulldozer are as follows:

- Excavate, cut, fill, backfill, and spread material
- Maintain haul roads and temporary roads through rocky or rough terrain
- Clear construction sites of debris, clear land, and demolish stumps and other buried items
- Tow accessory equipment or push other heavy equipment

Summary

The purpose of this training module is not to teach you how to use all of the equipment presented here but to introduce you to the types of equipment that you may encounter on the job as a pipeline maintenance technician. Before using any piece of equipment on the job, you must be properly trained in the correct use of the tool as to prevent injury to yourself or to your co-workers. Before using any tool, the tool must be inspected to ensure that it is in proper working condition and free of defects, excessive wear, and other damage that might make the tool inoperable or dangerous.

Review Questions

1. Never use a probe rod to locate a(n) _____ line.
   a. water
   b. electrical
   c. drain
   d. gas

2. The only wrench designed to be hit with a hammer is a _____ wrench.
   a. standard
   b. T-handle
   c. slug
   d. bolt-thru

3. A drift pin that is tapered on both ends is called a _____ pin.
   a. hole
   b. flange
   c. alignment
   d. barrel

4. Rotary pipe cutters will cut pipe up to _____ inches in diameter.
   a. 24
   b. 32
   c. 34
   d. 36

5. T-handle flange spreaders make _____ easier.
   a. pipe bending
   b. pipe removal
   c. gasket replacement
   d. maintenance

6. Hydraulic flange spreaders operate at pressures of _____ tons.
   a. 2 to 4
   b. 4 to 5
   c. 4 to 8
   d. 5 to 10

7. Most taps and connections are made with the main line _____.
   a. shut down
   b. in service
   c. removed
   d. disassembled

8. The T-101 series hand drilling and tapping machine is used to make _____ inch taps.
   a. ½ to 4
   b. 1 to 6
   c. 2 to 8
   d. 6 to 10½

9. Portable compressors have a safety value that is usually set _____ psig above the maximum allowable pressure.
   a. 5
   b. 10
   c. 15
   d. 20

10. Cleaning/priming machines are available for pipe up to _____ inches in diameter.
    a. 20
    b. 40
    c. 60
    d. 80

11. Air horns operate on _____.
    a. electricity
    b. compressed air
    c. gasoline
    d. hydraulics

12. To square the ends of a pipe, a _____ is used.
    a. beveling machine
    b. grinder
    c. pipe facing machine
    d. bandsaw

13. Handheld thickness gauges use _____ to check pipe wall integrity.
    a. light beams
    b. vibration
    c. sound waves
    d. infrared beams

14. Pipeline side boom tractors are sometimes called _____.
    a. trenches
    b. trackhoes
    c. pipe layers
    d. excavators

15. Bulldozers are capable of top speeds from _____ miles per hour.
    a. 8–10
    b. 10–12
    c. 13–14
    d. 14–16

GLOSSARY

Trade Terms Introduced in This Module

Backfilling: The technique for covering a completed pipeline so that adequate fill material is underneath and over the pipe to protect it from loose rocks, abrasion, shifting, and washouts.

Calibrate: To determine, by measurement or comparison with a standard, the correct value of each scale reading on a meter or other device.

Excavation: A man-made cavity or depression in the earth's surface formed by earth removal. This includes a trench, deep foundation, tunnel, shaft, or open excavation, but does not include borrow pits, gravel pits, and quarries unless specified by the local regulations or safety and health officer/inspector.

Hazardous atmosphere: An atmosphere that is identified as being explosive, flammable, poisonous, corrosive, oxidizing, irritating, oxygen deficient, toxic, or otherwise harmful, which may cause illness, injury, or death.

Holiday: An imperfection in the pipe coating that exposes unprotected surface to the environment.

Pipe coating: Dielectric coating on pipe to provide electrical insulation from other materials.

Skid: Pipeline skids are timbers used to support and stabilize the pipe off the ground during construction or maintenance work. These timbers, which typically measure 4" × 6" × 36", are usually cut from oak or other hardwood for strength and durability. Skids are built up one level at a time until the structure is sufficient for the task. Skids can be used as a single support or in multiples for larger jobs.

Sling: A wide rubber and fabric sling at the end of the boom cat's hoisting lines that is used for lowering in or handling coated or wrapped pipe.

Tap: A hole or opening in a line into which a gauge or valve may be inserted and screwed tight.

Tapping (hot): To tap into a pipeline that is in service and contains a hazardous liquid and/or vapors.

Trench: An excavation having a depth that exceeds its width measured at the bottom.

ACKNOWLEDGMENT

Figure Credit

Ridge Tool Company 104F07

NCCER CURRICULA — USER UPDATE

NCCER makes every effort to keep its textbooks up-to-date and free of technical errors. We appreciate your help in this process. If you find an error, a typographical mistake, or an inaccuracy in NCCER's curricula, please fill out this form (or a photocopy), or complete the online form at **www.nccer.org/olf**. Be sure to include the exact module ID number, page number, a detailed description, and your recommended correction. Your input will be brought to the attention of the Authoring Team. Thank you for your assistance.

Instructors – If you have an idea for improving this textbook, or have found that additional materials were necessary to teach this module effectively, please let us know so that we may present your suggestions to the Authoring Team.

NCCER Product Development and Revision

13614 Progress Blvd., Alachua, FL 32615

Email: curriculum@nccer.org
Online: www.nccer.org/olf

❑ Trainee Guide ❑ Lesson Plans ❑ Exam ❑ PowerPoints Other ______________________

Craft / Level: Copyright Date:

Module ID Number / Title:

Section Number(s):

Description:

Recommended Correction:

Your Name:

Address:

Email: Phone:

Module 62105-02

Pipeline Documentation

Introduction to the Pipeline Operator Qualification Rule and NCCER

It could be said that petroleum products keep America running. Petroleum products like oil and natural gas are the principal sources of energy in the United States and are used in many manufacturing and chemical processes. Petroleum products are primarily transported by pipeline. There are more than a million miles of liquid and natural gas pipeline in this country, ranging in diameter from two inches to four feet. These pipelines bring crude oil and natural gas from production fields to processing plants and refineries—and ultimately to the consumer.

Pipeline Operator Qualification

Petroleum products are considered hazardous, so pipeline personnel must take extra precautions and follow specific procedures when working with them. To help ensure the safe handling and transportation of petroleum products, the United States Department of Transportation (DOT) mandates safe practices as federal regulations.

One such regulation, the Operator Qualification (OQ) rule, requires pipeline operators to develop a written qualification plan to evaluate their personnel and their contract personnel (see *49 CFR Part 192 Subpart N—Gas*, and *49 CFR Part 195 Subpart G—Liquid*). The evaluation hinges on the operator's ability to perform certain covered tasks and to be able to recognize and respond to abnormal operating conditions that may be encountered while performing these covered tasks.

Covered Tasks

A pipeline performance task is deemed covered if the activity passes the following four-part test:

1. Is it performed on a pipeline facility?
2. Is it an operations or maintenance task?
3. Is it performed as a requirement of *49 CFR Part 195 (Liquid)* or *49 CFR Part 192 (Gas)*?
4. Does it affect the operation or integrity of the pipeline facility?

If the answer to all four questions is yes, then the activity is considered covered and must figure into the evaluation process of pipeline personnel.

Consortium on Operator Qualification

The Consortium on Operator Qualification (COOQ) became the guiding force for the identification and development of covered tasks for the liquid pipeline industry. Their work product, API publication 1161, *Guidance Document for the Qualification of Liquid Pipeline Personnel*, formed the basis for the resulting Pipeliner Training and Assessment Program.

The covered task list created in API publication 1161 has now evolved to include gas, as well as construction covered tasks. NCCER maintains a version of the covered task list, which can be accessed at **www.nccer.org**.

NCCER

With the guidance and support of the Consortium on Operator Qualification, NCCER was able to design and implement the Pipeliner Training and Assessment Program.

NCCER is a not-for-profit 501(c)(3) education foundation established in 1995 by the world's largest and most progressive construction and maintenance contractors and national contractor associations. NCCER was founded to address the severe workforce shortage facing the construction and maintenance industries and to develop a standardized training process and curricula. Today, NCCER is supported by hundreds of leading construction and maintenance companies, manufacturers, and national associations and developed more than 70 construction and maintenance titles.

NCCER also maintains a Registry, which contains the records of trainees and instructors who have received NCCER training and certification. All completion information for every trainee is sent to NCCER and filed within the Registry. The Registry can then confirm training and skills for workers as they move from company to company, state to state, or even within their own company.

NCCER Standardized Curricula

NCCER standardized curricula are developed using the expertise of industry subject matter experts (SMEs). The use of SMEs as an integral part of every new development or revision ensures that the curriculum is current, industry-driven,

MODULE 62105-02 CONTENTS

Figures

MODULE 62105-02

Pipeline Documentation

Objectives

When you have completed this module, you will be able to do the following:

1. Read alignment sheets.
2. Identify and explain documentation processes and forms.

Prerequisites

Before you begin this module, it is recommended that you successfully complete the following: Core Curriculum; Pipeline Core, Modules 66101 and 66102; Pipeline Maintenance Level One, Module 62103.

1.0.0 ◆ INTRODUCTION

The generation and maintenance of maps and records associated with the pipeline is necessary for a number of reasons. Below are just some of these:

- Maps are used in the design of the pipeline.
- Maps are used to show the location of the pipeline.
- Drawings and schematics are used to provide greater detail on the parts of the pipeline, such as pump stations and tank facilities.
- Records are used to document what has been done on the pipeline and to equipment.
- Records are required to plan personnel training and to document personnel qualifications.
- Records and maps are required by federal, state, and local regulations.

Specific forms and maps are needed for specific purposes. The forms needed for different purposes are identified in a company's procedure manual. Some of the most common ones are discussed here.

For documents to be useful, they must be controlled and maintained, so that they are current, accurate, and available. Document management makes this possible. Although each company may have its own form designs and document management system, they all have a common purpose. An overview of these is presented in this module. This module is divided into the following sections:

- Pipeline maps and drawings
- Documentation
- Document management

2.0.0 ◆ PIPELINE MAPS AND DRAWINGS

Maps of the pipeline route showing how the route is aligned with true north and its surroundings are called alignment maps. They include several types of these maps as well as diagrams and schematics:

- Maps
- Global positioning system (GPS)
- **Piping and instrumentation diagrams (P&IDs)**
- **Electrical schematics**

2.1.0 Maps

Maps are important documents throughout a pipeline's existence. Initially, maps are used to determine what the pipeline route should be. Maps showing area highways, railroads, political boundaries, and topographic features are studied. The locations of oil fields, refineries, other pipelines, river ports, and facilities that could be used are identified. Crossing certain topographic features, such as mountains, lakes, or major rivers must be analyzed for feasibility, and the most favorable sites chosen.

The pipeline route is imposed on the maps, and these are used throughout the pipeline's operation for locating the pipe when necessary and for defining the area along its route. Current maps of the pipeline are required by regulations as well.

Maps that may be used include the following:

- Survey maps/**alignment sheets**
- Topographic maps
- Profile maps
- Strip maps

2.1.1 Survey Maps/Alignment Sheets

When the route is decided, the proposed **right-of-way (ROW)** is laid out and permits are obtained. At that point, the proposed ROW is surveyed using design maps or aerial photographs, and alignment maps of the route are produced. As the pipeline is built, the line is imposed on the survey map, which includes angles of direction changes, property crossings, land survey lines, and features legally defining property limits (see *Figure 1*). The position of each feature such as fences, direction changes, property lines, ditches, creeks, and others is shown by its distance along the route from the company-designated starting point.

Station numbers are the number of feet from the company-designated starting or ending point. Special features such as rivers, highways, and railroads require additional information and planning. These maps/alignment sheets will be maintained and updated throughout the life of the pipeline.

2.1.2 Topographic Maps

Topographic maps, also called topo maps, show the ground's physical features and their elevations above sea level. The earth's shapes and forms are shown by contour lines spaced to show the most useful contours of features. Spacing may be different in different areas, depending on the rate of elevation changes. Each contour line connects points of equal elevation and contour lines cannot cross other lines or divide. Closely spaced lines indicate steep slopes, while contours spaced far apart represent gentle slopes. Contours trend up valleys, cross the streambed and return down the valley on the other side, forming a V which points upstream. *Figure 2* shows the relationship between topographic features and contour lines.

Numbers giving the elevation are incorporated into lines at appropriate intervals, generally every fifth line. These are called index contours and are shown with darker lines. Lines between the index contours represent specific increments that are indicated on the map. These increments can be from 5 to 200 feet, depending on the topography being mapped (see *Figure 3*). The elevation of a particular point is determined by its location relative to these lines.

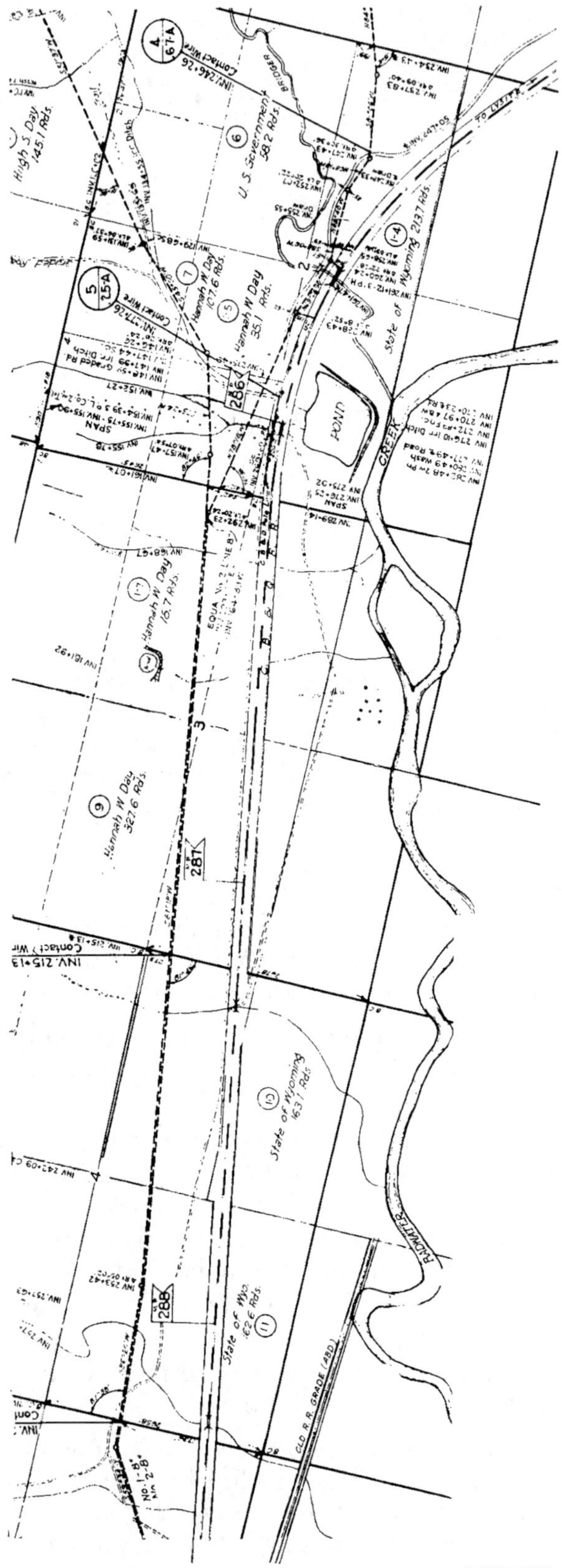

105F01.EPS

Figure 1 ◆ Pipeline survey map.

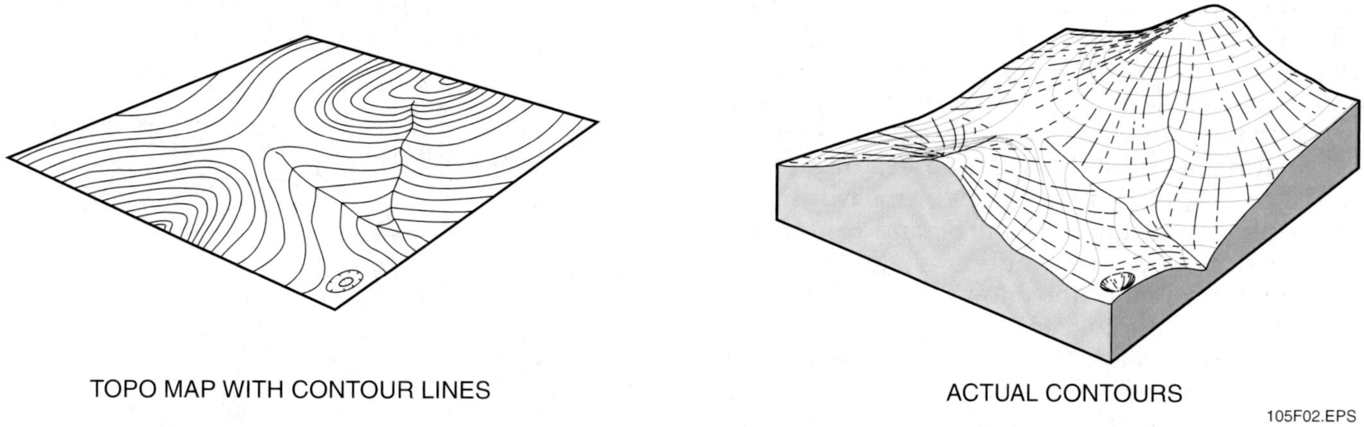

Figure 2 ◆ Relationship between topographic features and contour lines.

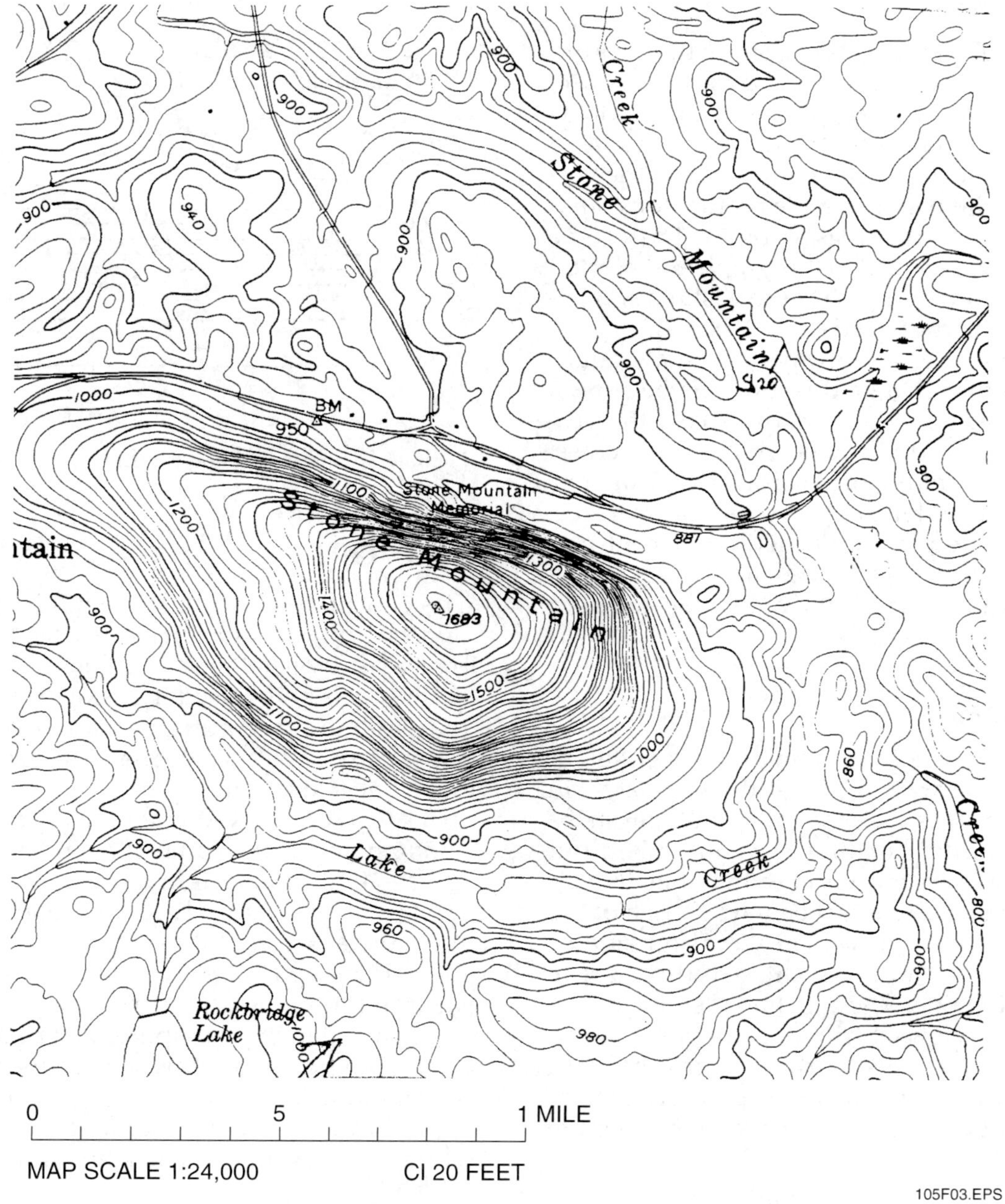

Figure 3 ◆ Sample topographic map.

Topographic maps are important for pipeline planning because they indicate extreme elevation changes, slope of the ground, and features that should be avoided. They can also be useful in field work by helping to identify locations from physical features. Gradients of land areas or streams can be calculated from the maps by measuring the distance, converting it to miles, and dividing it into the difference in elevation.

Topographic maps look down on the ground as from an aerial photograph; in fact, most modern ones have been made from photographs taken by planes or satellites. Topo maps are available for every area on the earth, including the ocean floors. The U.S. Geological Survey is a good resource for topographic maps.

2.1.3 Profile Maps

Profile maps show the elevation of the ground and pipeline from the side view or as a cross section. Hills and valleys are seen from the side as they normally would be. To produce a profile, a graph is drawn with an elevation scale on the vertical axis and the distance scale on the horizontal axis. The elevation readings along the line on the topo map are transferred to this graph and plotted at specific points along the line distances indicated by the horizontal axis (see *Figure 4*).

The pipeline does not always run in a straight line. Sections that change direction can be shown in profile by separate sections, or usually the profile is shown as a straight line all the way across. The vertical scale of the profile is usually exaggerated to provide better definition of land features. The depth of the pipeline is also plotted on the profile in the same way. Profile maps give a good visual indication of where the pipe is located both vertically and horizontally.

2.1.4 Strip Maps

Strip maps are usually combinations of different types of maps or drawings over the same area; in this case, the area is the pipeline route. The strips of each type of map or drawing can be laid out side by side on the same sheet so that they line up along the route. There may be views shown of the ground from an aerial or plan view and side or elevation views, showing the ground profile along a specific line. The pipeline may be superimposed on all views. Strip maps help to coordinate surface features with topography and pipeline depth, as well as other pipeline data at any point on the route. *Figure 5* shows a sample strip map.

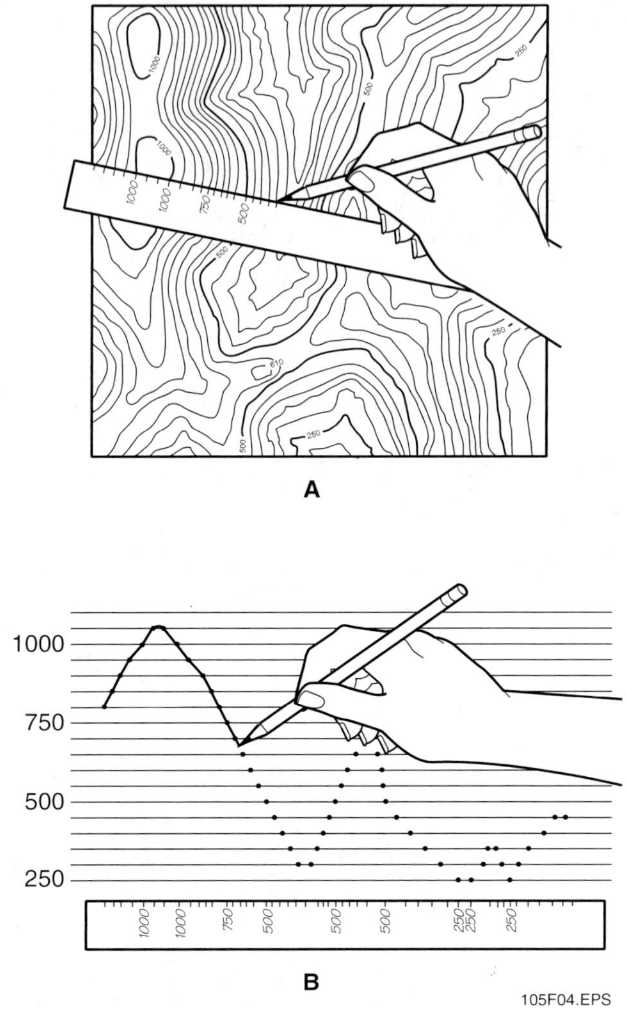

Figure 4 ◆ Simple method for constructing a profile.

2.2.0 GPS

The Global Positioning System (GPS) is a worldwide navigation system that uses radio signals broadcast by satellites. The GPS is included here because it is being used to make maps of the earth that are more accurate than any ever made and can be used with other maps to precisely determine one's location.

NAVSTAR, or NAVigation System using Time and Ranging, is the name of the navigational satellite system. The system uses 24 satellites in six orbits with an average height of about 12,550 miles. As many as eight of these satellites may be above the horizon when viewed from any point on earth. *Figure 6* shows the arrangement of these satellites.

For navigation, a computerized radio receiver in an airplane, ship, or other vehicle uses the satellite signals to calculate its own location. Hikers and others on foot may use small, portable receivers. The system has both civilian and military users.

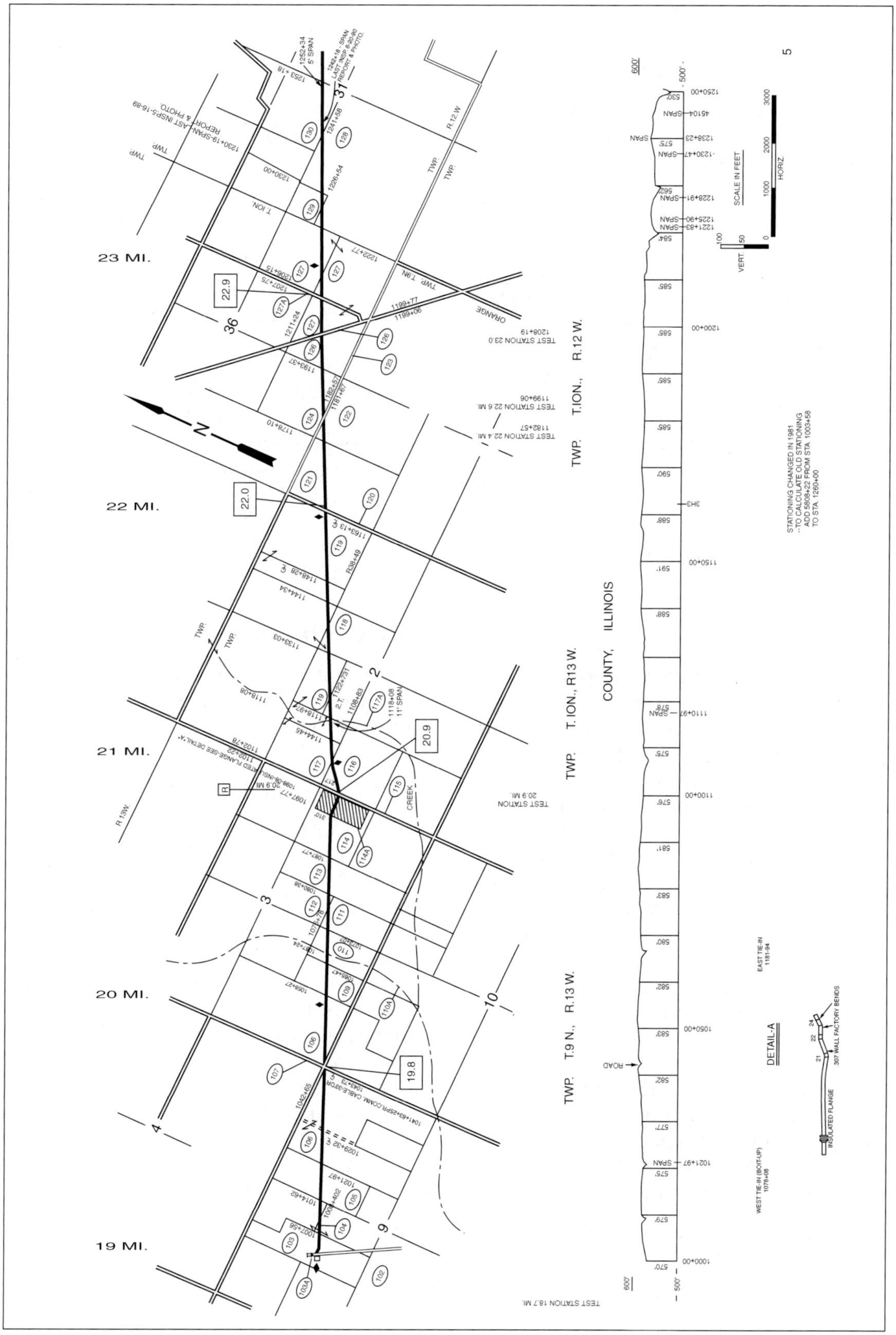

105F05.EPS

Figure 5 ◆ Sample strip map.

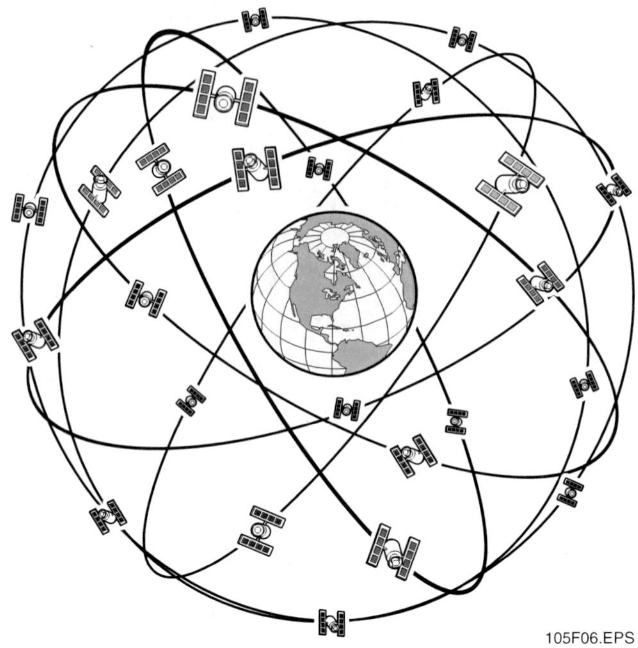

Figure 6 ◆ Arrangement of navigational satellites.

A large percentage of the world's base information maps (including topographic maps) display both geographic coordinates (latitude and longitude) and the Universal Transverse Mercator Grid (UTM). The portable receivers are designed to report positions in both of these coordinate systems, so users have maps showing where they want to go and receivers that give them the location where they are. With this and a compass, users should be able to find their target. For military use, some receivers are designed to use the Military Grid Reference System (MGRS) as well as the other coordinate systems. MGRS is also available on some topographic maps.

A receiver uses signals from at least four and often fourteen satellites. It will lock onto the best four signals. The signal indicates the precise time and the coordinates of the satellite. From these, the receiver determines its latitude and longitude. GPS is used for surveying, exploring, and tracking ground movements. In the newest systems, GPS can be used to locate positions within inches.

2.3.0 P&IDs

A P&ID depicts the functional layout of a liquid or gas system and its piping, valves, and instrumentation (see *Figure 7*). It is accurate to the extent that all components are connected to each other as shown in relation to flow path orientation. A P&ID does not always represent the actual physical layout of equipment; that is, a valve that may appear to be right of the discharge of a pump can actually be located some distance away or even on another floor. Many times a P&ID will use a broken line around groups of equipment to indicate that they are all in the same building.

The P&ID depicts all components and details of a particular system, including pipe sizes, flange sizes, valve sizes, flow direction, and references to other diagrams. Standardized symbols are used for these components instead of pictures. Industry standards for these symbols have been developed and adopted by organizations such as ANSI, ISA, and ASME to aid interpretation. *Figure 8* shows standard P&ID piping symbols, and *Figure 9* shows P&ID instrumentation symbols.

2.4.0 Electrical Drawings

Electrical drawings include several types, which are used for different applications. The most common of these are the blueprint and schematic.

2.4.1 Electrical Blueprints

Electrical **blueprints** are layouts of electrical circuits and their related components shown to scale on the structure (see *Figure 10*). The lines and symbols used on the blueprint may vary from one designer to another, so the legend must be reviewed in order to interpret it correctly.

ANSI has published and recommends standard symbols to be used on electrical drawings. An electrical symbol is a figure or mark that stands for a component used in the system. Many of the symbols use a similar form, but with minor differences which change their meaning. Each category in a system is usually represented by a basic distinguishing symbol. *Figure 11* shows standard electrical symbols.

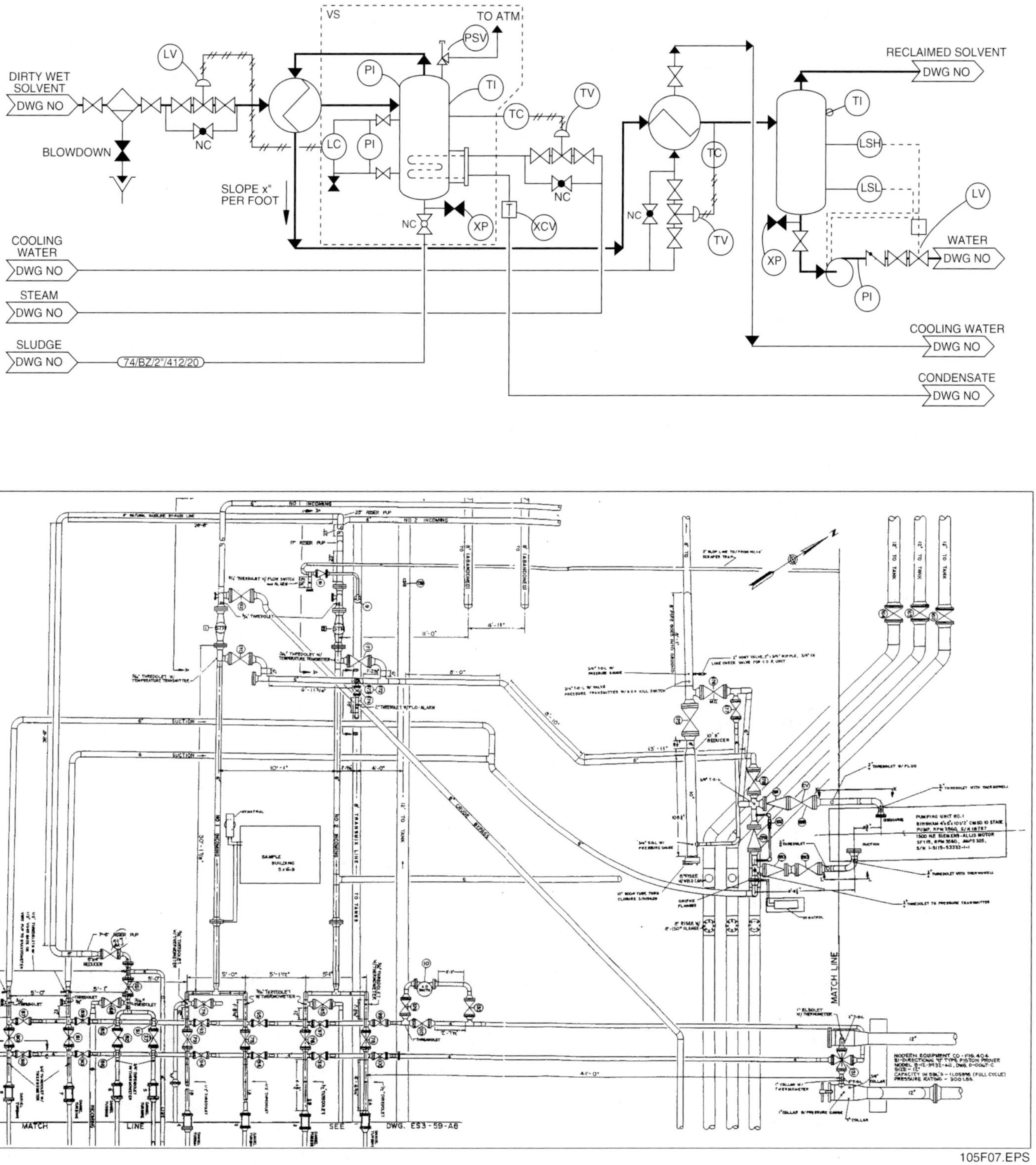

Figure 7 ◆ Sample P&IDs.

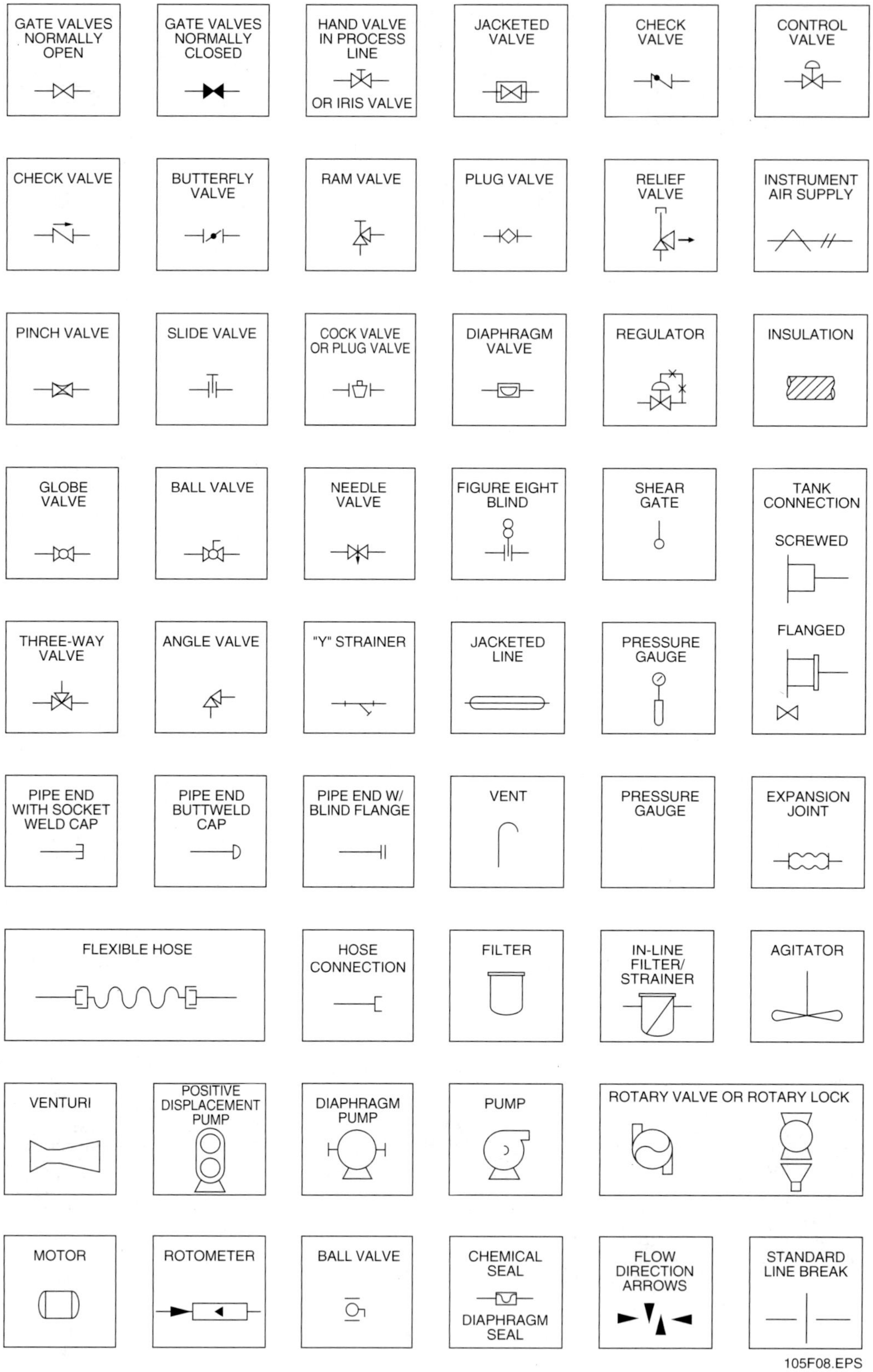

Figure 8 ◆ Standard P&ID piping symbols.

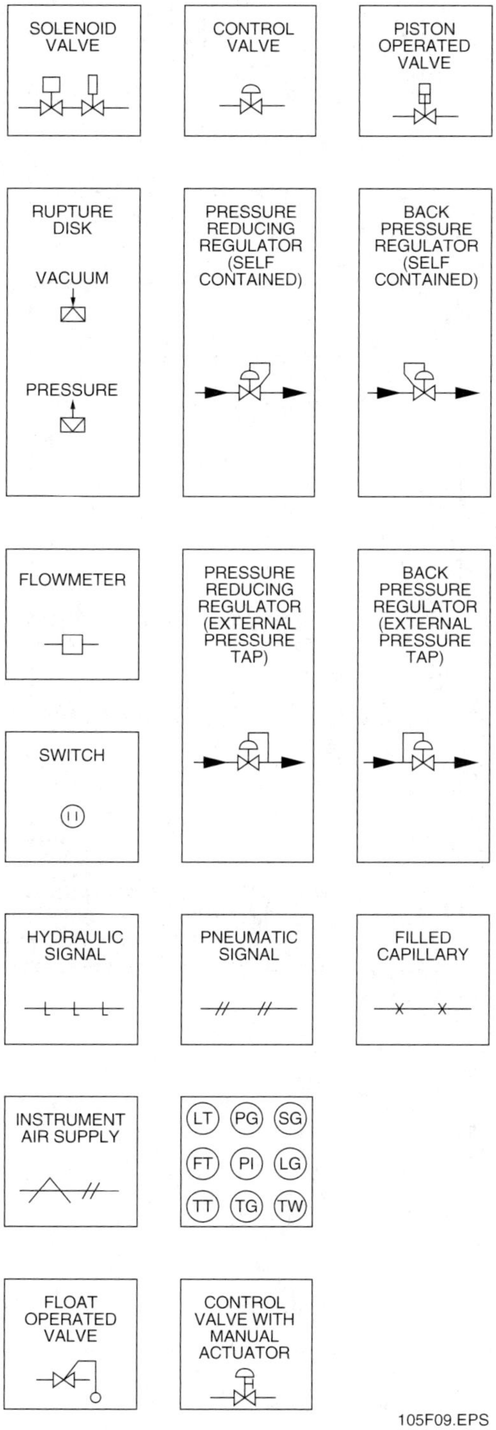

Figure 9 ◆ P&ID instrumentation symbols.

2.4.2 Electrical Schematics

Unlike an electrical blueprint, the schematic is seldom drawn to scale (see *Figure 12*). Its purpose is to show electrical components and their related connections. Schematics are normally used only in complicated electrical systems, such as control circuits. Components are represented by symbols and every wire is shown by itself or included in an assembly of wires, which appear as one line. Each wire is numbered when it enters an assembly and has the same number when it comes out again to be connected to some component.

3.0.0 ◆ DOCUMENTATION

Documentation is an important part of pipeline operations. Many documents are required by regulations, but many are also generated to ensure smooth operations.

Documents that may be required include the following:

- Valve inspection reports
- Exposed pipe reports
- Aerial patrol reports
- Aerial patrol followups
- Tank inspection reports
- Pipe repair/replacement reports
- Safety-related condition reports
- Accident reports
- Work orders
- Permits

3.1.0 Valve Inspection Reports

All valves that are necessary for safe operation of the pipeline shall be maintained in **good working order** at all times. In addition, valves must be protected from vandalism and unauthorized operation. Minimum inspection and maintenance schedules and general procedures for performing this work are given in the operations and maintenance manual.

Valve inspections are divided into main-line valves and station valves. Main-line valve maintenance and inspections are to be documented on a main-line valve inspection report form. Station valve maintenance and inspections are to be documented on a station valve inspection report form.

Both of these forms ask for similar information. This information to be provided includes service performed, condition of valve and site, deficiencies, action required, and action taken. This is documented with the date, valve tag number, serial number, and location. A main-line valve form might also include the site inspection checklist. This checklist includes markers, soil, fencing security, connections, trash and vegetation, component fittings, and other items connected with the condition of the site. The form is dated and signed by the inspector. *Figure 13* shows a valve inspection report.

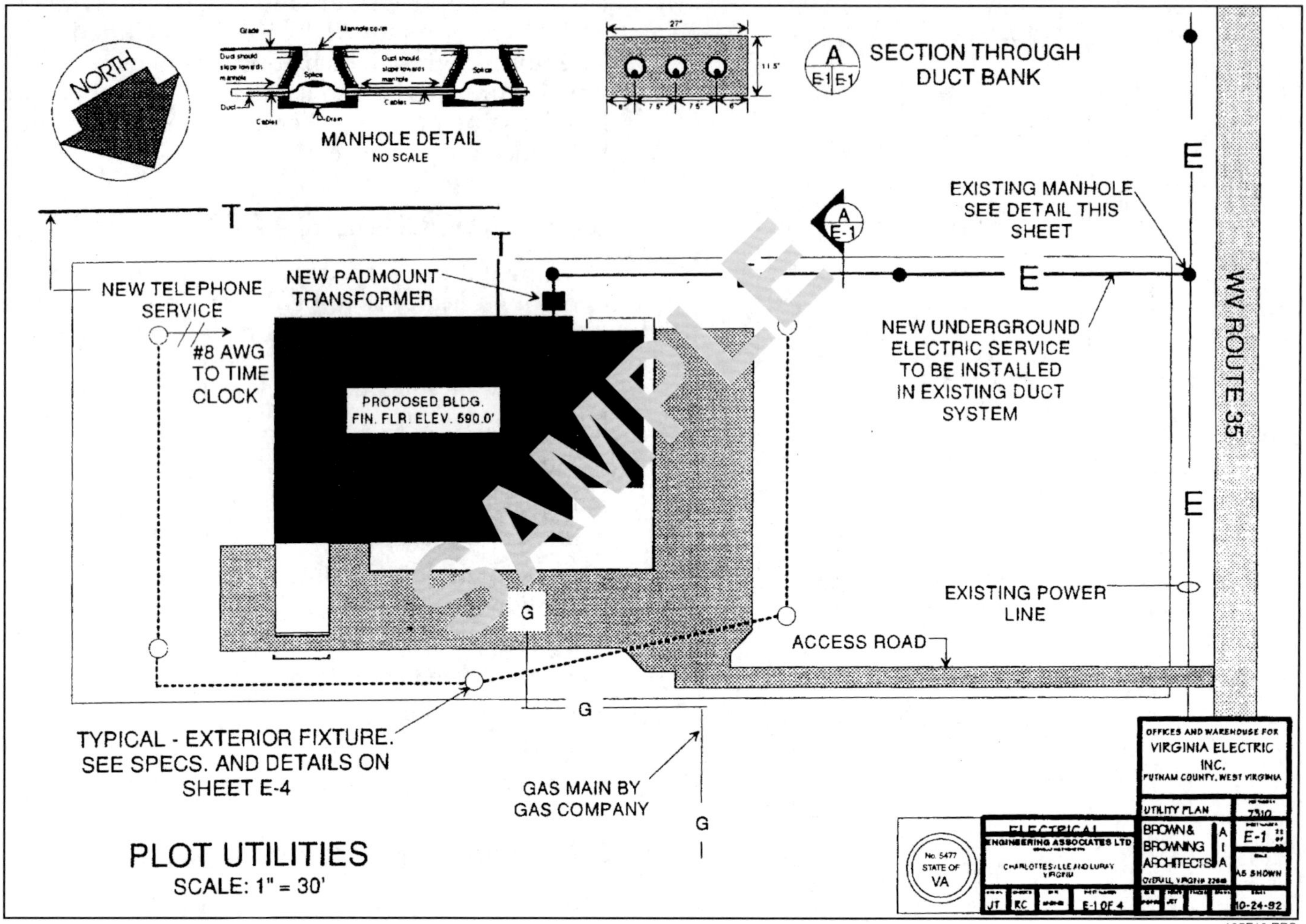

105F10.EPS

Figure 10 ◆ Electrical blueprint.

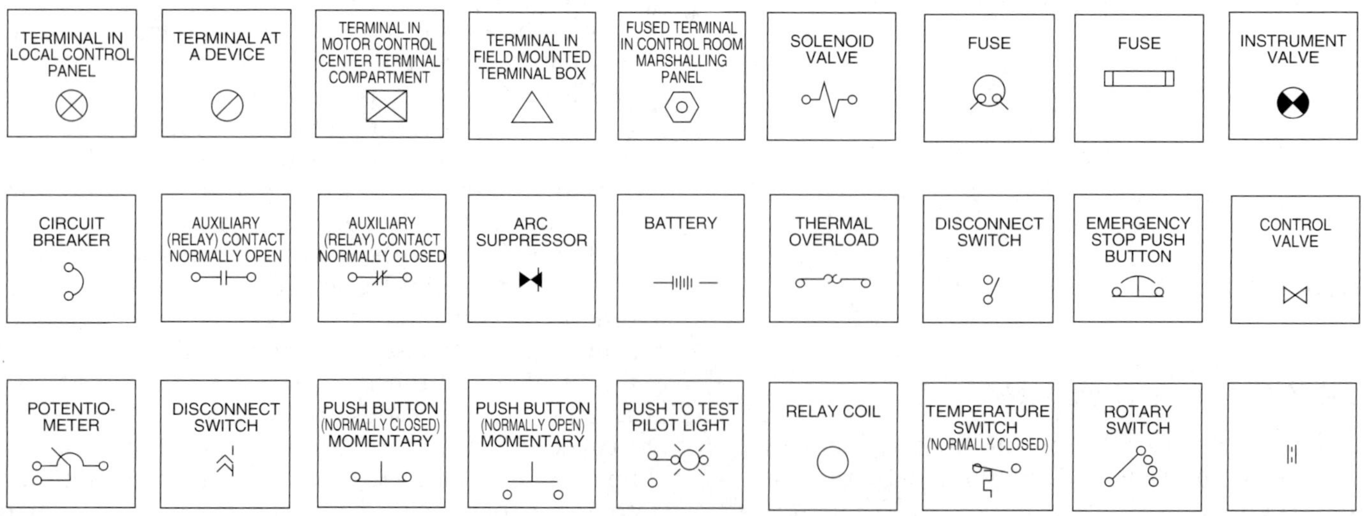

105F11.EPS

Figure 11 ◆ Standard electrical symbols.

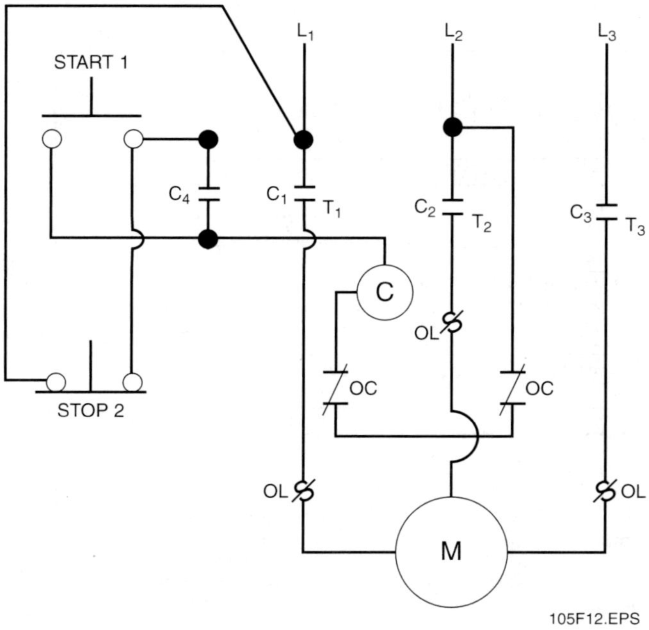

Figure 12 ◆ Electrical schematic.

3.2.0 Exposed Pipe Reports

When exposed pipe is discovered, it must be inspected for evidence of external damage, corrosion, and the extent of damage or corrosion. Corrective action is taken as required. The inspection and action are recorded on a pipeline inspection report form, a form that asks for information on the reason for inspection, the pipe and material conditions, general corrosion, defects or damage, and defect, damage, or corrosion measurements (see *Figure 14*). A sketch of the damaged pipe is also required. The form is dated and signed by the inspector.

Rev. 12/97

Service Performed	Condition of Valve/Site	Deficiency	Action Required	Action Taken
VI - Visual Inspection MO - Mechanical Operation LO - Local Mode Operation RO - Remote Mode Operation RM - Routine Maintenance AS - All Services Listed	AC - Acceptable Condition NR - Needs Repair / Work RR - Replacement Required OS - Out of Service	DO - Does Not Operate LK - Leaking PM - Part Missing DP - Defective Paint NS - Security WO - Wash Out SU - Site Upkeep OT - Other (describe)	RP - Replace Part RR - Repair Required PR - Paint Required SS - Secure Site DW - Dirt Work Required SC - Site Clean Up OT - Other (describe)	PR - Part Replaced RP - Repair Performed RE - Repainted SP - Security Provided DP - Dirt Work Performed SM - Site Cleaned-Up OT - Other (describe)

Date	Print Order	Valve Tag No.	Serial Number	Location / Service	Service Performed	Condition of Valve/Site	Deficiency	Action Required	Action Taken
11/01/01	111	42-37B	BR549	#3 Pump Room	VI	NR	DP	PR	RE
11/13/01	112	54-18C	XJ242	Pump Station 14	MO	RR	LK	RP	PR

SAMPLE

REMARKS: ______________________________

DIVISION: ______________ STATION: ______________ INSPECTED BY: ______________

105F13.EPS

Figure 13 ◆ Valve inspection report.

PIPELINE INSPECTION REPORT

Frequency: At Time of Inspection

PART A — LOCATION

Reason For Inspection ______________________________

1. Location: Division ______________ County ______________ State ______ Date of Inspection __________

A) **Mainline**
Alignment Sheet ________ Pipeline Size/Segment ________
Beg. M.P. ____________ End M.P. ____________
Beg. Sta. No. ____________ End Sta. No. ____________

Note: Station numbers shall be measured from a known landmark (Road Crossing, etc.)

B) **Company Facility: Pump Station, Tank Farm, etc.**
Location Name ______________________
Physical Location of Pipe Inspected ______________________
____
Milepost ______________________

PART B — MATERIAL/CONDITIONS

2. Piping:
Pipe Size ____________ In. O.D. Nom. W.T. __________ In. Actual W.T. __________ Grade __________

3. External Pipe Coating:

A) Coating
☐ Bare
☐ LAC 450
☐ Taped
☐ Insulated
☐ Other ____________

B) Application
☐ Mill Coated
☐ Yard Coated
☐ Field Coated
☐ Other

C) Coating Thickness
Factory Applied____ mile
Field Applied ____mile

D) Condition of Coating
☐ Good ☐ Fair
☐ Poor

E) Bond
☐ Good
☐ Fair
☐ Poor

F) Cause of Coating Failure
☐ Damage
☐ Disbondment
☐ Defective Application
☐ Decomposition
☐ Other ________

4. Environment
☐ Clay ☐ Loam ☐ Sand ☐ Gumbo ☐ Caliche ☐ Rock ☐ Other ________
☐ Wet ☐ Dry ☐ Moist Depth of Cover (Inches) ____________
Soil Resistivity Measurement ______________ (ohm-cm) Measurement Method ____________

5. Coating Repaired ☐ No ☐ Yes
Repair System ______________________

6. Pipe-to-Soil Potential at Inspected Location (At Grade) __________ (volts) (At Pipe/Soil Interface) __________ (volts)

7. Were any Non-Destructive Tests Completed? ☐ No ☐ Yes Type __________

8. Evidence of: ☐ Corrosion ☐ Defect ☐ Damage ☐ Stress Corrosion Cracking (S.C.C.) ☐ Calcareous Deposits

9. Test Station Added ☐ No ☐ Yes I.D. No. ______________

NOTES:
1) If S.C.C. or Corrosion is Noted, Complete Below and Back of Report (External Only, See Note 4)
2) If Defect or Damage is Noted, Complete Back of Report
3) If Repairs are Required, AAPL Form 502 "REPAIR REPORT FOR LEAK DAMAGE, CORROSION OR TEST FAILURE" Must be Attached.
4) AAPL Form 508 "INTERNAL CORROSION INSPECTION REPORT" Should be Completed When Internal Corrosion Is Detected.

SAMPLE

PART C — EXTERNAL CORROSION

10. General Corrosion Information:

A. DESCRIPTION
☐ Less than 10% of Nom. Wall Thickness, Complete Front of Report
☐ Greater than 10% of Nom. W.T., Complete Front and Back of Report
☐ Pitting ☐ General ☐ Localized
Max Depth ________ (In.) Longitudinal Length ________ (In.)

B. STATUS
☐ Active
☐ Inactive
☐ Unknown

C. CAUSE
☐ Galvanic
☐ Stray Current
☐ Unknown
☐ Other ________

11. Cathodic Protection: ☐ Impressed ☐ Galvanic

12. Pipe-to-Soil Potential:
a) Last Pipe-to-Soil Measurement at Nearest Point Each Side of Corroded Area
Upstream __________ (volts) Downstream __________ (volts)
b) Distance from Corroded Area to Measurement Points:
Upstream __________ (feet) Downstream __________ (feet)
c) Date of Measurements: Upstream __________ Downstream __________

13. pH of Soil Near Corroded Area __________ pH of Moisture Under Coating __________

14. Pipe Surface Temperature if Fluid is Heated __________ (deg F)

Remarks ______________________________
____

Distribution: (Original) Corr. Ctrl./Eng (Copy) Division Office (Copy) Other __________

______________ *Technician or Inspector* ___/___/___ *Date* ______________ *Supervisor* ___/___/___ *Date*

105F14A.EPS

Figure 14 ◆ Pipeline inspection report (Page 1 of 2).

PIPELINE INSPECTION REPORT

Frequency: At Time of Inspection

PART D — DEFECT, DAMAGE OR CORROSION INVESTIGATION

Notes:
1) Tabulate all Dents, Gouges, Slivers, Corrosion, S.C.C., etc.
2) List Reference Point - Valve, Tee, Center Road etc., as shown on Alignment Sheet with appropriate Station Number, if available.
3) Indicate Location of Longitudinal Seam. If Pipe is Vertical Substitute Compass direction for o'clock Position. For Spiral Weld Pipe, Note as Such.
4) Defects Such as Dents, Slivers, Gouges, etc., shall be tabulated individually. Corrosion Cracks and/or Corrosion Pits Closely Grouped May be Tabulated as One Defect.
5) Defect Width and Length Will be Measured Along the Girth and Longitudinal Axis Respectively.

Reference Point: ☐ Upstream End (Station No.) ________________ Description ________________________________

☐ Downstream End (Station No) ________________ Description ________________________________

Upstream End **Downstream End**

SKETCH DEPICTING PIPE CUT AT BOTTOM AND LAID FLAT

NOTE: Indicate Girth Weld Location If Known

________ O'Clock Position →

← Length of Inspected Area →

PART E — DEFECT, DAMAGE OR CORROSION MEASUREMENTS AND LOCATION

No.	Type	Distance from Girth Weld	Distance from Long Seam	Width (Circumferencial)	Length (Longitudinal)	Maximum Depth	Number of Anomalies
1.							
2.							
3.							
4.							
5.							
6.							
7.							
8.							
9.							
10.							
11.							
12.							
13.							
14.							

If Repair is Required, send to Division Manager for Approval.

Repair Required ☐ No ☐ Yes Type of Repair ________________________________

Approved by ________________________________ Title ________________________________

Remarks: ________________________________

105F14B.EPS

Figure 14 ◆ Pipeline inspection report (Page 2 of 2).

3.3.0 Aerial Patrol Reports

As the aerial patrol flies the route of line, the pilot notes any abnormal conditions of the flight log, which is typed up on the return to base. This log is distributed to all division and operations offices. The aerial patrol report form or log contains the following information:

- Date
- Location
- Time
- Remarks
- Report number

A report number is assigned by the pilot to each abnormal sighting when it is reported. There is also a space on the form to indicate if the sighting is a new one or an old one.

3.4.0 Aerial Patrol Followups

The person(s) who investigate the reported abnormalities complete an ROW investigation report, a hard copy of which is given to the originating pilot to check against the log (see *Figure 15*). The original ROW investigation report goes to the division compliance records coordinator and then to division DOT files. The report details the reported information, the date and result of the investigation, and the action taken. The form is signed and dated by the inspector.

3.5.0 Tank Inspection Reports

Breakout tanks are inspected externally once a month for stability, physical condition, venting and safety valves, drains, containment structures, equipment, grounding, and alarms. A more detailed inspection by a qualified person is also required at least once per calendar year, not to exceed 15 months between inspections. A visual inspection is made, and any problems are recorded and corrected as necessary. Tanks are inspected externally by API 653-certified personnel every five years. They are inspected internally at one half the corrosion rate of the bottom or every 20 years, whichever is less.

Inspection results are recorded on the Inspection Report for Oil Storage Tanks, with internal inspection noted if performed (see *Figure 16*). The form asks for the location of the tank and then a description of the condition of the physical parts of the tank and all of the associated equipment. Repairs that are required and/or made must also be described. The form is dated and signed by the inspector.

3.6.0 Pipe Repair/Replacement Reports

Pipe changes involve replacing portions of pipe, installing fittings, moving pipe, and making permanent repairs to pipe. These changes may be required for relocation or rearrangement of the line, repairs, changes in pressure requirements, or for other reasons.

Pipe change events must be planned and controlled to avoid spills and interruption in operations. Emergency pipe changes are planned ahead as much as possible, based on events that could occur. All pipe changes must be explained and documented on a pipe report.

3.7.0 Safety-Related Condition Reports

Safety-related conditions must be reported if they occur on an active pipeline or facility that is:

- Within 220 yards of any building intended for human occupancy or outdoor place of assembly
- Within the right-of-way of an active railroad, paved road, street, or highway
- Offshore
- At locations where an oil spill could reasonably be expected to pollute any stream, river, lake, reservoir, or other body of water

Safety-related conditions involving pipelines in service that must be reported are:

- General corrosion that has reduced wall thickness to less than that of maximum operating pressure (*OSHA Part 195.55*)
- Unintended movement or abnormal loading of a pipeline
- Any material defect or physical damage impairing serviceability
- Any malfunction or operating error causing pressure to rise above MOP
- A leak constituting an emergency
- Any condition that could lead to a hazard, a reduction in pressure, or shutdown

All employees are expected to learn how to recognize safety-related conditions and report these to the manager. The manager will initiate action and file the report.

RIGHT-OF-WAY INVESTIGATION REPORT

Frequency: As Required by Aerial Patrol Report

Report Date ________

Note: Right-of-Way personnel should be consulted before confronting enroachment situations for advice pertaining to right-of-way restrictions and agreements.

PART A — REPORTED INFORMATION

1. Right-of-Way Investigation Report Number: ________
2. Date Condition Reported: ________
3. Milepost Location Reported: ________
4. Descriptive Location Reported: ________
5. Problem Reported: ________

PART B — INVESTIGATION INFORMATION

1. Date Condition Investigated: ________
2. Actual Milepost Location: ________
3. Actual Location of Condition: ________
4. Actual Problem Found: ________

☐ Encroachment Advisory (AAPL Form 202) was given to encroaching party

Name of Encroaching Party: ________

PART C — ACTION TAKEN

____

PART D — DISTRIBUTION INFORMATION

(Original)	Division Compliance Records Coordinator
(Copy)	Forward to Pilot at the Arizona Division Office
(Copy)	Right-of-Way

Technician or Inspector ________ *Date* ________ *Supervisor* ________ *Date* ________

Retention: Two Years

105F15.EPS

Figure 15 ◆ ROW investigation report.

LEAK/ACCIDENT REPORT

Frequency: When ANY leak or accident occurs

PART A — TIME AND LOCATION OF ACCIDENT

1. Date: *(Month, Day, Year)* ____________________
2. Hour: *(24 hour clock)* ____________________
3. Give County or City: ____________________
4. Did accident occur on Federal Land? ☐ Yes ☐ No
5. Specify location. *(If location is near buildings, other land marks such as highways, waterways, or railroads, attach a sketch or drawing showing relationship to these landmarks).*

____

____

____

PART B — ORIGIN OF RELEASE OF LIQUID *(Check all applicable items)*

1. Part of system involved:
   ☐ Mainline ☐ Tank Farm ☐ Pump Station
2. Item Involved:
   ☐ Pipe ☐ Valve ☐ Scraper Trap ☐ Pump ☐ Welded Fitting
   ☐ Girth Weld ☐ Tank ☐ Bolted Fitting ☐ Longitudinal Weld
   Other *(Specify)* ____________________
   ____
3. Year Item Installed ____________________
4. Depth of Cover ____________________ inches

PART C — CAUSE OF ACCIDENT

☐ Corrosion ☐ Failed Pipe ☐ Failed Weld ☐ Outside Force Damage
☐ Malfunction of Control Relief Equipment ☐ Incorrect Operation by Operator Personnel
☐ Other *(Specify)* ____________________
____

PART D — DEATH OR INJURY

1. Number of persons killed
   ____________ Operator Employees ____________ Non-Employees
2. Number of persons injured and requiring medical treatment other than on-site first aid.
   ____________ Operator Employees ____________ Non-Employees

PART E — ESTIMATED TOTAL COST

$ ____________________

PART F — AMOUNT OF OIL INVOLVED

1. Estimated amount of oil spilled
   ________ barrels spilled ____________ barrels recovered
2. Was there an explosion?
   ☐ Yes ☐ No
3. Was there a fire?
   ☐ Yes ☐ No
4. Was there a rupture?
   ☐ Yes ☐ No
5. * Area covered by spill
   ________ Feet X ________ Feet
6. * Depth of Soil Saturation
   ________ Inches
   * (Attach Sketch)
7. Was contaminated soil removed?
   ☐ Yes ☐ No

____________________ ___/___/___ ____________________ ___/___/___

Technician or Inspector *Date* *Supervisor* *Date*

Retention: Life of System

105F17A.EPS

Figure 17 ◆ Accident report (Page 1 of 2).

LEAK/ACCIDENT REPORT

Frequency: When ANY leak or accident occurs

INSTRUCTIONS: *Answer Sections G, H or I only if it applies to the particular accident being reported.*

PART G — OCCURRED IN LINE PIPE

1. Nominal diameter (inches) ____________
2. Wall Thickness (inches) ____________
3. SMYS (psi) ________
4. Type joint: ☐ Welded ☐ Flanged ☐ Threaded ☐ Coupled ☐ Other
5. Pipe was ☐ Below Ground ☐ Above Ground
6. Maximum operating pressure ____________
7. Pressure at time and location of accident *(psig)* ____________
8. Had there been a pressure test on the system? ☐ Yes ☐ No
9. Duration of test *(hrs)* ____________
10. Maximum test pressure *(psig)* ____________
11. Date of last test ____________

PART H — CAUSED BY CORROSION

1. Location of Corrosion ☐ Internal ☐ External
2. Facility Coated? ☐ Yes ☐ No
3. Facility Under Cathodic Protection? ☐ Yes ☐ No
4. Type of Corrosion? ☐ Galvanic ☐ Other *(specify)* ____________

PART I — CAUSED BY OUTSIDE FORCE

1. ☐ Damage by operator or by AAPL Contractor
   ☐ Damage by Others
   ☐ Damage by Natural Forces
   ☐ Landslide
   ☐ Subsidence
   ☐ Washout
   ☐ Earthquake
   ☐ Mudslide
2. Was damage prevention program in effect?
   ☐ Yes ☐ No
3. If yes, was the program
   ☐ "one-call" ☐ Other ________
4. Did excavator call?
   ☐ Yes ☐ No
5. Was the pipeline location temporarily marked?
   ☐ Yes ☐ No

PART J — ACCOUNT OF ACCIDENT

____

____

____

____

____

____

____

____

____

____

____

Retention: Life of System

105F17B.EPS

Figure 17 ◆ Accident report (Page 2 of 2).

NOTICE: This report is required by 49 CFR Part 195. Failure to report can result in a civil penalty not to exceed $25,000 for each violation for each day that such violation persists except that the maximum civil penalty shall not exceed $500,000 as provided in 49 USC 60122

Form Approved
OMB No. 2137-0047

U.S. Department of Transportation
Research and Special Programs Administration

ACCIDENT REPORT – HAZARDOUS LIQUID PIPELINE SYSTEMS

Report Date ______________
No. ______________
(DOT Use Only)

INSTRUCTIONS

Important: *Please read the separate instructions for completing this form before you begin. They clarify the information requested and provide specific examples. If you do not have a copy of the instructions, you can obtain one from the Office Of Pipeline Safety Web Page at http://ops.dot.gov.*

PART A – GENERAL REPORT INFORMATION Check: ☐ **Original Report** ☐ **Supplemental Report** ☐ **Final Report**

1. a. Operator's OPS 5-digit Identification Number (if known) / / / / / /
2. b. If Operator does not own the pipeline, enter Owner's OPS 5-digit Identification Number (if known) / / / / / /
   c. Name of Operator ______________
   d. Operator street address ______________
   e. Operator address ______________
   City, County, State and Zip Code

IMPORTANT: IF THE SPILL IS SMALL, THAT IS, THE AMOUNT IS AT LEAST 5 GALLONS BUT IS LESS THAN 5 BARRELS, COMPLETE THIS PAGE ONLY, UNLESS THE SPILL IS TO WATER AS DESCRIBED IN 49 CFR §195.52(A)(4) OR IS OTHERWISE REPORTABLE UNDER §195.50 AS REVISED IN CY 2001.

2. Time and date of the accident

/ / / / / (hr.) / / / (month) / / / (day) / / / (year)

3. Location of accident
   (If offshore, do not complete a through d. See Part C.1)

   a. Latitude: ________ Longitude: ________
   (if not available, see instructions for how to provide specific location)

   b. ______________
   City, and County or Parish

   c. ______________
   State and Zip Code

   d. Mile post/valve station O or survey station no. O
   (whichever gives more accurate location)

   ____

4. Telephone report

/ / / / / / / / (NRC Report Number) / / / (month) / / / (day) / / / (year)

5. Losses *(Estimated)*

Public/Community Losses reimbursed by operator:

Public/private property damage $__________
Cost of emergency response phase $__________
Cost of environmental remediation $__________
Other Costs $__________
(describe) ______________

Operator Losses:

Value of product lost $__________
Value of operator property damage $__________
Other Costs $__________
(describe) ______________

Total Costs $__________

6. Commodity Spilled OYes O No
(If Yes, complete Parts a through c where applicable)
   a. Name of commodity spilled ______________
   b. Classification of commodity spilled:
      O HVLs /other flammable or toxic fluid which is a gas at ambient conditions
      O CO_2 or other non-flammable, non-toxic fluid which is a gas at ambient conditions
      O Gasoline, diesel, fuel oil or other petroleum product which is a liquid at ambient conditions
      O Crude oil

c. Estimated amount of commodity involved :
O Barrels
O Gallons (check only if spill is less than one barrel)

Amounts:
Spilled : __________
Recovered: __________

CAUSES FOR SMALL SPILLS ONLY (5 gallons to under 5 barrels) : ***(For large spills [5 barrels or greater] see Part H)***

O Corrosion **O Natural Forces** **O Excavation Damage** **O Other Outside Force Damage**
O Material and/or Weld Failures **O Equipment** **O Incorrect Operation** **O Other**

PART B – PREPARER AND AUTHORIZED SIGNATURE

______________ (type or print) Preparer's Name and Title ______________ Area Code and Telephone Number

______________ Preparer's E-mail Address ______________ Area Code and Facsimile Number

______________ Authorized Signature ______________ (type or print) Name and Title ________ Date ______________ Area Code and Telephone Number

Form RSPA F 7000-1 (01-2001) *Page 1 of 4*

Figure 18 ◆ DOT accident report form (page 1 of 4).

105F18A.PDF

PART C – ORIGIN OF THE ACCIDENT (Check all that apply)

1. Additional location information
   a. Line segment name or ID ____________
   b. Accident on Federal land other than Outer Continental Shelf O Yes O No
   c. Is pipeline interstate? O Yes O No

Offshore: O Yes O No *(complete d if offshore)*
d. Area ____________ Block # ____________
State / / / or Outer Continental Shelf ☐

2. Location of system involved *(check all that apply)*
   ☐ Operator's Property
   ☐ Pipeline Right of Way
   ☐ High Consequence Area (HCA)? Describe HCA____________

3. Part of system involved in accident
   O Above Ground Storage Tank
   O Cavern or other below ground storage facility
   O Pump/meter station; terminal/tank farm piping and equipment, including sumps
   O Other *Specify:* ____________
   O Onshore **pipeline**, including valve sites
   O Offshore **pipeline**, including platforms

   *If failure occurred on **Pipeline**, complete items a - g:*

a. Type of leak or rupture
   O Leak: O Pinhole O Connection Failure *(complete sec. H5)*
   O Puncture, diameter *(inches)* ____________
   O Rupture: O Circumferential – Separation
   O Longitudinal – Tear/Crack, length *(inches)* ____________
   Propagation Length, total, both sides *(feet)* ____________
   O N/A
   O Other ____________

b. Type of block valve used for isolation of immediate section:
   Upstream: ☐ Manual ☐ Automatic ☐ Remote Control ☐ Check Valve
   Downstream: ☐ Manual ☐ Automatic ☐ Remote Control ☐ Check Valve

c. Length of segment isolated ______ ft

d. Distance between valves ______ ft

e. Is segment configured for internal inspection tools? O Yes O No

f. Had there been an in-line inspection device run at the point of failure? O Yes O No O Don't Know
   O Not Possible due to physical constraints in the system

g. If *Yes*, type of device run (*check all that apply*)
   ☐ High Resolution Magnetic Flux tool Year run: ______
   ☐ Low Resolution Magnetic Flux tool Year run: ______
   ☐ UT tool Year run: ______
   ☐ Geometry tool Year run: ______
   ☐ Caliper tool Year run: ______
   ☐ Crack tool Year run: ______
   ☐ Hard Spot tool Year run: ______
   ☐ Other tool Year run: ______

4. Failure occurred on
   O Body of Pipe O Pipe Seam O Scraper Trap
   O Pump O Sump O Joint
   O Component O Valve O Metering Facility
   O Repair Sleeve O Welded Fitting O Bolted Fitting
   O Girth Weld
   Other *(specify)* ____________

   Year the component that failed was installed: / / / / /

5. Maximum operating pressure *(MOP)*
   a. Estimated pressure at point and time of accident: ________ *PSIG*
   b. MOP at time of accident: ________ *PSIG*
   c. Did an overpressurization occur relating to the accident? O Yes O No

PART D – MATERIAL SPECIFICATION

1. Nominal pipe size *(NPS)* / / / / / in.
2. Wall thickness / / / / / in.
3. Specification ____________ SMYS / / / / / / /
4. Seam type ____________
5. Valve type ____________
6. Manufactured by ____________ in year / / / / /

PART E – ENVIRONMENT

1. Area of accident
   O Under pavement O In open ditch
   O Underground O Above ground
   O Inside/under building O Under water
   O Other ____________

2. Depth of cover: ____________ inches

PART F – CONSEQUENCES

1. Consequences *(check and complete all that apply)*

a.

	Fatalities	Injuries
Number of operator employees:	______	______
Contractor employees working for operator:	______	______
General public:	______	______
Totals:	______	______

b. Was pipeline/segment shutdown due to leak? O Yes O No
   If Yes, how long? ______ days ______ hours ______ minutes

c. Product ignited O Yes O No d. Explosion O Yes O No

e. ☐ Evacuation *(general public only)* / / / / / people
   Reason for Evacuation:
   O Precautionary by company
   O Evacuation required or initiated by public official

f. Elapsed time until area was made safe:
   / / / hr. / / / min.

2. Environmental Impact

a. Wildlife Impact: Fish/aquatic O Yes O No
   Birds O Yes O No
   Terrestrial O Yes O No

b. Soil Contamination O Yes O No
   If Yes, estimated number of cubic yards: ________

c. Long term impact assessment performed: O Yes O No

d. Anticipated remediation O Yes O No
   If Yes, check all that apply: ☐ Surface water ☐ Groundwater ☐ Soil ☐ Vegetation ☐ Wildlife

e. Water Contamination: O Yes O No *(If Yes, provide the following)*
   Amount in water ________ barrels
   Ocean/Seawater O No O Yes
   Surface O No O Yes
   Groundwater O No O Yes
   Drinking water O No O Yes *(If Yes, check below.)*
   O Private well O Public water intake

Form RSPA F 7000-1 (01-2001) *Page 2 of 4*

105F18B.PDF

Figure 18 ◆ DOT accident report form (page 2 of 4).

PART G – LEAK DETECTION INFORMATION

1. Computer based leak detection capability in place? O Yes O No

2. Was the release initially detected by? (check one): O CPM/SCADA-based system with leak detection O Static shut-in test or other pressure or leak test O Local operating personnel, procedures or equipment O Remote operating personnel, including controllers O Air patrol or ground surveillance O A third party O Other (specify) ________________

3. Estimated leak duration days ____ hours ____

PART H – APPARENT CAUSE

Important: *There are 25 numbered causes in this Part H. Check the box corresponding to the primary cause of the accident. Check one circle in each of the supplemental categories corresponding to the cause you indicate. See the instructions for guidance.*

H1 – CORROSION

1. ☐ External Corrosion

2. ☐ Internal Corrosion

(Complete items a – e where applicable.)

a. Pipe Coating O Bare O Coated

b. Visual Examination O Localized Pitting O General Corrosion O Other ________________

c. Cause of Corrosion O Galvanic O Atmospheric O Stray Current O Microbiological O Cathodic Protection Disrupted O Stress Corrosion Cracking O Selective Seam Corrosion O Other ________________

d. Was corroded part of pipeline considered to be under cathodic protection prior to discovering accident? O No O Yes, Year Protection Started: /__/__/__/__/

e. Was pipe previously damaged in the area of corrosion? O No O Yes ⇒ Estimated time prior to accident: /__/__/ *years* /__/__/ *months* *Unknown* ☐

H2 – NATURAL FORCES

3. ☐ Earth Movement ⇒ O Earthquake O Subsidence O Landslide O Other ________________
4. ☐ Lightning
5. ☐ Heavy Rains/Floods ⇒ O Washouts O Flotation O Mudslide O Scouring O Other ________________
6. ☐ Temperature ⇒ O Thermal stress O Frost heave O Frozen components O Other ________________
7. ☐ High Winds

H3 — EXCAVATION DAMAGE

8. ☐ Operator Excavation Damage (including their contractors/Not Third Party)
9. ☐ Third Party *(complete a-f)*

a. Excavator group
O General Public O Government O Excavator other than Operator/subcontractor

b. Type: O Road Work O Pipeline O Water O Electric O Sewer O Phone/Cable O Landowner-not farming related O Farming O Railroad O Other liquid or gas transmission pipeline operator or their contractor O Nautical Operations O Other ________________

c. Excavation was: OOpen Trench O Sub-strata (boring, directional drilling, etc…)

d. Excavation was an ongoing activity (Month or longer) OYes O No If Yes, Date of last contact /___/___/___/

e. Did operator get prior notification of excavation activity?
O Yes; Date received: /__/__/ mo. /__/__/ day /__/__/__/__/ yr. O No
Notification received from: O One Call System O Excavator O Contractor O Landowner

f. Was pipeline marked as result of location request for excavation? O No O Yes *(If Yes, check applicable items i - iv)*
i. Temporary markings: O Flags O Stakes O Paint
ii. Permanent markings: O
iii. Marks were (check one) : O Accurate O Not Accurate
iv. Were marks made within required time? O Yes O No

H4 – OTHER OUTSIDE FORCE DAMAGE

10. ☐ Fire/Explosion as primary cause of failure ⇒ Fire/Explosion cause: O Man made O Natural
11. ☐ Car, truck or other vehicle not relating to excavation activity damaging pipe
12. ☐ Rupture of Previously Damaged Pipe
13. ☐ Vandalism

Form RSPA F 7000-1 (01-2001) *Page 3 of 4*

105F18C.PDF

Figure 18 ◆ DOT accident report form (page 3 of 4).

H5 – MATERIAL AND/OR WELD FAILURES

Material

14. ☐ Body of Pipe ⇒ O Dent O Gouge O Bend O Arc Burn O Other ______
15. ☐ Component ⇒ O Valve O Fitting O Vessel O Extruded Outlet O Other ______
16. ☐ Joint ⇒ O Gasket O O-Ring O Threads O Other ______

Weld

17. ☐ Butt ⇒ O Pipe O Fabrication O Other ______
18. ☐ Fillet ⇒ O Branch O Hot Tap O Fitting O Repair Sleeve O Other ______
19. ☐ Pipe Seam ⇒ O LF ERW O DSAW O Seamless O Flash Weld
O HF ERW O SAW O Spiral O Other ______

*Complete a-g if you indicate **any** cause in part H5.*

a. Type of failure:
O Construction Defect ⇒ O Poor Workmanship O Procedure not followed O Poor Construction Procedures
O Material Defect

b. Was failure due to pipe damage sustained in transportation to the construction or fabrication site? O Yes O No

c. Was part which leaked pressure tested before accident occurred? O Yes, *complete d-g* O No

d. Date of test: /__/__/__/__/ yr. /__/__/ mo. /__/__/ day

e. Test medium: O Water O Inert Gas O Other ______

f. Time held at test pressure: /__/__/ hr.

g. Estimated test pressure at point of accident: ______ *PSIG*

H6 – EQUIPMENT

20. ☐ Malfunction of Control/Relief Equipment ⇒ O Control valve O Instrumentation O SCADA O Communications
O Block valve O Relief valve O Power failure O Other ______
21. ☐ Threads Stripped, Broken Pipe Coupling ⇒ O Nipples O Valve Threads O Dresser Couplings O Other ______
22. ☐ Seal Failure ⇒ O Gasket O O-Ring O Seal/Pump Packing O Other ______

H7 – INCORRECT OPERATION

23. ☐ Incorrect Operation

a. Type: O Inadequate Procedures O Inadequate Safety Practices O Failure to Follow Procedures
O Other ______

b. Number of employees involved who failed a post-accident test: drug test: /__/__/__/ alcohol test /__/__/__/

H8 – OTHER

24. ☐ Miscellaneous, *describe:* ______
25. ☐ Unknown

O Investigation Complete O Still Under Investigation *(submit a supplemental report when investigation is complete)*

PART I – NARRATIVE DESCRIPTION OF FACTORS CONTRIBUTING TO THE EVENT *(Attach additional sheets as necessary)*

Form RSPA F 7000-1 (01-2001) *Page 4 of 4*

105F18D.PDF

Figure 18 ◆ DOT accident report form (page 4 of 4).

Pipeline Work Permit

No:

Effective From: Time: Date:	Effective To: Time: Date:
Work Location:	
Specific Work To Be Done:	

1. OCC notification required?	☐ Yes ☐ No
2. Alyeska representative required on site when work is being performed?	☐ Yes ☐ No
3. P&CM required on site when work is being performed?	☐ Yes ☐ No
4. All required government permits in hand (Title 16,C.O.E.)?	☐ Yes ☐ No
5. All required Company Work Permits in hand?	☐ Yes ☐ No
6. Type of work being performed has potential to adversely affect the buried Fiber Optics Cable? **Note:** If answer is "Yes" contact GCI, 1-888-560-2022 for cable locate.	☐ Yes ☐ No
7. Responsible Work Supervisor informed of the need to avoid damage to TAPS Survey Monuments/Markers, and that any damage (pre-existing or otherwise) should be immediately reported?	☐ Yes ☐ No
8. Mining and/or materials required from Company Operations Material Sites (OMS)? **NOTE:** Mineral material used for any work will require tracking and reporting on a daily basis to the local Maintenance Coordinator utilizing the "Daily Mineral Material Hauling Voucher," Company Form #10164.	☐ Yes ☐ No

Specifications/Procedures To Be Utilized:

Special Instructions, Re-notification Milestones, etc.:

Responsible Work Supervisor:	Date:	Maintenance Coordinator Signature:	Date:
Job Completed: Time: Date:	Initial:	Permit Cancelled: Time: Date:	Initial:

SAMPLE

3459 (03/01)

105F19.EPS

Figure 19 ◆ Work permit form.

4.1.3 Operation and Maintenance Maps and Records

Each operator shall maintain current maps and records of its pipeline systems, including location and identification of facilities, all crossings, maximum operating pressures, and specifications of all pipe, until these are revised.

4.1.4 Daily Operating Records

Each operator shall maintain for at least three years daily operating records that indicate discharge pressure at each pump station and any emergency or abnormal operation.

4.1.5 Miscellaneous Records

There are many types of miscellaneous records. These records vary in how long they must be retained. The following shows types of records and how long they must be retained:

- Retained for the life of the system, facility, pipe, or part:
  - Pipe repair description
  - Pressure testing
  - Weld qualification tests
  - Welder qualification tests
  - Non-destructive testing of welds
- Retained for three years or until the next inspection:
  - Liaison with fire, police and other official
  - Review of personnel training
  - Supervisor knowledge records
- Retained for two years or until the next inspection, whichever is greater:
  - ROW inspections
  - Cathodic protection testing reports
  - Rectifier inspections
  - Inspections of exposed buried pipe
  - Investigation of corrosive effects
  - Internal corrosion monitoring tests
  - Inspections of internal pipe surface
  - Station valve inspections
  - Main-line valve inspections
  - Overpressure safety device inspections
  - Fire-fighting equipment inspections and tests
  - Breakout tank inspections
  - Continuing public education programs
  - Review of training effectiveness
  - Procedure review record
- Retained until the next inspection or test:
  - Navigable waterway inspection
  - Corrosion testing of bare pipe
  - Records of each inspection and test

4.2.0 Document Control

All control documents generated are assigned numbers according to the company numbering system. Documents are distributed within the company according to specific distribution lists. Documents published in the company are numbered and signed out to authorized persons. Revisions to these documents are recorded on revision sheets with descriptions, dates, and names and distributed to all document holders. Documents sent to DOT and other regulatory agencies are logged as sent. Logs and copies are kept on file.

4.3.0 Liquid Inspection Intervals

Inspection intervals are regulated by DOT Inspection Requirements, *Part 195, Subpart F*. This requirement sets the following intervals:

- Not exceeding fifteen months, but at least once per calendar year:
  - Operations and maintenance manual reviews
  - Public education material reviews
  - Procedure reviews
  - Training reviews of personnel
  - Training program revisions
  - Potential surveys of cathodic protection adequacy
  - Cathodic protection test leads inspections
  - Aboveground pipe/component coating inspections
  - Station valve inspections
  - Overpressure safety device inspections (non-HVL)
  - Breakout tanks (in service)
- Not exceeding seven and a half months, but at least twice per calendar year:
  - Corrosion inhibitor monitor examinations
  - Mainline valve inspections
  - Halon or inert gas system inspections
  - Overpressure safety device inspections (HVL)
- Not exceeding two and a half months, but at least six times per calendar year:
  - Rectifier inspections and readings

GLOSSARY

Trade Terms Introduced in This Module

Alignment sheets: Maps and drawings that identify the pipeline and its route in relation to direction, physical features, and political features.

Blueprints: Construction and engineering drawings showing the actual layout and measurements of the pipeline and facilities.

Good working order: All valves are required by *49 CFR Part 195 Paragraph 195.420(a)* to be in good working order, which can be interpreted to mean that a valve's performance, based on in-service criteria, meets all necessary functions of the valve.

Piping and instrumentation diagrams (P&IDs): Diagrammatic layouts of the pipe and equipment to show the flow through the line, using standard piping and instrumentation symbols.

Right-of-way (ROW): Area specifically set aside to contain the pipeline and provide access for maintenance and inspection.

Station number: The distance from the starting point on a survey map, shown as xxx + 00.

NCCER CURRICULA — USER UPDATE

NCCER makes every effort to keep its textbooks up-to-date and free of technical errors. We appreciate your help in this process. If you find an error, a typographical mistake, or an inaccuracy in NCCER's curricula, please fill out this form (or a photocopy), or complete the online form at **www.nccer.org/olf**. Be sure to include the exact module ID number, page number, a detailed description, and your recommended correction. Your input will be brought to the attention of the Authoring Team. Thank you for your assistance.

Instructors – If you have an idea for improving this textbook, or have found that additional materials were necessary to teach this module effectively, please let us know so that we may present your suggestions to the Authoring Team.

NCCER Product Development and Revision

13614 Progress Blvd., Alachua, FL 32615

Email: curriculum@nccer.org

Online: www.nccer.org/olf

❑ Trainee Guide ❑ Lesson Plans ❑ Exam ❑ PowerPoints Other ______________________

Craft / Level: Copyright Date:

Module ID Number / Title:

Section Number(s):

Description:

Recommended Correction:

Your Name:

Address:

Email: Phone:

Module 67102-02

Basic Pipeline Pneumatics and Equipment

Introduction to the Pipeline Operator Qualification Rule, API, and NCCER

It could be said that petroleum products keep America running. Petroleum products like oil and natural gas are the principal sources of energy in the United States and are used in many manufacturing and chemical processes. Petroleum products are primarily transported by pipeline. There are more than a million miles of liquid and natural gas pipeline in this country, ranging in diameter from two inches to four feet. These pipelines bring crude oil and natural gas from production fields to processing plants and refineries—and ultimately to the consumer.

Pipeline Operator Qualification

Petroleum products are considered hazardous, so pipeline personnel must take extra precautions and follow specific procedures when working with them. To help ensure the safe handling and transportation of petroleum products, the United States Department of Transportation (DOT) mandates safe practices as federal regulations.

One such regulation, the Operator Qualification (OQ) rule, requires pipeline operators to develop a written qualification plan to evaluate their personnel and their contract personnel (see *49 CFR Part 192 Subpart N—Gas,* and *49 CFR Part 195 Subpart G—Liquid*). The evaluation hinges on the operator's ability to perform certain covered tasks and to be able to recognize and respond to abnormal operating conditions that may be encountered while performing these covered tasks.

Covered Tasks

A pipeline performance task is deemed *covered* if the activity passes the following four-part test:

1. Is it performed on a pipeline facility?
2. Is it an operations or maintenance task?
3. Is it performed as a requirement of *49 CFR Part 195 (Liquid)* or *49 CFR Part 192 (Gas)*?
4. Does it affect the operation or integrity of the pipeline facility?

If the answer to all four questions is *yes*, then the activity is considered *covered* and must figure into the evaluation process of pipeline personnel.

American Petroleum Institute

The OQ rule and its related covered tasks provided the catalyst to develop a unified pipeline training and assessment program—so that pipeline operators and owners have ready access to the assessment and training tools they need. This effort was the direct result of the vision and leadership provided by The American Petroleum Institute (API) and API members.

Efforts to unify the industry are nothing new to API. Indeed, the impetus for forming the Institute in 1919 was the need to standardize engineering specifications for drilling and production equipment. Today, the most pressing issues revolve around public perceptions and government policies toward the industry. Speaking with one voice on these issues has become as essential as having interchangeable parts in the field. API's members determine the petroleum industry's positions on public policy as well as the standards that govern its day-to-day operations worldwide.

Consortium on Operator Qualification

API assembled a focus group whose primary purpose was to identify covered tasks and provide compliance recommendations to pipeline operators.

The Consortium on Operator Qualification (COOQ) became the guiding force for the identification and development of covered tasks for the liquid pipeline industry. Their work product, API publication 1161, *Guidance Document for the Qualification of Liquid Pipeline Personnel*, formed the basis for the resulting Pipeliner Training and Assessment Program.

The covered task list created in API publication 1161 has now evolved to include gas covered tasks and is maintained by NCCER. The most current covered task list can be downloaded at **www.nccer.org**.

NCCER

With the guidance and support of API, its members, and the Consortium on Operator Qualification, NCCER was able to design and implement the Pipeliner Training and Assessment Program.

NCCER is a not-for-profit 501(c)(3) education foundation established in 1995 by the world's largest and most progressive construction and maintenance contractors and national contractor associations. NCCER was founded to address the severe workforce shortage facing the construction and maintenance industries and to develop a standardized training process and curricula. Today,

NCCER is supported by hundreds of leading construction and maintenance companies, manufacturers, and national associations and developed more than 35 construction and maintenance titles.

NCCER also maintains a National Registry, which contains the records of trainees and instructors who have received NCCER training. All completion information for every trainee is sent to the NCCER and filed within the Registry. The Registry can then confirm training and skills for workers as they move from company to company, state to state, or even within their own company.

NCCER Standardized Curricula

NCCER standardized curricula are developed using the expertise of industry subject matter experts (SMEs). The use of SMEs as an integral part of every new development or revision ensures that the curriculum is current, industry-driven, and indicative of industry practices. The development and/or revision process is facilitated from start to finish by NCCER project managers, who work closely with the SMEs and a select team of technical writer(s) on the development and presentation of the curriculum.

Pipeline: Five Training Paths

The pipeline curriculum focuses on essential skills and competencies within five skill areas:

- Corrosion Control
- Electrical and Instrumentation
- Maintenance
- Mechanical
- Operations (Field and Control Center)

SMEs from the pipeline industry have determined the skills and competencies needed based on API's covered tasks and the consensus of the group/industry. Specific company information is generally not incorporated into the curriculum unless the SMEs determine that the information is the industry standard.

Training Modules and Covered Tasks

The smallest unit of instruction in the curriculum is a *module*. Each curriculum is composed of modules (the equivalent of stand-alone chapters) that are logically sequenced into complete *levels* of instruction. Every module is task-oriented and requires the successful completion of a written (knowledge-based) examination and a performance (hands-on) test. Modules may be individually selected and/or combined for efficient task training. Modules in the pipeline curricula, to the extent possible, will align with API-developed covered tasks. While not all modules will relate to covered tasks, all covered tasks will relate to at least one module.

Benefits of Training to the Pipeline Marketplace

Applying NCCER's standardized training process allows pipeline owners and contractors in the United States to establish the benchmark for acceptable skill and competency levels within the pipeline industry. This process will assist training organizations in developing and/or improving standardized pipeline training programs that effectively match their resources and operating conditions while responding to regulatory demands and to the needs of the industry. The process will also recognize training organizations that meet the minimum standards of competence, warranting regulatory, public, and professional confidence. This in turn will foster national unity within the pipeline industry for the mutual benefit of individual technicians, owners, contractors, and ultimately pipeline-users and society.

Additional Resources

To review the Pipeliner Training and Assessment Program Implementation and Resource Guide and the latest version of pipeline covered tasks, visit NCCER's website at **www.nccer.org**, or contact:

NCCER
13614 Progress Boulevard
Alachua, FL 32615
(888) 622-3720
Web site: **www.nccer.org**

For more information on API, please contact:

American Petroleum Institute
1220 L Street, Northwest
Washington, D.C. 20005
(202) 682-8000
Web site: **www.api.org**

To order *API Pub 1161*, contact Global Engineering Documents by calling 1-800-854-7149 or visiting their Web site at www.global.his.com.

Additional sources of information on the pipeline operator qualification regulation include:

DOT – Office of Pipeline Safety
www.ops.dot.gov

Transportation Safety Institute
www.tsi.dot.gov

COURSE MAP

This course map shows all of the modules in the Gas Pipeline Operations curriculum. The suggested training order begins at the bottom and proceeds up. Skill levels increase as you advance on the course map. The local Training Program Sponsor may adjust the training order.

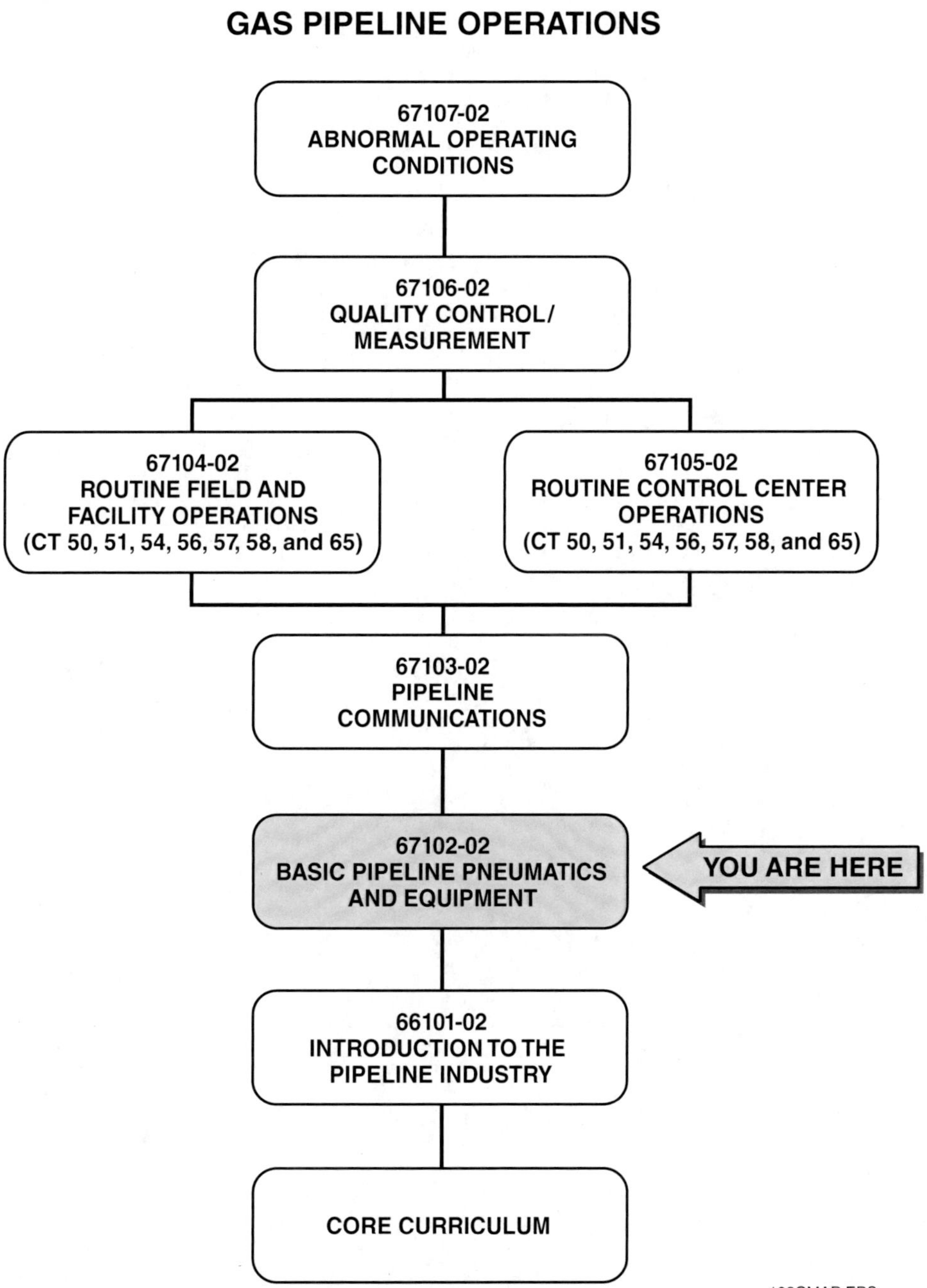

102CMAP.EPS

MODULE 67102-02 CONTENTS

Figures

Tables

Basic Pipeline Pneumatics and Equipment

Objectives

When you have completed this module, you will be able to do the following:

1. Explain pneumatic safety.
2. Explain the physical characteristics of gas.
3. Explain compressing gases.
4. Explain the pneumatic transmission of energy.
5. Explain the principles of compressor operation.
6. Identify and explain types of gas pipeline equipment.
7. Identify and explain pneumatic system components and symbols.
8. Demonstrate an understanding of the design limits of pipelines.

Prerequisites

Before you begin this module, it is recommended that you successfully complete the following: *Core Curriculum; Pipeline Core,* Module 66101-02.

1.0.0 ◆ INTRODUCTION

Pneumatic systems use a gas, usually air, under pressure to create energy and motion. The earliest and simplest type of pneumatic tool is the bellows. Air is compressed inside the accordion-shaped bellows and forced out at greater speed. A **compressor,** which is also a pneumatic device, is a more complex version of a bellows. Compressors are used on the gas pipeline to boost the speed of flowing gas.

When natural gas is released from an underground well, it is under a lot of pressure, which causes the gas to flow rapidly into a pipeline; however, as the gas flows through the pipeline, friction slows its movement, and the pressure drops. Compressors are used to bring the gas pressure back up and keep the gas moving. Because gas pipelines cover thousands of miles, many compressor stations are located along the pipeline. The gas gets a pressure boost as it passes through these stations. Pipeline personnel must carefully calculate compression requirements to properly size the compressors and determine the number and spacing of compressors along the line.

This module introduces you to the basics of working with pneumatic systems and equipment. You will learn about some of the science that governs how gases behave under different conditions. You will also be introduced to some of the equipment used on the gas pipeline and acquire a basic understanding of pipeline design.

2.0.0 ◆ PNEUMATIC SAFETY

Any element placed under pressure develops a certain amount of force. This force is what moves gas through a pipeline. However, if not properly handled, this force can be dangerous. Even low-pressure systems can cause injuries or damage equipment. In addition, working around pneumatic equipment may expose you to toxic or flammable gases. Your company has specific safety procedures for working around pneumatic equipment. You must study these procedures and follow them to ensure your safety and the safety of your co-workers. You must understand and follow all safety procedures before attempting to troubleshoot or work on pneumatic equipment or systems. A list of general safety guidelines follows:

- Always wear the appropriate personal protective equipment.
- Learn and follow company safety procedures.

- Study the **material safety data sheet (MSDS)** if appropriate.
- Study the appropriate technical manuals for proper procedures and specific safety guidelines.
- Ensure that the equipment or system is **locked out** and **tagged out.**
- Ensure that the equipment or system is de-energized both electrically and mechanically.
- Check pressure gauges to ensure that the pressure reading is at zero. Be aware that pressure may be trapped in the equipment or the system.
- Study the possible hazards before opening or closing a **valve** mechanically or electrically.
- Ensure that machines are in a down or safe position.
- Run breathing air through a dehydrator and a carbon dioxide monitor.
- Check the area with a gas detection instrument to ensure that oxygen has not been displaced by another gas.

> **NOTE**
>
> Gas detection instruments are available in a wide variety of styles and applications. Some are programmed to test for a single gas. Others can continuously monitor for several gases at one time. There are both portable units for fieldwork and fixed units for indoor installations. Many include audible and visual alarms.

- Replace defective parts with identical parts. Pay special attention to sizes and pressure ratings.
- Check your work to ensure that all repairs have been completed properly.

3.0.0 ◆ CHARACTERISTICS OF NATURAL GAS

Natural gas is a hydrocarbon mixture. The main gases found in this mixture are methane and ethane. Other hydrocarbon gases in the mixture include propane, butane, pentane, and hexane. Natural gas may also contain some nonhydrocarbon gases, such as nitrogen, carbon dioxide, hydrogen, and hydrogen sulfide. Some water is also present in natural gas.

The main ingredient of natural gas—methane—is colorless, odorless, and flammable. Some natural gas can be used without any processing or refinement. However, most natural gas is processed to remove unwanted elements, such as water vapor, carbon dioxide, and hydrogen sulfide. The resulting refined gas is almost totally methane. Natural gas may be transported through **gathering lines, transmission lines,** or **distribution lines.**

Like any other substance, gas is made up of molecules. Gas molecules do not stick together like molecules in some other substances. For example, if you pour water from a container, the water molecules stick together to form puddles. Gas molecules, however, scatter or disperse into the atmosphere. When gas is contained in a pipeline, the molecules race along, constantly colliding with each other and with the walls of the pipeline as they try to disperse. A pressure gauge interprets these billions of collisions as a single pressure reading. All that activity produces heat, and the greater the compression, the greater the heat. The hotter the gas is, the faster the molecules move and the more heat energy is generated. However, the heat cannot be allowed to build indefinitely. Gas flowing through the pipeline must be diverted to cooling towers where its temperature is brought down to acceptable levels before it is further transported. This is done to protect the pipeline from damage.

3.1.0 Compression

When a gas fills an open container, the pressure and temperature inside the container is the same as the pressure and temperature outside of it. If the container has a close-fitting piston, gas volume is reduced, or compressed, when the piston is pushed in. As the gas is compressed, the speed and temperature of the gas molecules increase. Pressure also increases because the same number of molecules occupies a smaller space, and the molecular collisions are more frequent.

Pipeline personnel must control three variables—pressure, volume, and temperature—to compress and move gas through a pipeline. These variables have a close relationship to one another. If one variable remains constant, the other two change either directly or inversely (opposite to each other). The relationships between these variables are explained with three gas laws: Boyle's law, Amontons' (Gay-Lussac's) law, and Charles' law. These gas laws apply to any gaseous element.

3.2.0 Boyle's Law

Boyle's law was developed by Irish chemist Robert Boyle in 1662. This law explains the relationship between pressure and volume if temperature remains constant. If volume decreases, pressure increases. If volume increases, pressure decreases. This is an example of an inverse relationship; when one value goes up, the other value always goes down.

Following is an example of how Boyle's law works in a pipeline when temperature is constant. When volume is greater than pressure, gas molecules have farther to travel, so they hit the pipeline walls less often. Gas pressure then decreases because there are fewer molecule hits over time. When volume is less than pressure, gas molecules have a shorter distance to travel. Therefore, they strike the pipeline walls more often. Pressure therefore increases because there are more molecule hits over time.

3.3.0 Amontons' (Gay-Lussac's) Law

Amontons' law was developed by French physicist Guillaume Amontons in 1699. Because Joseph Gay-Lussac built the apparatus to prove the theory, this law is sometimes given his name. Amontons' law is an example of a direct relationship. When there are two connected values, whatever happens to one value also happens to the other. This law explains the relationship between pressure and temperature when volume remains constant. When temperature increases, pressure also increases. If temperature decreases, pressure also decreases.

Following is an example of how Amontons' law works in a pipeline when volume remains constant. When the temperature increases, gas molecules move faster. Faster-moving molecules hit the pipeline walls more often, causing the gas pressure inside the pipeline to increase. When the temperature decreases, gas molecules move more slowly. Slower-moving molecules hit the pipeline walls less often, causing the pressure inside the pipeline to decrease.

3.4.0 Charles' Law

Charles' law was developed by French scientist Jacques Charles in 1787. This law also defines a direct relationship. It gives the relationship between volume and temperature when pressure remains constant. When volume increases, temperature increases. If volume decreases, temperature decreases.

Following is an example of how Charles' law works in a pipeline when pressure remains constant. When temperature increases, gas molecules move faster, hitting the pipeline walls more often. The gas pressure inside the pipeline increases as well, but only for an instant. The greater pressure on the inside of the pipeline walls pushes the walls outward, increasing the volume. When this happens, the gas molecules have to travel farther to hit the walls. The lower number of hits drops the pressure back to its constant value.

4.0.0 ◆ PNEUMATIC TRANSMISSION OF ENERGY

Pneumatic systems use compressed air or other gases to produce energy. Compressed air is commonly used to power such tools as impact wrenches and sandblasting equipment. Flexible hoses or tubes carry the compressed air from the compressor to the tool.

A compressed air transmission system carries air through pipes from the compressor to an air-powered device (see *Figure 1*). Most air transmission systems consist of a main line, branch lines, and feeder lines. The feeder lines connect to individual pieces of air-powered equipment. Sometimes air distribution is more efficient if the ends of the lines are tied together to form a loop system. This loop allows two-way airflow to a point of heavy demand.

In the pipeline industry, field personnel may use bottled high-pressure gas to charge relief valves and flow-control devices. Bottled gas may also be used to purge small sections of piping or small tanks. The gases used for these smaller jobs include nitrogen, argon, and helium.

The pressure of the gas transported in the pipeline can be reduced to open or close valves. The main-line compressors for gas pipeline compressor stations usually are started with compressed air.

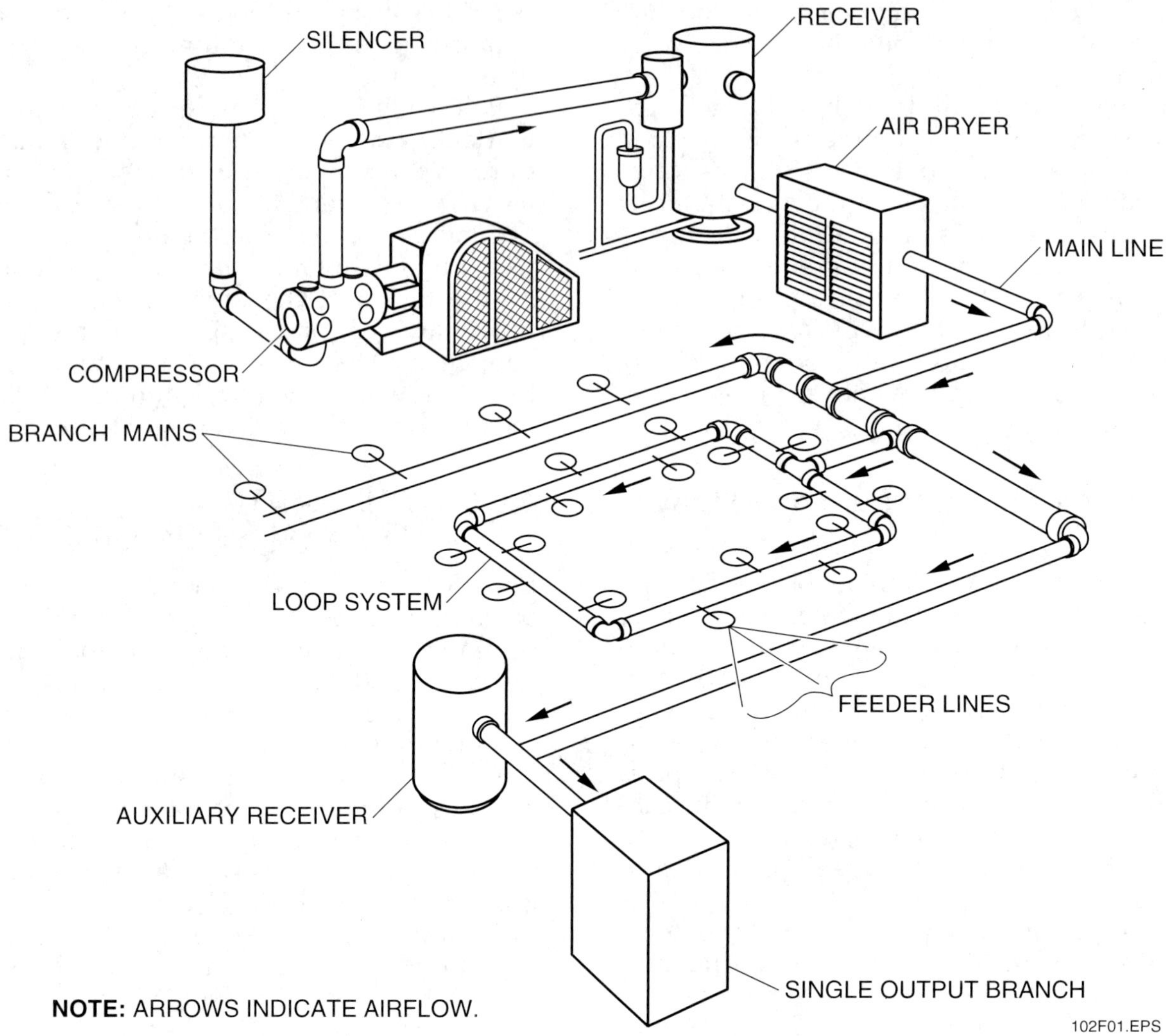

Figure 1 ◆ Compressed air transmission system.

Review Questions

Sections 2.0.0–4.0.0

1. Before attempting to troubleshoot or work on pneumatic equipment or systems, you must ensure that the equipment or system is _____.
   a. locked out and tagged out
   b. marked with a color-coded sign
   c. marked with an authorized work order
   d. redlined as a service item

2. The main gases found in natural gas are _____.
   a. butane and propane
   b. butane and ethane
   c. methane and ethane
   d. methane and pentane

3. As gas is compressed in a pipeline, it _____.
   a. liquifies
   b. separates into butane and ethane
   c. cools off
   d. heats up

4. _____ explains the relationship between pressure and volume if temperature remains constant.
   a. Boyle's law
   b. Amontons' law
   c. Guy-Lussac's law
   d. Charles' law

5. In the pipeline industry, field personnel may use _____ to charge relief valves and flow-control devices.
   a. pressure-bleeding units
   b. bottled low-pressure gas
   c. bottled high-pressure gas
   d. a pneumatic unit

5.0.0 ◆ COMPRESSORS

A compressor increases the pressure of gas by reducing its volume. Compressor stations are generally placed at approximately 100-mile intervals along transmission lines. Compressors may be located at more frequent intervals along distribution lines. Some compressors are powered by natural gas from the pipeline, and some are powered by electrically operated motors.

Every compressor is made up of one or more basic components. A single-stage compressor may be made up of a single component or a group of components. Single-stage compressors may not be able to handle all operating conditions. For example, too great a **compression ratio** may raise the **discharge temperature** too much. To solve this problem, components are combined to form a multistage unit with two or more stages of compression. The gas is cooled between the stages to reduce its temperature and volume before it enters the next stage. Each stage of a multistage unit contains an individual compressor. Each compressor is sized to operate with the other compressors in the series, and all of the compressors in the series operate from one power source.

The two main types of compressors are positive-displacement and continuous-flow. There are several different versions of compressors within these two main types (see *Figure 2*).

5.1.0 Positive-Displacement Compressors

In positive-displacement compressors, air is forced into an enclosed space. Pistons, gears, or pumps then increase the pressure within that space. Positive-displacement compressors are classified as either reciprocating or rotary.

In the reciprocating compressor, a piston compresses the air (see *Figure 3*). The piston rests in a cylinder that has both **suction** and **discharge** valves. The term *suction* (sometimes called the inlet or intake) refers to the gas flow coming into the compressor. The term *discharge* (sometimes called the outlet) refers to the gas flow leaving the compressor. The suction valve opens when the pressure in the cylinder is slightly below the **suction pressure.** The discharge valve opens when the pressure in the cylinder is slightly above the **discharge pressure.** This type of compressor may be single acting or double acting. In the single-acting unit, gas is compressed on only one side of the piston. In the double-acting unit, gas is compressed first on one side of the piston and then the other.

In the rotary compressor, gears or an electric pump move confined air to smaller and smaller spaces. As the volume of air decreases, the pressure increases. It is a simpler unit than the reciprocating compressor. There are no valves or pistons. Instead, rotors that fit together like gears compress air flowing through the unit. Following are common types of rotary compressors:

- Sliding-vane compressors
- Liquid-piston compressors
- Twin-impeller, straight-lobe compressors
- Helical, or spiral-lobe, compressors

5.1.1 Sliding-Vane Compressors

A sliding-vane compressor consists of a case that holds a rotor assembly and vanes. The compression cycle of a sliding-vane compressor is shown in *Figure 4*. Gas flows in through the suction port. The rotor turns and traps the gas in pockets formed by the vanes. Once the pockets between the vanes are filled, the suction port closes. The pocket volume decreases as the rotor turns, and the gas is compressed. The compressed gas is then pushed out of the discharge port.

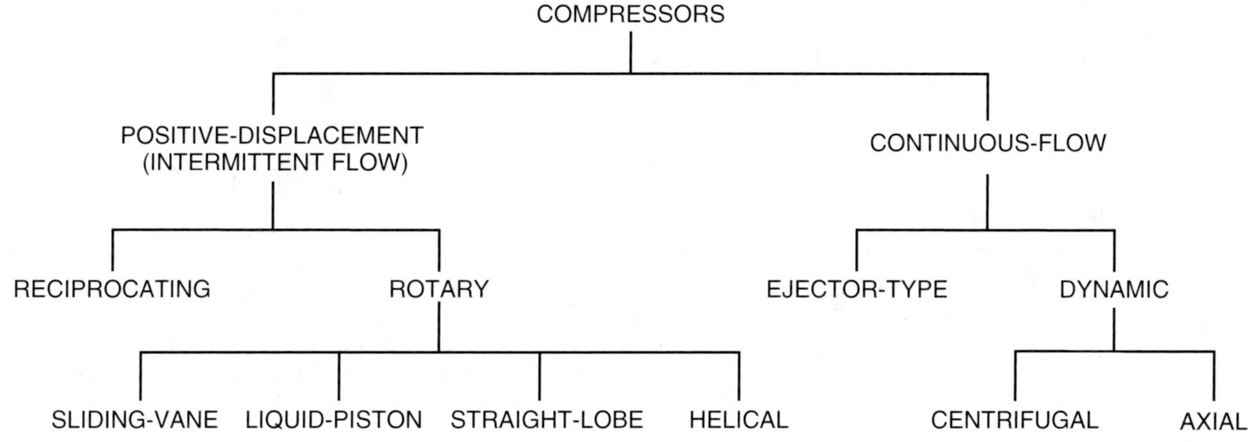

Figure 2 ◆ Various types of compressors.

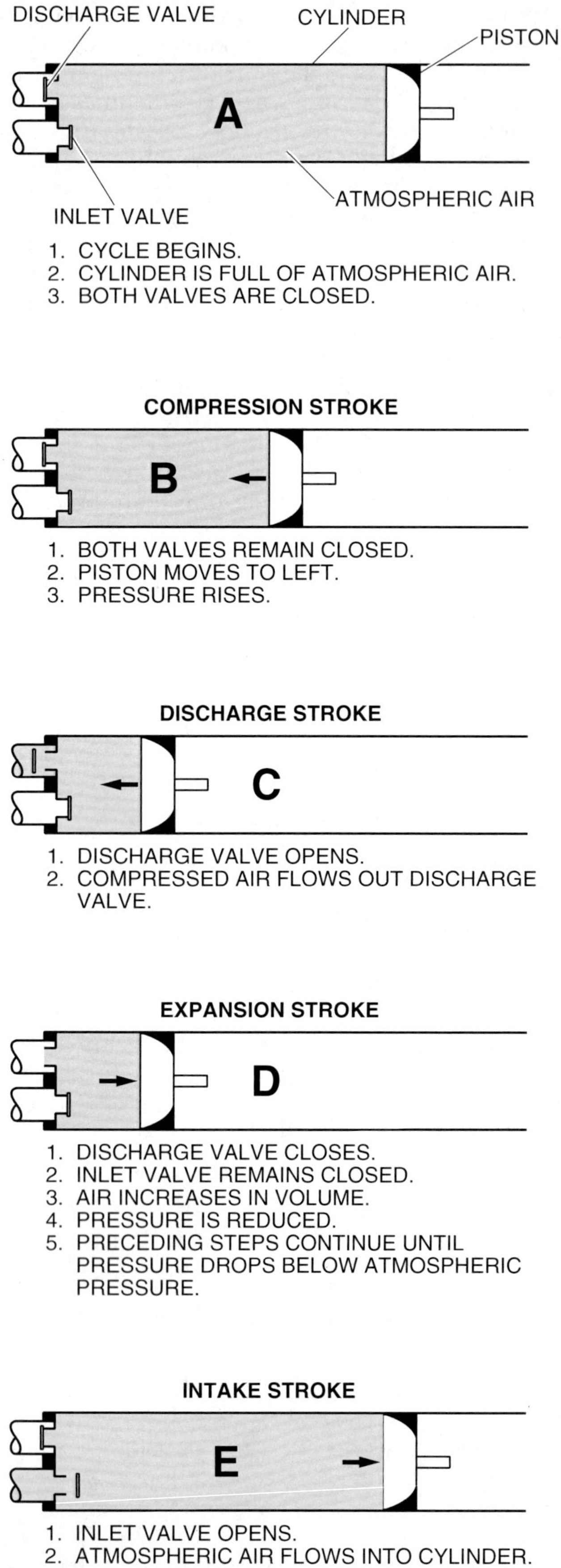

Figure 3 ◆ Steps in a reciprocating compressor cycle.

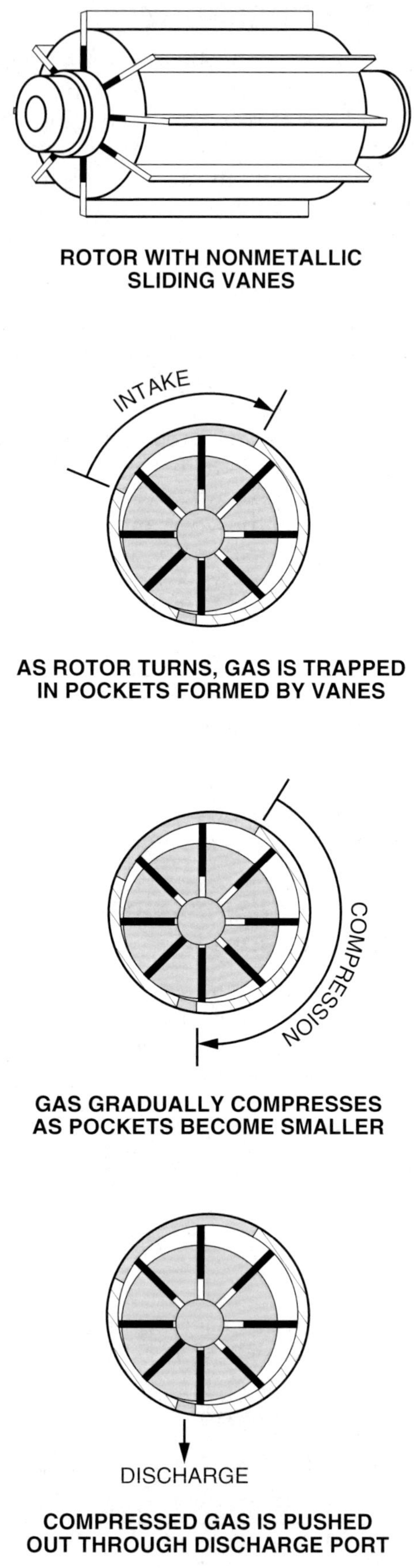

Figure 4 ◆ Compression cycle of a sliding-vane compressor.

5.1.2 Liquid-Piston Compressors

A liquid-piston, or liquid-ring, compressor uses a rotor with forward-turned blades. These blades turn around a central cone that holds suction and discharge ports (see *Figure 5*). The blades drive a ring of liquid compressant around the inside of the casing. The liquid acts as a lubricant, sealer, and coolant. Some liquid is trapped in the space between the blades, and as the rotor turns, it moves in and out of the space, creating a liquid piston. Lubrication is required only in the bearings located outside the casing for this type of compressor; therefore, the compressed gas is oil-free.

5.1.3 Twin-Impeller, Straight-Lobe Compressors

The casing of the twin-impeller, straight-lobe compressor holds two identical impellers, or rotors. When viewed in cross section, each impeller looks like a figure eight (see *Figure 6*). The impellers rotate in opposite directions and mesh somewhat like gears. As they rotate, the impellers compress the gas from the suction port against the casing. External timing gears keep the impellers moving smoothly against one another.

5.1.4 Helical (Spiral-Lobe) Compressors

The casing of the helical, or spiral-lobe, compressor holds two rotors: a driven rotor and a gate rotor (see *Figure 7*). This unit compresses gas between the rotor lobes and the housing chamber. Unlike the lobes on the twin-impeller compressor, the lobes of the helical compressor are not identical. The male, or driven, rotor fits into a pocket formed by the female, or gate, rotor. In addition, the driven rotor uses about 90 percent of the power in this unit.

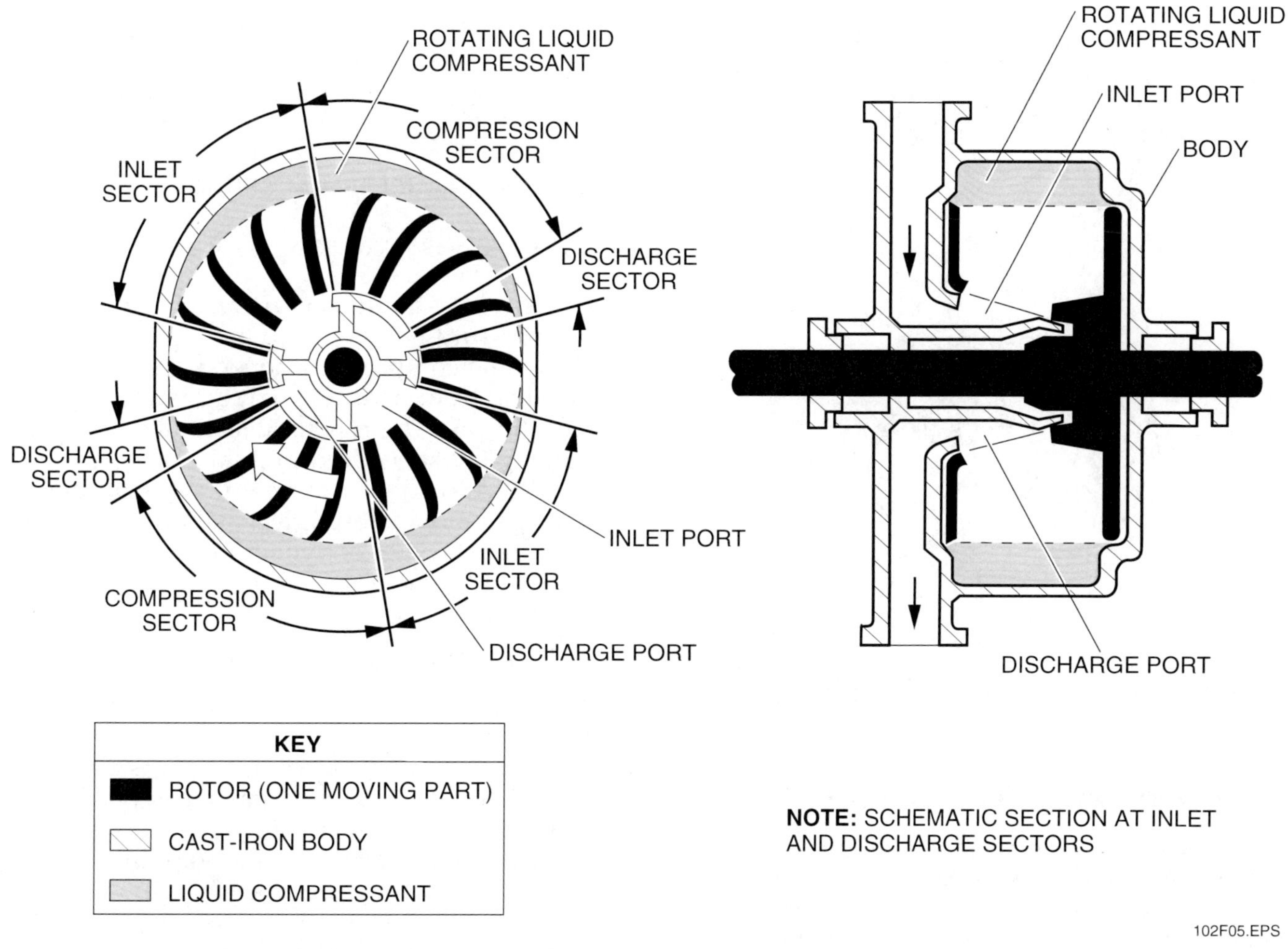

Figure 5 ◆ Typical liquid-piston rotary compressor.

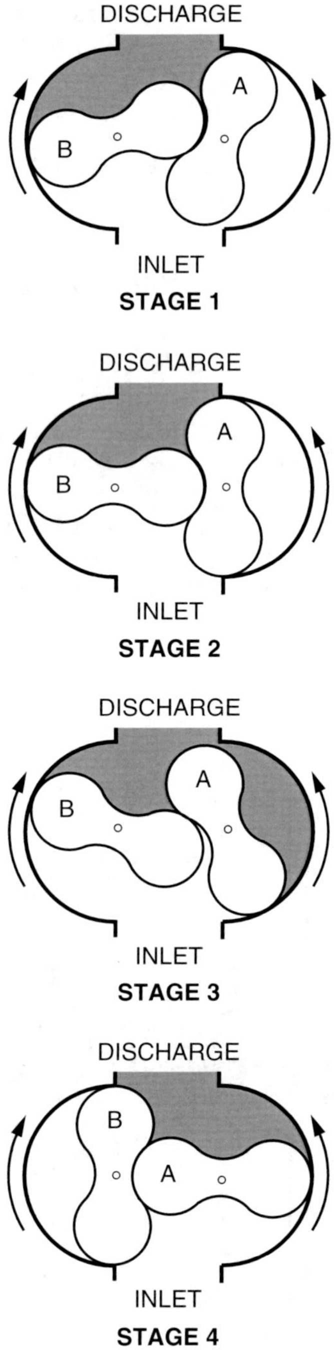

Figure 6 ◆ Twin-impeller, straight-lobe compressor operating cycle.

5.2.0 Continuous-Flow Compressors

Continuous-flow compressors contain a rapidly rotating element that boosts the speed of the gas as it passes through the compressor. Pressure is created partly in the rotating element and partly in the stationary **diffusers** or blades. There are two main types of continuous-flow compressors: ejector-type and dynamic.

Ejector-type compressors use a high-speed gas or steam jet to capture the flowing gas. The gas is

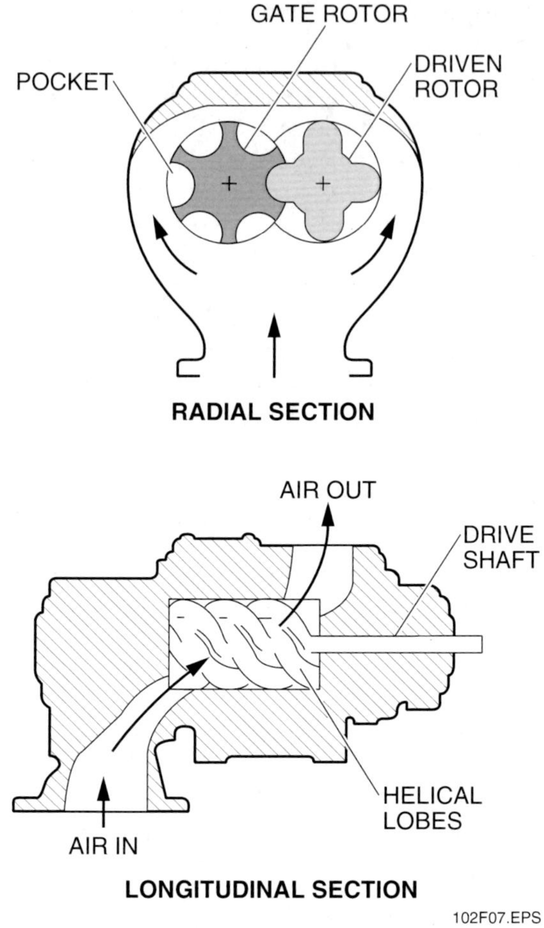

Figure 7 ◆ Sectional views of the helical compressor.

then carried to a diffuser where speed is converted to pressure.

Dynamic compressors depend on the transfer of energy from a set of rotating blades to compress the gas. The rotor transfers this energy by changing the speed and pressure of the gas. The speed is converted into useful pressure energy as the gas moves through a stationary diffuser. There are two types of dynamic compressors: centrifugal and axial.

5.2.1 Centrifugal Compressors

In a cross-section view, the centrifugal compressor looks somewhat like a French horn. A single-stage centrifugal compressor is shown in *Figure 8*. This compressor has an impeller with radial, or backward-leaning, vanes. Gas is forced through the impeller by the rapidly rotating vanes. This rotation produces speed, which is converted into pressure partly in the impeller and partly in the radial diffuser. The impeller forces the gas through the radial diffuser and into the volute (spiral-shaped) collector. The pressurized gas circulates around the volute collector and flows away from the compressor, through the volute diffuser, and into the pipeline.

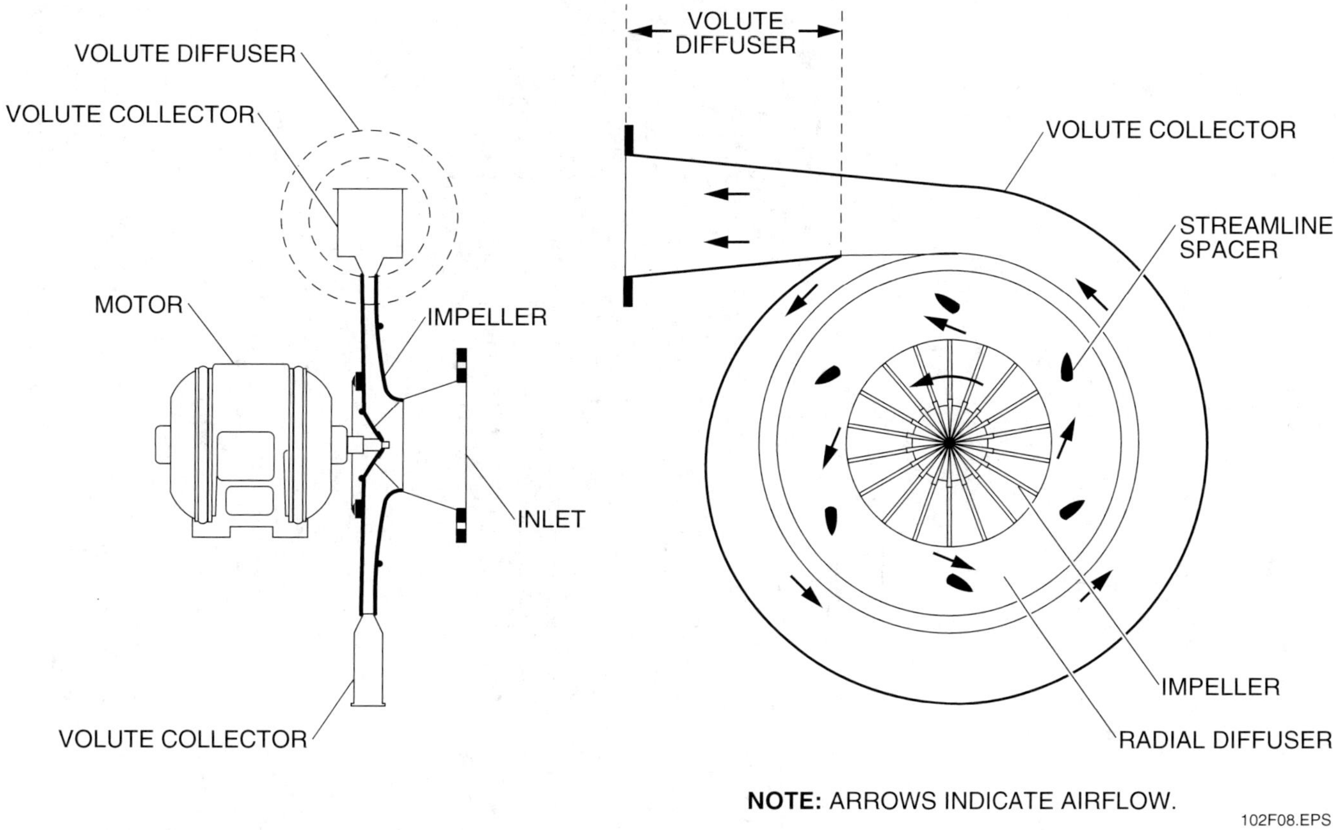

Figure 8 ◆ Single-stage centrifugal compressor.

5.2.2 Axial Compressors

An axial compressor is a large-capacity, high-speed machine (see *Figure 9*). Each stage in this compressor includes two rows of vanes: one rotating, one stationary. The rotating vanes, or rotors, increase gas speed. The stationary vanes, or stators, decrease gas speed. The speed produced by the rotor vanes is converted into pressure. This compressor does not create much swirling action, so gas flows in a straight line (also called axial direction).

5.3.0 Compressor Startup and Shutdown

Compressors in the gas pipeline must be started and stopped carefully. Gas compressor units are complicated and dangerous if not handled properly. You must carefully read and fully understand the manufacturer's startup and shutdown information.

> **WARNING!**
> Gas and air mixtures under pressure can explode. You can be severely injured or killed. Always follow the manufacturer's instructions for purging a compressor unit of any explosive mixture.

5.3.1 Startup

Pipeline personnel must perform a number of checks before starting up compressors. Most companies have a complete checklist that must be followed to ensure that the compressor is started safely. A wide variety of compressors are available, and different operating instructions go with each type. Therefore, you must always read and understand the manufacturer's technical and operations manual before starting a compressor. This manual includes proper installation instructions, startup checklists, and operating instructions. In general, startup checklists include the following operations:

- Ensure that the correct manufacturer's technical manuals are available at the work site for easy reference.
- Check the design limitations for the compressor, such as the operating pressure, discharge temperature, and recommended operating speeds.
- Review the design operating conditions.
- Ensure that all moving parts are working properly.
- Ensure that the gas high-discharge temperature shutdown device is working properly and is set at the correct percentage above the anticipated

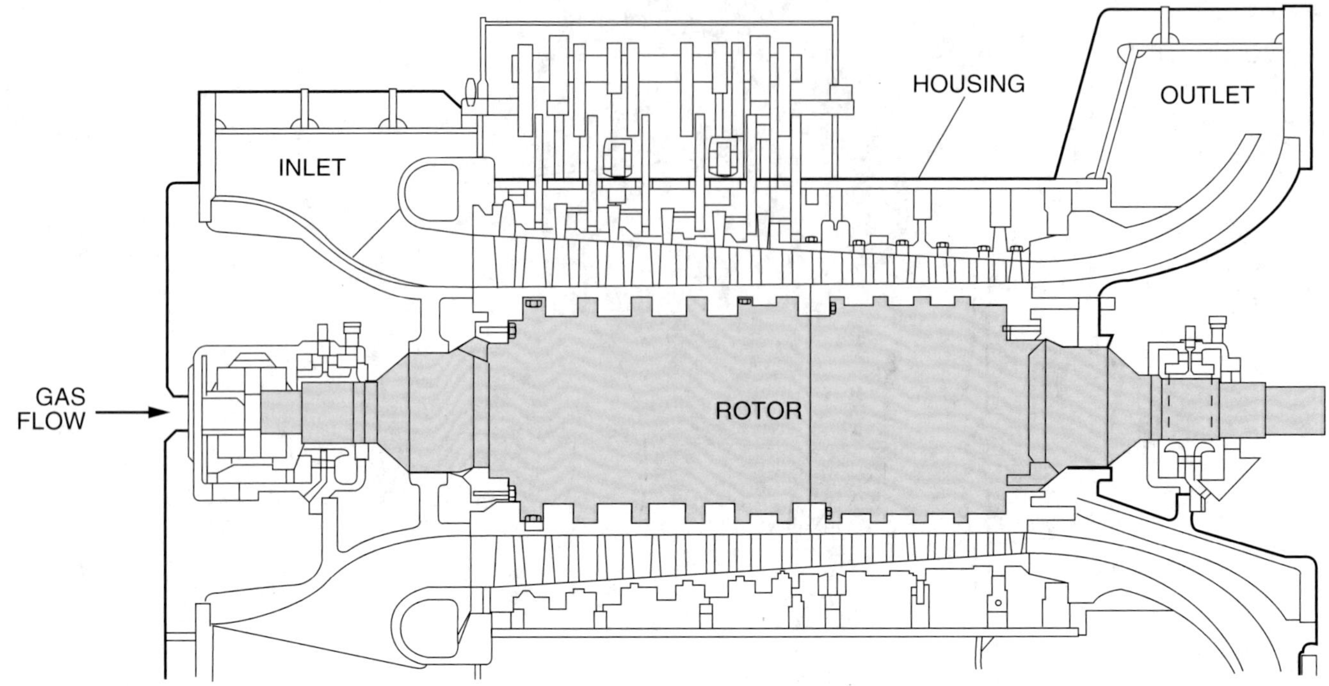

Figure 9 ◆ Axial compressor components.

temperature. This device monitors gas temperature at the discharge area and shuts down the compressor if the temperature exceeds a preset limit.

- Ensure that the strainer screen is installed **upstream** of compressor suction.
- Ensure that the compressor bearings and mechanical seal are properly lubricated.

> **CAUTION**
>
> Mechanical seals on compressors are designed to work with an oil film. Therefore, the seals must be lubricated before startup and be kept lubricated during operation. If the seals are allowed to dry out, they wear out more quickly.

- Ensure that the driver rotation matches the compressor rotation.
- Ensure that the piping and the compressor have been purged to remove all air and contaminants.

Many manufacturers also provide a log sheet on which details about maintenance can be recorded. Note that many manufacturers require that the log sheet be completed to maintain the warranty.

5.3.2 Remotely Controlled Shutdown Devices

Remotely controlled shutdown devices (RCSDs) may be operated by a signal sent from the pipeline's **control center.** A simplified example of how these devices work follows. A meter sends a pressure reading to a sensor in the control center. This reading is converted into information that is stored in a database. The system can then compare stored normal operating pressure against the current operating pressure. If the system detects a problem, a signal is sent to the RCSD to **shut in,** or isolate, part of the pipeline.

5.3.3 Emergency Shutdown Devices

By monitoring the line carefully and using pressure-relief valves, pipeline companies can protect the pipeline from too much pressure buildup and still maintain a safe flow of gas. In these situations, emergency shutdown devices (ESDs) may be used.

ESDs are heavy-duty valves designed to handle any pressure imbalance in the pipeline. Most are installed between a pressure-reducing regulator and **downstream** equipment. These devices allow pipeliners to quickly and safely shut in a section of the pipeline. ESDs shut down when a sensor detects that pressure has gone over a preset limit. An ESD may be designed to shut down in response to only low-pressure conditions, only high-pressure conditions, or both.

The activation of an ESD does not always signal an emergency condition. Generally, the following three conditions must be met:

- People may be harmed or killed.
- There may be extensive property damage.
- There may be extensive environmental damage.

Emergencies usually involve actions outside business as usual. For example, a buildup of pressure may be routinely handled by activating a relief valve. If the relief system fails, however, pressure may build up to emergency levels. Following are examples of conditions that may trigger an emergency situation:

- Pressure imbalance in the system
- Uncontrolled release
- Fire or explosion near or on the pipeline
- Bad weather conditions such as floods, tornadoes, hurricanes, and earthquakes
- Riots or violent protests on or near the pipeline
- Criminal or terrorist actions

In the event of an emergency, it is important to follow company policy for notifying local emergency response agencies. Many situations can be safely handled without causing undue alarm. Therefore, pipeline employees follow these general steps:

Step 1 Evaluate the situation to decide whether a true emergency exists.

Step 2 Notify the appropriate company personnel.

Step 3 Carry out established company procedures to protect people, property, and the environment.

Step 4 Notify the local response emergency agency if appropriate.

5.4.0 The Operation of Gas Compressors

It is important to understand the major parts of gas compressors. These include distance pieces, pistons, rings, and valves.

5.4.1 Distance Pieces

As gas is being compressed, some of it may get by the compressor's piston rings and packing. If the gas gets into the crankcase, a hazardous condition can result. Therefore, devices called distance pieces are installed to prevent excess gas, moisture, or lubrication from building up in the compressor area. A **stuffing box** keeps the gas from getting into the crankcase.

5.4.2 Compressor Pistons and Rings

Compressor pistons are made from cast steel, cast iron, aluminum, various alloys of machined steel, and steel alloys. Each pair of pistons should be weight balanced to reduce vibration. In single-stage compressors, achieving a complete balance is not possible. To offset the vibration, the counterweights on the crankshaft can be adjusted to improve performance, and a heavier crankcase and frame can be added.

In every reciprocating compressor, a set of rings seals the gas. The rings also keep the piston from rubbing against the cylinder wall and wearing out too fast. In piston-type compressors, lubrication must be pumped into the compression cylinder to lubricate the piston rings. Providing some of the same lubrication to the packing extends its life.

5.4.3 Compressor Valves

Suction and discharge valves in the head of the compressor control the gas intake and discharge of the compressor cylinder. The valve area (total open area when the valve is open) affects the compressor's efficiency. The proper valve design minimizes pressure drop across the valve and helps the unit to operate efficiently. Selecting valves carefully also increases the efficiency of the compressor. Valves with a lower spring tension open more easily and operate more efficiently.

6.0.0 ◆ COMPRESSED-AIR TREATMENT

Air entering a compressor is often hot and contaminated, which reduces compressor efficiency. Therefore, the air must be treated. The following devices are used to remove contaminants and regulate temperature:

- Intake filters
- Intercoolers
- Aftercoolers
- Separators
- Chemical dryers
- Receivers
- In-line filters
- Pressure regulators
- Lubricators
- Filter-regulator-lubricators

6.1.0 Intake Filters

Compressors work best when the air flowing into them is cool, dry, and clean. However, air contains dirt and dust, which, if allowed to get into the compressor, cause it to overheat and lose efficiency. Intake filters, which must be cleaned or replaced on a regular schedule, solve this problem. Follow the manufacturer's instructions and recommendations for cleaning and replacing filters. The two main types of intake filters are dry and wet.

The dry intake filter works a lot like an air filter on a car. As air flows through the filter, it traps dirt and dust (see *Figure 10*).

In the wet intake filter, air passes over oil before it enters the filter (see *Figure 11*). The oil traps dirt and dust and helps remove moisture from the air before it enters the compressor. Wet intake filters are rarely used on modern equipment, but some may still be found on older equipment.

> **NOTE**
> Filters are rated by the size of the openings in the material. This rating is based on the size of the largest particle that can pass through the filter. Filtration is measured in microns. One micron is 39 millionths of an inch (0.0000039). The size of filter to use for a particular application depends on the micron size and the flow volume.

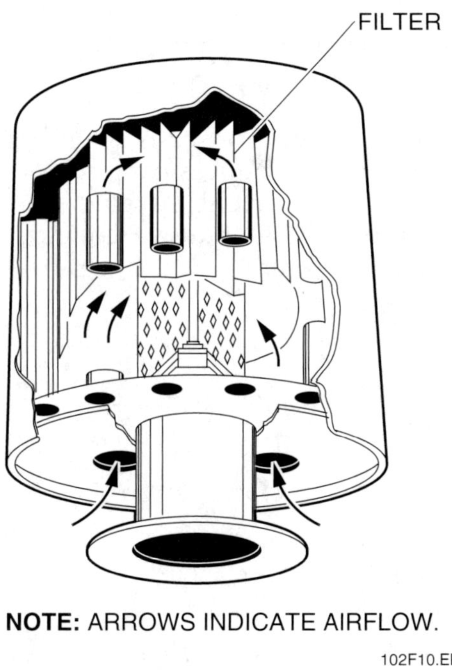

Figure 10 ◆ Dry intake filter.

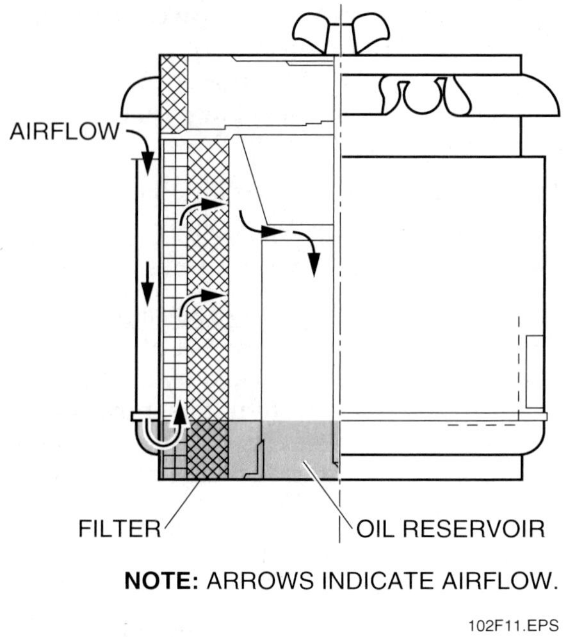

Figure 11 ◆ Wet intake filter.

6.2.0 Intercoolers

Compressing a gas produces heat, and heat reduces compressor efficiency. To solve this problem, the gas is cooled between compression stages with intercoolers. The two types of intercoolers used in multistage compressors are air and liquid.

The air intercooler has finned tubing that runs between the stages of the compressor (see *Figure 12*). As the gas hits the tubing, the heat is scattered, and the gas cools off.

The liquid intercooler uses water to cool the air. The water circulates through tubing inside a tank that contains the compressed gas.

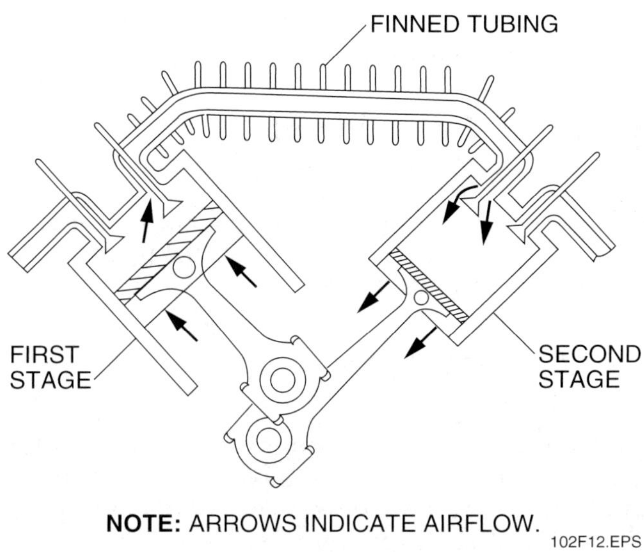

Figure 12 ◆ Air intercooler.

6.3.0 Aftercoolers

Aftercoolers lower the discharge temperature of compressed gas and remove oil vapor and water vapor. The main types of aftercoolers are air-cooled and water-cooled types and refrigeration dryers.

In the air-cooled aftercooler, compressed gas passes over finned tubing on its way to the air receiver. A fan-type flywheel draws the gas over the tubes, cooling it and condensing the moisture. The moisture passes on to a receiver and is drained either manually or by an automatic drain trap.

In the water-cooled aftercooler, a steel shell encases copper tubing and cooling fins (see *Figure 13*). Headers on each end of the aftercooler provide piping connections for water, air, and condensate piping.

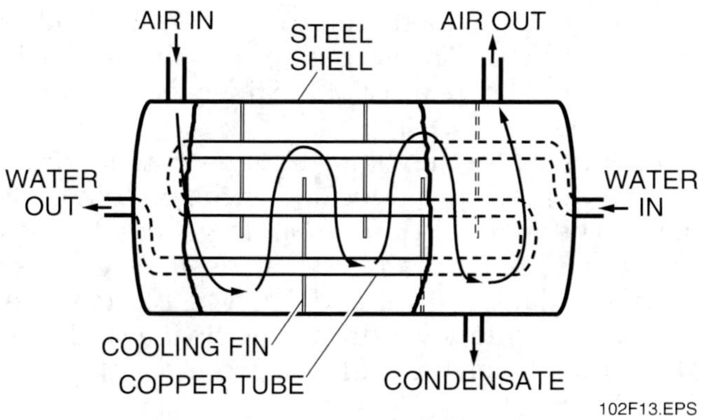

Figure 13 ◆ Water-cooled aftercooler.

In the refrigeration dryer, humidity is lowered to dry the gas. The refrigeration system captures moisture from the compressed gas as it passes through the heat exchanger and discharges it through a drain (see *Figure 14*).

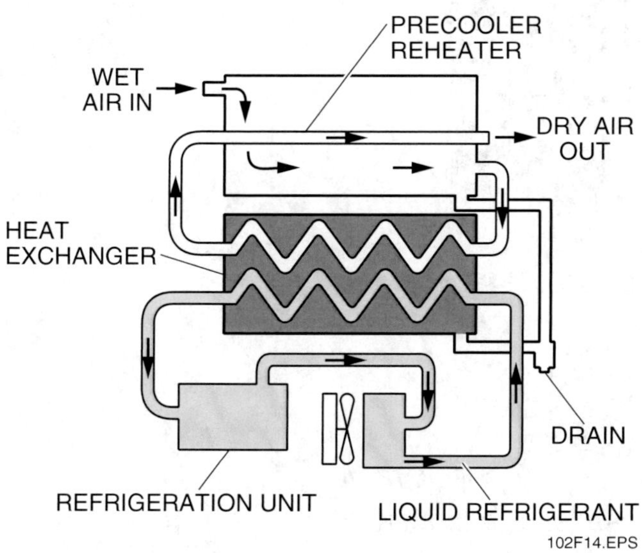

Figure 14 ◆ Refrigeration dryer.

6.4.0 Separators

Natural gas may contain other hydrocarbons and water that condense during transmission. Separators, which contain scrubbers or filters, remove these elements from the gas at the compressor station (see *Figure 15*). The gas usually enters the separator before entering the compressor station.

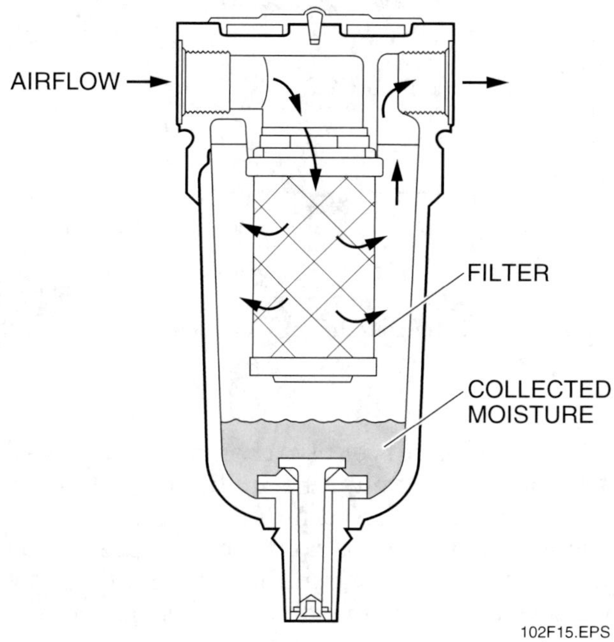

Figure 15 ◆ Separator.

6.5.0 Chemical Dryers

Chemical dryers clean and dry the air. They generally are placed in the pipeline following the separator. The two main types of chemical dryers are the deliquescent and the desiccant. The deliquescent dryer is an **absorption** dryer; the desiccant dryer is an **adsorption** dryer.

The deliquescent dryer uses a chemical to absorb water from the system (see *Figure 16*). As it absorbs the water, this chemical slowly melts and dissolves; therefore, it must be replaced on a regular basis. Some absorption dryers use two-stage systems with heaters to help the water evaporate. The two-stage system allows one side of the dryer to dry out while the other side is in use. Wet air flows into the dryer through a four-way valve that controls which side of the dryer is used. Heating elements dry the air, which then flows back through the four-way valve into the line. At the same time, check valves control the flow of condensate that comes out of the bottom of the dryer.

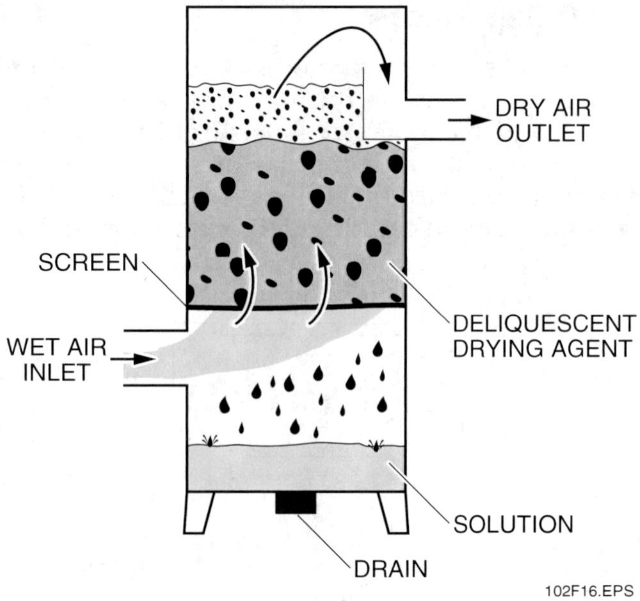

Figure 16 ◆ Deliquescent dryer.

The desiccant dryer uses chemicals, such as silica gel or activated alumina, in the adsorption process to dry compressed air (see *Figure 17*). These chemicals dry the air by trapping water in tiny pores. Adsorption systems have no moving parts, and the desiccant does not have to be replaced. Generally, however, an adsorption system costs more to run than an absorption system does.

6.6.0 Receivers

A receiver is a tank-like structure that stores compressed air (see *Figure 18*). It removes heat from the compressed air and collects moisture. The receiver acts as a buffer and allows air to flow smoothly and quietly from the compressor to the pneumatic system.

6.7.0 In-Line Filters

In-line filters (see *Figure 19*) remove most solid and liquid particles from compressed air. As air enters the filter unit, louvers force it into a swirling pattern. Centrifugal force separates the coarse, solid particles, which then drop to the bottom of the filter's bowl. The air then passes through a filter element, where most of the finer solid particles are trapped.

The filter unit may have either a manual drain or an automatic drain located at the bottom of its bowl. The manual drain is a single small valve. In the automatic drain, a float rises in response to accumulated liquid and opens a bleed valve. When the liquid level drops, the float also drops; at a certain level, the draining process is stopped.

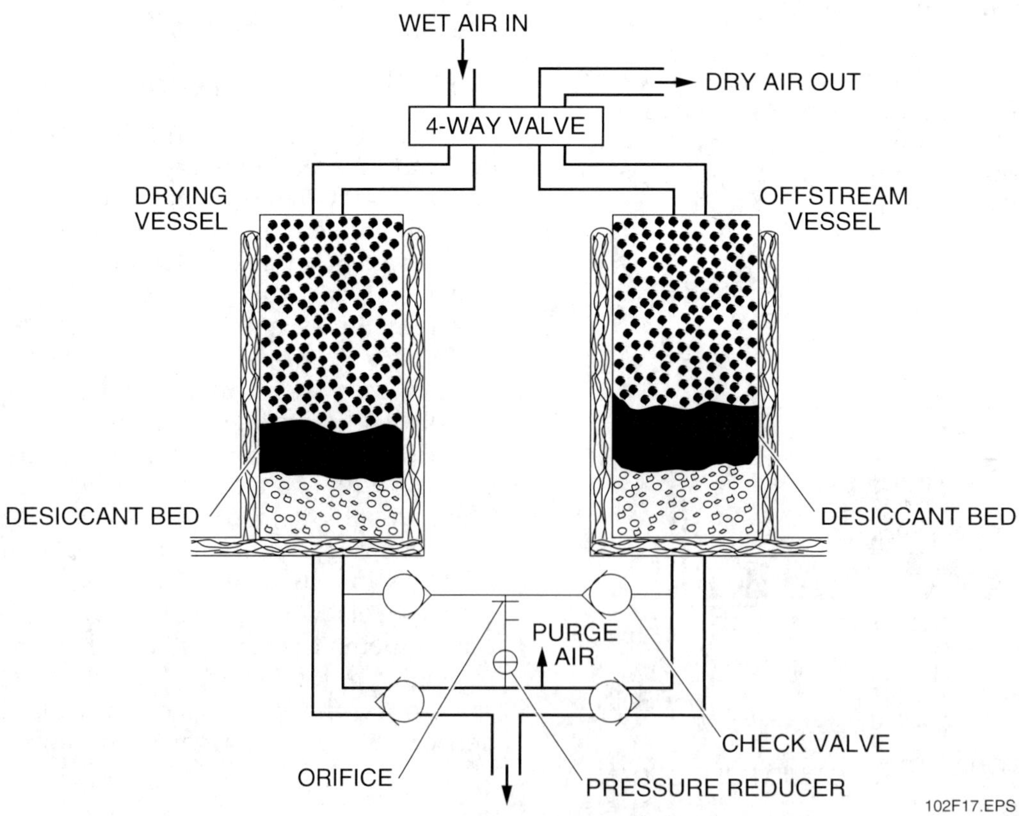

Figure 17 ◆ Desiccant dryer.

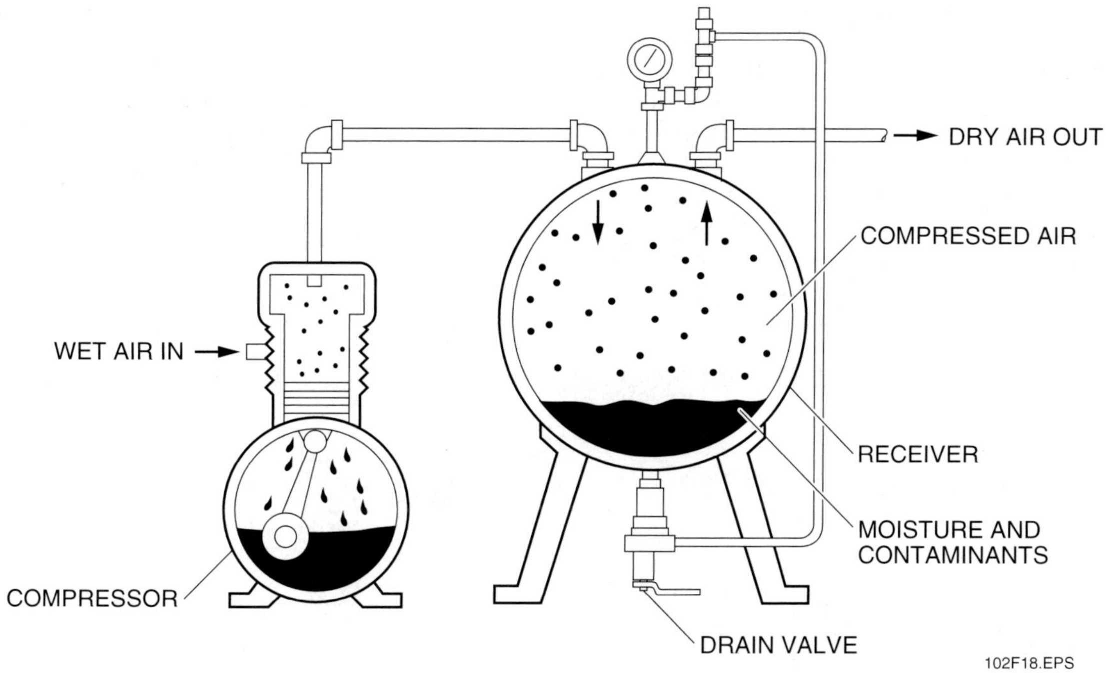

Figure 18 ◆ Receiver.

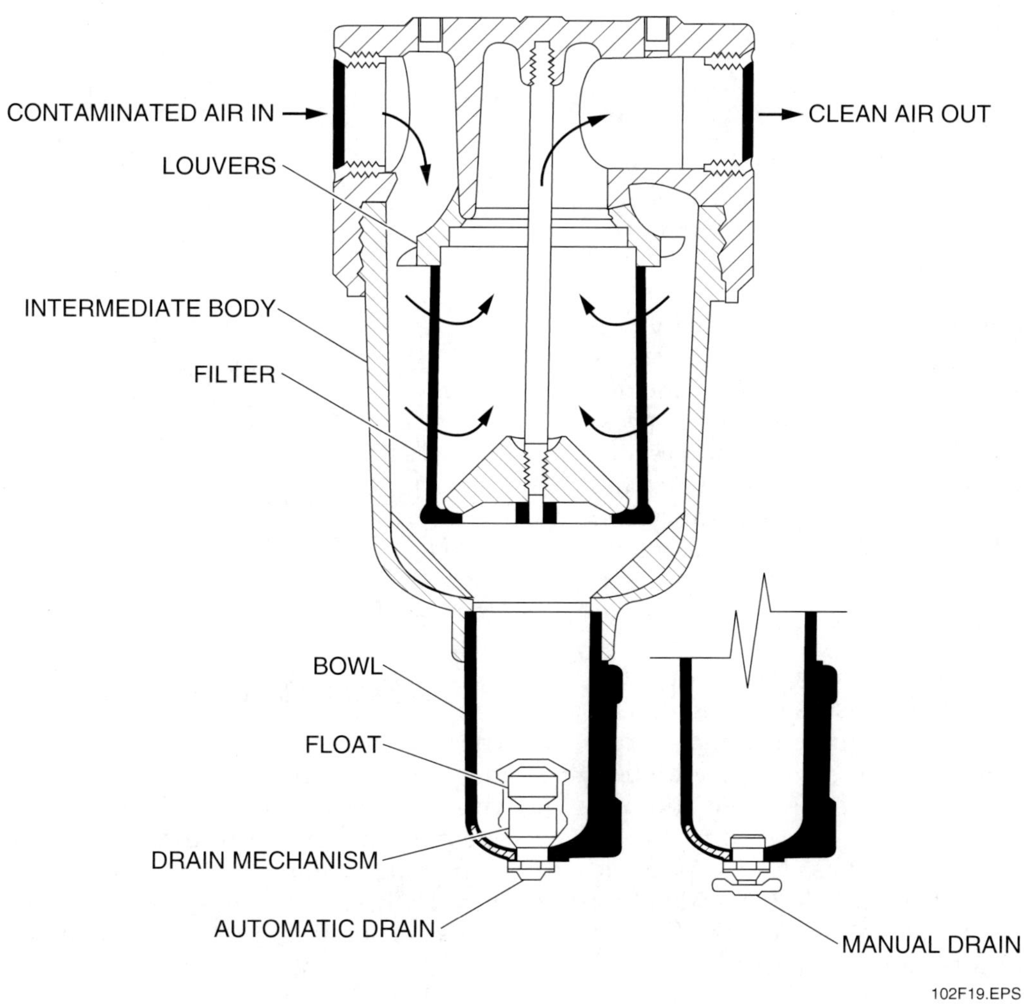

Figure 19 ◆ In-line filter.

6.8.0 Pressure Regulators

A pressure regulator controls the operation of a valve that either opens or closes to control pressure in the line. The two main types of regulators are diaphragm and piston.

In a diaphragm regulator, pressure either increases or decreases against a thin diaphragm (see *Figure 20*). As the downstream pressure increases, the diaphragm closes the valve. As the pressure drops, the diaphragm opens the valve.

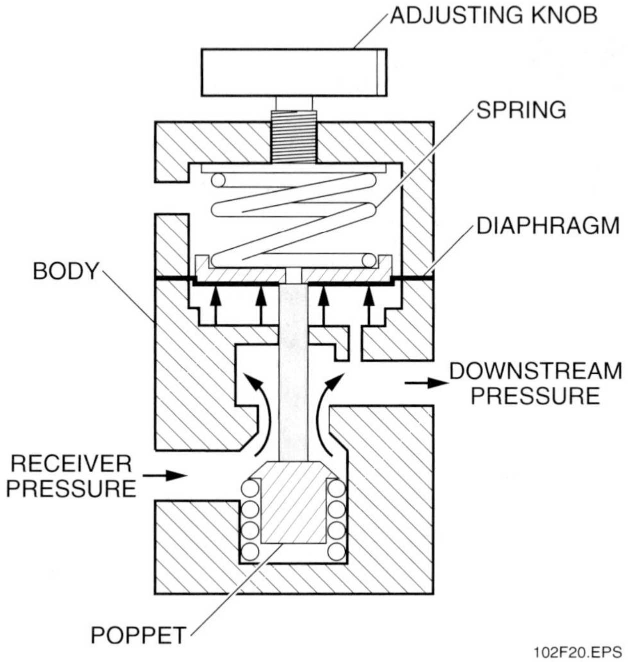

Figure 20 ◆ Diaphragm regulator.

In a piston regulator, a spring-loaded valve responds to changes in pressure (see *Figure 21*). A pressure increase is sensed in an area near the underside of the piston. When the downstream pressure nears a preset level, the piston moves upward and pulls the **poppet** into its **seat.** Once the poppet is seated, a constant source of compressed air is made available to an **actuator** downstream.

6.9.0 Lubricators

Lubricators add a small amount of oil into the air flowing through compressors (see *Figure 22*). The air breaks the oil into a fine mist and carries it downstream to the compressor. Oil is added to compressed air to lubricate seals, keep moving parts from sticking together, and reduce wear. Lubricators must be installed as close as possible to the components to keep the oil from traveling as a liquid. Incorrect installation can stop flow in the system. You must always refer to the manufacturer's manual to determine the amount and type of lubrication to use.

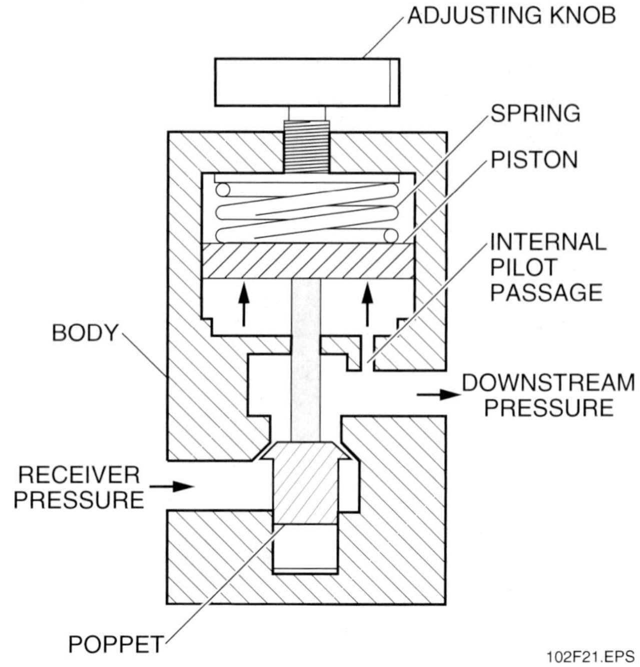

Figure 21 ◆ Piston regulator.

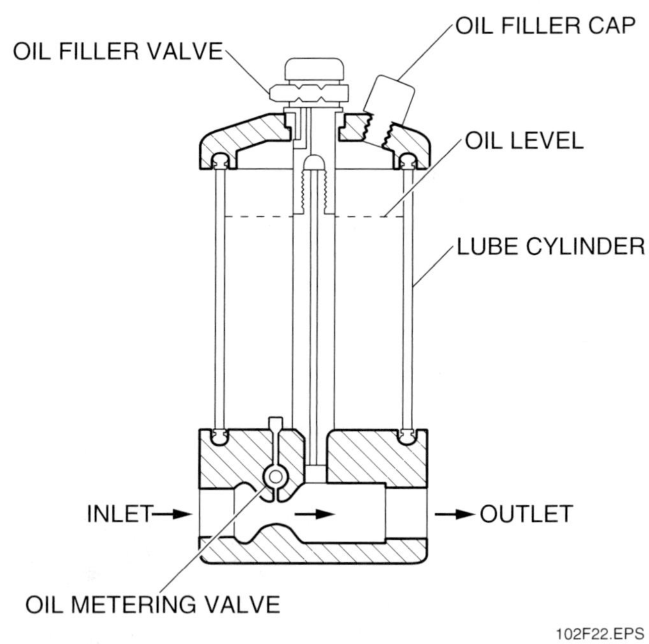

Figure 22 ◆ Lubricator.

6.10.0 Filter-Regulator-Lubricators

The filter-regulator-lubricator (F-R-L) unit is used in place of three separate units (see *Figure 23*). This unit allows you to take the system apart in one place instead of in three; thus, the chance of contaminants getting into the system is greatly reduced. The F-R-L unit must be connected in the correct order: filter first, then the regulator, and finally the lubricator. The order of connection must be followed to prevent damage resulting from contamination of the pneumatic system.

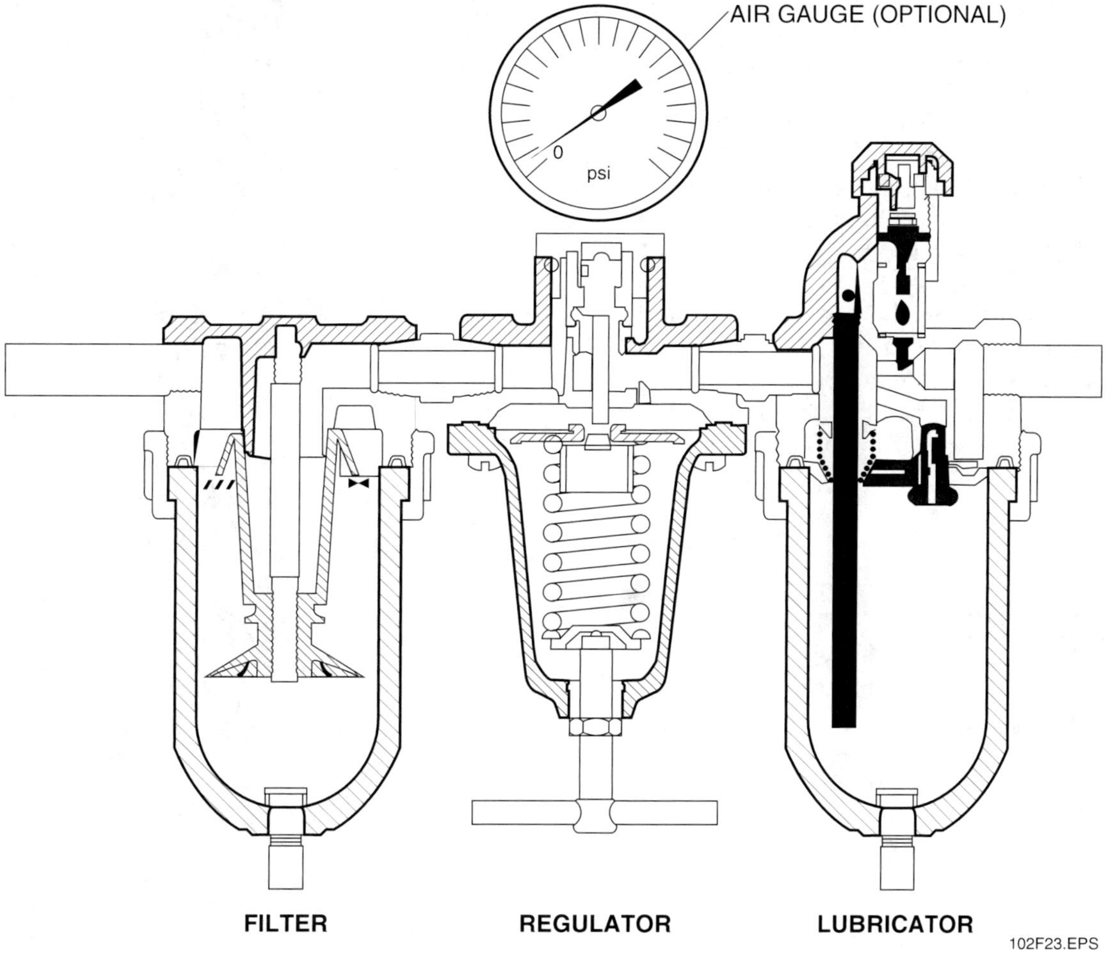

Figure 23 ◆ F-R-L assembly.

7.0.0 ◆ PNEUMATIC SYSTEM COMPONENTS

Pneumatic system components are designed to control flow, operate system equipment, and muffle noise. The main pneumatic system components are as follows:

- Valves
- Actuators
- Mufflers

7.1.0 Valves

Valves are grouped by how they control airflow. Valves may direct, check, block, or restrict airflow. They may also control or relieve pressure in the line. Following are the five most common types of valves:

- Directional-control valves
- Check valves
- Flow-control valves
- Pressure-control valves
- Pressure-relief valves

NOTE

Each valve that might be needed during an emergency must be inspected and operated on a regular schedule. Government regulations state that these inspections must take place at intervals not greater than 15 months, but at least once each calendar year (*49 CFR 192.745*).

7.1.1 Directional-Control Valves

Directional-control valves allow or prevent airflow through the pipeline and come in a wide range of sizes and types. This type of valve may be controlled manually, mechanically, or with a pilot device. Manual controls include push buttons, switches, push-pull levers, or foot pedals. In valves that are mechanically controlled, an electrical signal trips a **solenoid,** which causes another solenoid to operate. A push button, limit switch, or sequencing relay may activate the electrical signal. Pilot devices are activated by a pressure change in the circuit that makes the valve open or close.

7.1.2 Check Valves

Check valves (see *Figure 24*) allow air to flow in one direction and block it from flowing upstream or back through the line. This type of valve has suction and discharge ports at either end of the valve body and a movable part, or seal, that is activated by spring pressure. The seal might be a flapper, plunger, ball, or poppet. When the pressure at the suction port is higher than the pressure of the spring that holds the seal in place, the seal moves off its seat. Air then flows freely through the valve. When air tries to return through the discharge port, the air pushes the seal back onto its seat, and airflow through the valve is blocked.

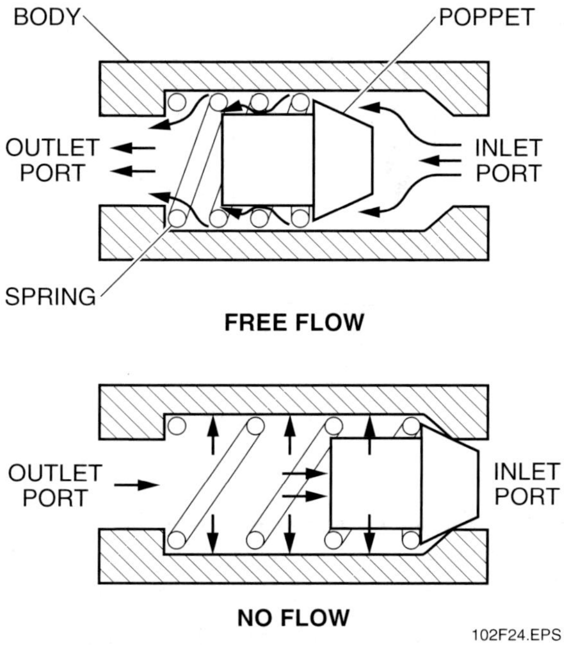

Figure 24 ◆ Check valve.

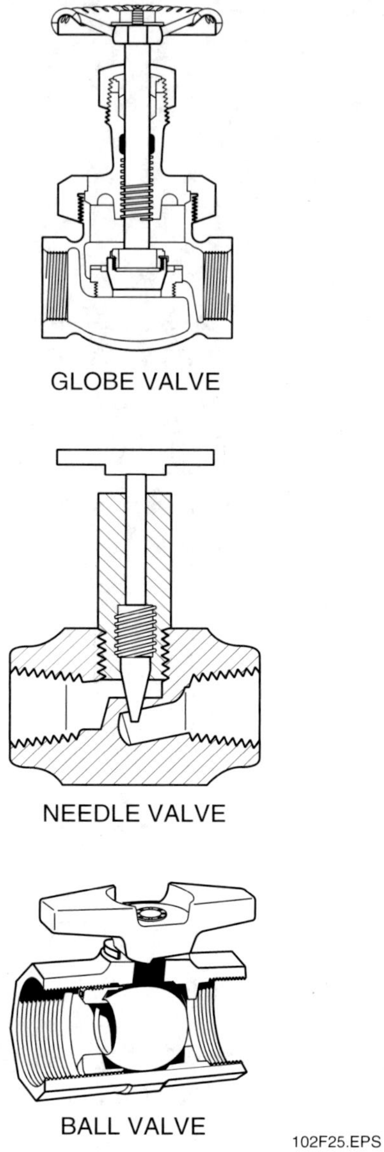

Figure 25 ◆ Flow-control valves.

7.1.3 Flow-Control Valves

Flow-control valves restrict airflow. Controlling how much air flows through the valve helps control the operating speed of the actuator. A flow-control valve might be fixed or adjustable. Three common flow-control valves are globe, needle, and ball valves (see *Figure 25*).

Air does not flow in a straight line through a globe valve. It must turn 90 degrees and pass through the orifice, or opening, that is the seat of the globe. The orifice size is adjustable and determines the rate of flow.

A common type of flow-control valve in a pneumatic system is a combination needle valve and bypass check valve. Air flowing through the needle valve must turn 90 degrees to pass through the orifice. The orifice is the seat for a cone-shaped rod, and the position of the rod in relation to the seat determines the orifice size. Fine threads on the valve stem allow gradual changes in the orifice size. The bypass check valve allows reverse airflow in the system.

A ball valve allows air to flow straight through its center. A handle is used to change the orifice size. As the handle is turned, a ball with a cross-drilled hole is positioned across the airflow path. The size of the hole determines the flow rate.

7.1.4 Pressure-Control Valves

Compressed air held in a compressed-air storage tank must be portioned out to the actuators downstream. Pressure-control valves control the pressure at two points in the system: after the compressor and after the air receiver tank.

Pressure-sensing devices automatically control the amount of pressure produced by the compressor. To manage the compressor's operating range, the system signals the compressor to turn on or off. If this system fails, a safety-relief valve vents some pressure to the atmosphere. This venting process allows the compressor to return to a safe operating level.

7.1.5 Pressure-Relief Valves

Pressure-relief valves (see *Figure 26*) are sometimes called unload or pop-off valves. These valves are safety devices that protect the line from a buildup of too much pressure. Pressure-relief valves should be installed close to the compressor or the system part for which they relieve pressure. The valve is preset to a specific pressure point. Pressure above this point overcomes a low-pressure spring that keeps the valve closed. When the valve opens, it vents air or gas to the atmosphere and reduces the pressure buildup. When the pressure decreases to the preset point, the spring closes the valve.

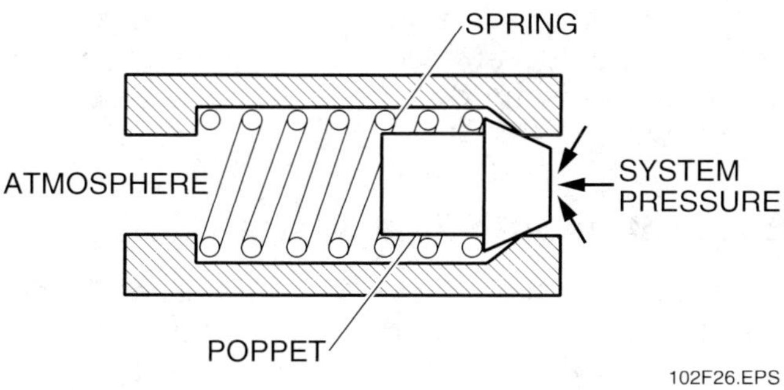

Figure 26 ◆ Pressure-relief valve.

7.2.0 Actuators

Actuators are mechanical devices that move or control the operating parts of a pneumatic system. Common examples of actuators include an air-driven dentist's drill and a pneumatic lift at a service station. Two types of actuators in pneumatic systems are pneumatic cylinders and pneumatic motors.

7.2.1 Pneumatic Cylinders

Pneumatic cylinders channel air pressure in a straight line. Air pressure in the cylinder pushes a piston forward to an area where there is less pressure. Pneumatic cylinders are found in jackhammers, drill presses, and other impact-type tools.

Pneumatic cylinders fall into two categories: single acting and double acting. The single-acting cylinder applies force to move the piston in one direction. A weight or spring pushes the piston back again. The double-acting cylinder applies force in both directions.

7.2.2 Pneumatic Motors

Pneumatic motors convert the force of compressed air into mechanical energy. The vane motor (see *Figure 27*) is the most common type of pneumatic motor. This motor is made up of vanes, a rotor, a shaft, a ring, and suction and discharge ports. Compressed air enters the suction port and pushes against the vanes, which causes the motor to produce an output **torque.** As air pressure against the vane increases, rotation also increases.

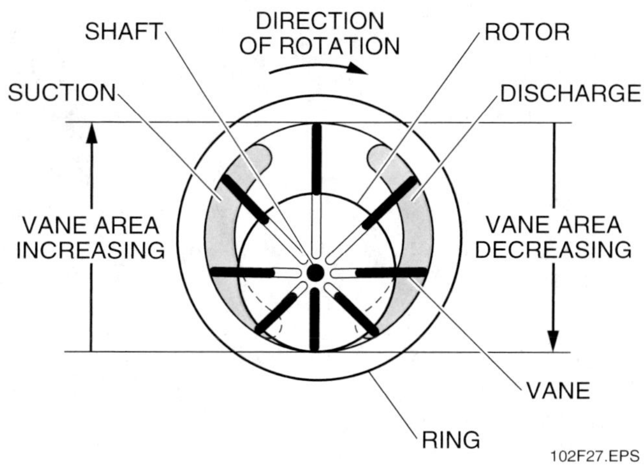

Figure 27 ◆ Vane motor.

7.3.0 Mufflers

When pressure is released from a pneumatic system, it makes a lot of noise. Much like mufflers on cars, mufflers are used on air systems to reduce noise. However, mufflers also restrict airflow, which can reduce system efficiency. Only mufflers that are compatible with the system should be used. If the muffler is not compatible, backlash is created in the system, which could damage the compressor. You must follow the manufacturer's technical specifications to ensure the right match between the muffler and the system.

CAUTION

Mufflers may not reduce noise levels enough to protect hearing. In these cases, earplugs must be worn. Follow company safety rules regarding the use of proper hearing protection.

Review Questions

Sections 5.0.0–7.0.0

1. A compressor increases the pressure of gas by _____.
   a. reducing its volume
   b. increasing its volume
   c. heating it up
   d. cooling it down

2. The two main types of compressors are _____.
   a. helical and axial
   b. reciprocating and sliding vane
   c. positive displacement and continuous flow
   d. negative displacement and intermittent flow

3. In a liquid intercooler, the air is cooled with _____.
   a. oil
   b. silica gel
   c. nitrogen
   d. water

4. _____ lower the discharge temperature of compressed gas and remove oil vapor and water vapor.
   a. Intercoolers
   b. Aftercoolers
   c. Separators
   d. Receivers

5. _____ valves allow air to flow in one direction and block it from flowing upstream, or back through the line.
   a. Gate
   b. Block
   c. Check
   d. Needle

8.0.0 ◆ METERS

Three basic types of meters are used on the gas pipeline: turbine, ultrasonic flow, and orifice. Meters are the cash registers of the pipeline industry. They record how much gas goes from one point to another, allowing the pipeline company to track deliveries and bill customers. Therefore, meters must be calibrated before use and checked on a regular basis to ensure accurate measurement. The three main costs associated with providing gas to customers are as follows:

- *Transmission cost*—The cost to move the gas from its source, through the pipeline, to a local gas distributor
- *Distribution cost*—The cost to move the gas from the local gas distributor to individual customers
- *Commodity cost*—The cost of the gas itself

8.1.0 Turbine Meters

Gas turbine meters are used for the **custody transfer** of natural gas (see *Figure 28*). They are also used in distribution and supply stations. These meters are designed to operate with high-energy efficiency at low gas speed. A turbine rotor guides the gas flow through the meter to an exit channel, which is designed to minimize pressure loss.

102F28.EPS

Figure 28 ◆ Gas turbine meter.

Turbine meters measure speed and convert it to volume. Rotors in the meter have a sensor that sends a signal to an electronic counter. The total of the signals is divided by the frequency per unit of volume.

8.2.0 Ultrasonic Flow Meters

An ultrasonic flow meter is also called a transit-time meter. It measures the amount of time it takes for an ultrasonic sound wave to transit, or move, from one sensor to another. The transit time is measured in both the upstream and downstream directions. Ultrasonic flow meters come in two basic types: single path and multipath (see *Figure 29*).

102F29.EPS

Figure 29 ◆ Multipath ultrasonic flow meter.

In the single-path meter, the transit time is measured with a single set of sensors. The transit time (stated in feet per second) is calculated by dividing distance by time.

A multipath ultrasonic meter works exactly like a single-path meter. However, the multipath meter uses several pairs of sensors to send and receive signals. Because more information is collected, multipath readings tend to be more accurate than single-path readings.

8.3.0 Orifice Meters

Orifice meters (see *Figure 30*) restrict the flow of gas, thus producing a difference in pressure. That pressure difference is measured to calculate actual flow. The orifice meter comes in a wide variety of sizes and types. Some models are designed for repair in the field, which makes them highly efficient for pipeline use.

102F30.EPS

Figure 30 ◆ Orifice meter.

Orifice plates (see *Figure 31*) allow precise flow of natural gas through the meter. These plates are set in the body of the meter housing, and you can easily remove them for maintenance. The plates of some meters may be removed without stopping the flow of gas. This feature saves time and money.

102F31.EPS

Figure 31 ◆ Orifice plates.

8.4.0 Meter Protection

Meters are sensitive units that must be kept in good working condition. Meters are equipped with protection and control equipment, including alarms, flow controllers, pressure controllers, strainers, and air ventilation equipment.

Flow controllers manage the flow of gas through the meter, which helps to protect the meter from a buildup of pressure. Flow controllers also prevent gas from flowing in the wrong direction, or back through the meter. Pressure gauges are designed to detect a change in pressure caused by a clog in the strainer. A sensor on the pressure gauge sends an alarm to the control center if the strainer clogs. Air ventilation equipment is installed upstream of meters to vent excess air. If excess air starts to build up in a meter, an alarm is sent to the control center.

9.0.0 ◆ FLOW AND TEMPERATURE CONTROLS

It is important to monitor and regulate gas flow and temperature as the gas moves through the pipeline. Faster-moving gas heats up more quickly and must be directed to cooling towers located between compressors before being returned to the main pipeline. In addition, gas may be heated for certain processes.

9.1.0 Flow Control

Flow controllers (see *Figure 32*) monitor and measure flow through the pipeline around the clock. Some flow controllers are stand-alone units. Others are networked to a data collection system. The information collected from the flow controller is processed and stored in the unit. Personnel can read this information in the field and send it by telephone, radio, or satellite transmission to other locations. Flow controllers are installed at various places along the line, including compressor stations, custody transfer stations, and gate stations. These controllers may use turbine meters, ultrasonic meters, or orifice plates to monitor the flow of gas.

102F32.EPS

Figure 32 ◆ Gas flow controller.

9.2.0 Temperature Control

As you have learned, compression causes gas to become hotter. In the transmission line, gas can be cooled to acceptable levels in several ways. On land, the gas moves through cooling stations where water or air is used to bring the temperature down to acceptable levels. The gas also may be cooled when the cooler ground temperature or cooler water temperature surrounding the pipeline draws off heat.

Gas may have to be heated for process control, such as to operate a valve, compressor, or other line equipment. The temperature is increased on the upstream side to power the equipment and then decreased on the downstream side. A programmable logic controller in the control center is used to control the process and the compressors.

Review Questions

Sections 8.0.0–10.0.0

1. The three basic types of meters used on gas pipelines are _____.
   a. turbine, ultrasonic flow, and orifice
   b. intercoolers, aftercoolers, and separators
   c. master, reciprocating, and sliding vane
   d. positive displacement, continuous flow, and axial

2. The three main costs associated with providing gas to customers are _____.
   a. research, development, and compliance
   b. transmission, distribution, and commodity
   c. research, transportation, and development
   d. transmission, transportation, and maintenance

3. Turbine meters measure speed and convert it to _____.
   a. energy
   b. heat
   c. volume
   d. temperature readings

4. Orifice meters _____ the flow of gas.
   a. redirect
   b. restrict
   c. measure
   d. monitor

5. Flow controllers may use _____ to monitor the flow of gas.
   a. turbine meters, ultrasonic meters, or orifice plates
   b. pressure gauges, check valves, or gate valves
   c. check valves, gate valves, or needle valves
   d. compressor pressure or flow volume

10.0.0 ◆ ODORIZATION EQUIPMENT

Natural gas must be odorized in distribution lines so that if a leak occurs, anyone with a normal sense of smell can detect it. Generally, **odorants** are not required in transmission lines, offshore lines, or lines that run through sparsely populated areas. Odorant, usually a chemical called mercaptan, is added to natural gas, which is odorless, to make it smell like rotten eggs.

There are three main types of odorization equipment: bypass, wick, and electronic. Each diverts the flow of gas across a sensor that triggers the flow of odorant into the gas stream. Personnel use monitors inserted into the gas stream to monitor the gas for correct odorant levels.

Generally, it is the responsibility of the local distribution company to add odorant to the distribution line. Computerized units are available that allow responsible personnel to do the following tasks:

- Calculate the correct amount of odorant to add to the gas stream
- Safely inject the odorant into the distribution line
- Monitor the odorant
- Calibrate the odorization equipment
- Generate reports regarding the odorization operation

11.0.0 ◆ PNEUMATIC SYMBOLS

Pneumatic symbols are used on blueprints of pneumatic systems. These symbols are a shorthand way of showing the location of items in the system, such as compressors and valves (see *Figure 33*). The pipeline mechanic, maintenance technician, and engineer use these symbols to plan, install, locate, troubleshoot, and maintain the system.

12.0.0 ◆ PIPELINE DESIGN

Pipeline used for transmission lines is made of carbon steel that meets standards set by the American Petroleum Institute. Distribution pipelines generally are made of plastic, but the material used may vary depending on location. Pipes fall into two broad size classifications: large-diameter piping (generally between 16 and 48 inches in diameter) and small-diameter piping (which can be as small as 0.05 inches in diameter). Large-diameter pipes, used in transmission lines, are formed by rolling the steel into a tube and welding the edges. Small-diameter pipes, used mainly in gathering and distribution lines, can be made seamlessly. To produce seamless piping, a hole is punched through a heated metal bar to create a tube.

Piping must be strong enough to endure many forces. On the inside, the pipeline must hold up under the heat and pressure from the flowing, compressed gas. Some lines also must be treated to rid them of bacteria that can form harmful acid deposits. Once a pipeline is buried, however, it is impossible for personnel to inspect the interior without digging it up again. Therefore, pipeline companies use sophisticated robots called pigs that can travel through the pipeline to inspect it. Pigs can perform the following tasks:

- Check the pipe wall thickness
- Detect corrosion

NOTE

Corrosion is the degradation of a material (in this case carbon or low alloy steel) through an interaction with its environment (called an electrolyte). Two electrochemical reactions are necessary for corrosion to occur: an oxidation reaction commonly called an anodic reaction and a reduction reaction commonly called a cathodic reaction.

For corrosion on a pipeline to occur, the following four conditions must be present:

- There must be an anode
- There must be a cathode
- There must be a metallic path to connect the anode and cathode (the pipeline itself can serve as this metallic path)
- There must be an electrolyte, such as moist soil

- Detect leaks
- Look for any problem that could interrupt gas flow
- Look for any problem that could cause a safety risk

On the outside, the pipeline must withstand the effects of weather, erosion, and corrosion. A special epoxy coating that fuses to the pipeline is applied to protect it from corrosion. In addition, **cathodic protection** is used on some lines.

The following material summarizes the government regulations that cover pipeline design. For specific design details, it is best to refer to *49 CFR 192, Subpart C.* The following pipeline design factors are covered by this regulation:

- **Specified minimum yield strength**
- **Nominal** outside pipe diameter (see *Appendix*)
- Nominal pipe wall thickness (see *Appendix*)
- Design factor, a number based on the pipeline's geographic location
- Longitudinal joint factor, a number based on the **American Society for Testing and Materials (ASTM)** classification and pipe class (see *Table 1*)
- Temperature **derating** factor, a number based on the gas temperature (see *Table 2*)

Each of these factors is considered in a formula used by pipeline design engineers to calculate design pressure in pounds per square inch gauge.

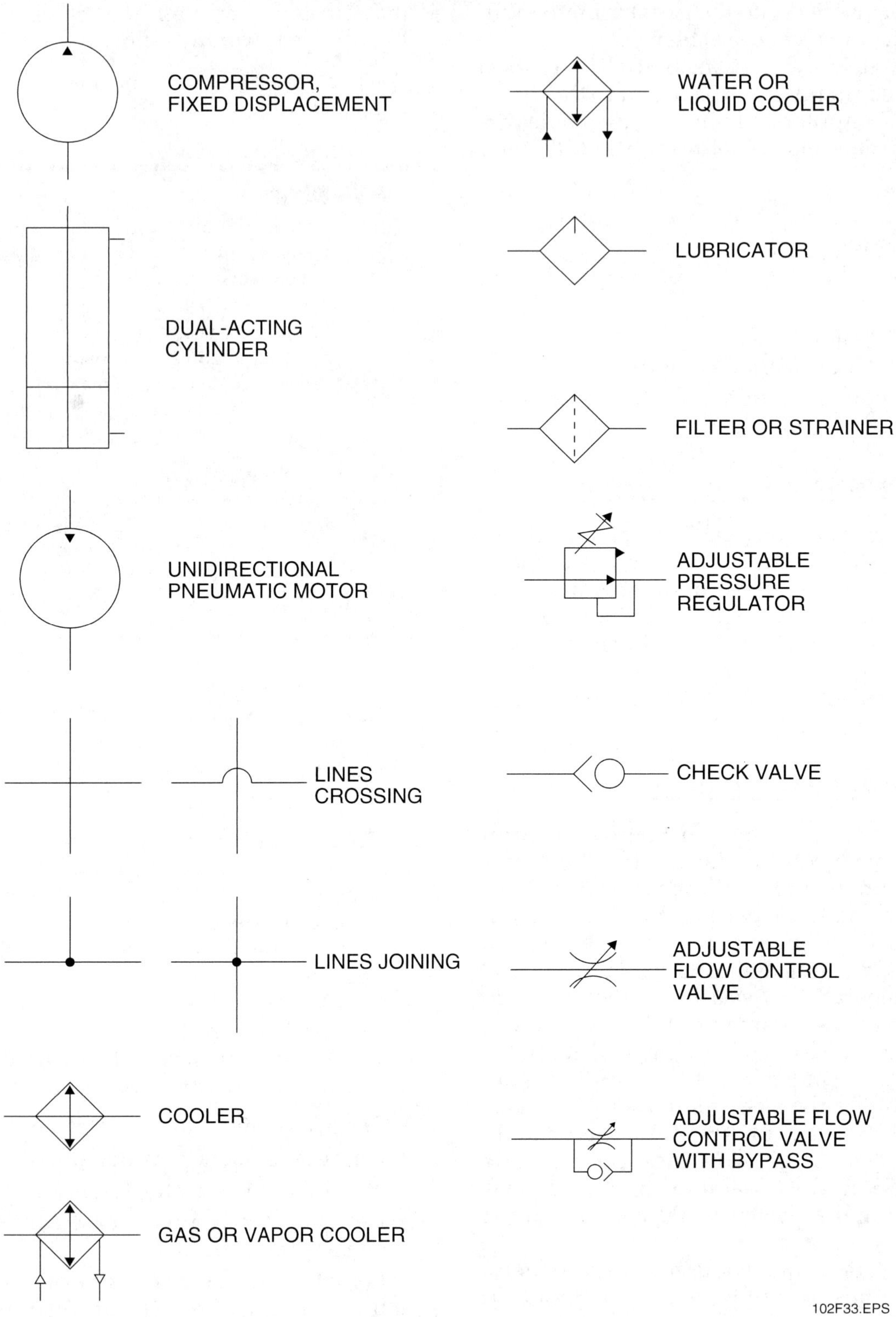

Figure 33 ◆ Basic pneumatic symbols.

Table 1 Longitudinal Joint Factor

ASTM Specification	Pipe Class*	Joint Factor
ASTM A 53	Seamless	1.00
ASTM A 53	Electric resistance welded	1.00
ASTM A 53	Furnace butt welded	0.60
ASTM A 106	Seamless	1.00
ASTM A 333/A 333M	Seamless	1.00
ASTM A 333/A 333M	Electric resistance welded	1.00
ASTM A 106	Seamless	1.00
ASTM A 381	Double welded	1.00
ASTM A 671	Electric fusion welded	1.00
ASTM A 672	Electric fusion welded	1.00
ASTM A 691	Electric fusion welded	1.00
API 5-L	Seamless	1.00
API 5-L	Electric resistance welded	1.00
API 5-L	Electric flash welded	1.00
API 5-L	Submerged arc welded	1.00
API 5-L	Furnace butt welded	0.60
Other	Pipe more than 4 inches (102 millimeters)	0.80
Other	Pipe 4 inches (102 millimeters) or less	0.60

* The type of longitudinal joint is determined by whether the pipe is welded and the type of weld used. If the type of longitudinal joint cannot be determined, the joint factor used must not be greater than that shown for "Other."

ASTM—The American Society for Testing and Materials

API—The American Petroleum Institute

Source: *49 CFR 192.113*

Table 2 Temperature Derating Factor

Gas Temperature in Degrees Fahrenheit (Celsius)*	Derating Factor
250°F (121°C) or less	1.000
300°F (149°C)	0.967
350°F (177°C)	0.933
400°F (204°C)	0.900
450°F (232°C)	0.867

* For intermediate gas temperatures, design engineers use interpolation to find the derating factor.

Source: *49 CFR 192.115*

12.1.0 Maximum Allowable Operating Pressure (MAOP)

The **maximum allowable operating pressure (MAOP)** for pipelines that carry natural gas is defined by government regulation *49 CFR 192.619.* If pipeline pressure goes higher than the lowest pressure that the weakest part of the pipeline can handle, the pipeline will fail. This regulation is designed to ensure that pipelines do not fail under operating pressure. It covers the following areas:

- Type of piping material used in the pipeline industry
- Areas where natural gas is transported by pipeline (onshore and offshore)
- Factors used by the pipeline industry to determine the MAOP

Trade Terms Introduced in This Module

Absorption: The penetration of a substance into the body of another.

Actuator: A mechanism that indirectly moves or controls a part.

Adsorption: The condensation of gases, liquids, or dissolved substances on the surfaces of solids.

American Society for Testing and Materials (ASTM): A nonprofit organization that works with volunteers in the industry to set standards.

Cathodic protection: A method used to prevent the destructive electrochemical process of corrosion of metal by using a sacrificial anode.

Compression ratio: The ratio of the absolute discharge pressure to the absolute intake pressure. The ratio usually applies to a single stage of compression, but it may be applied to multiple stages of compression.

Compressor: A machine that compresses natural gas to reduce its volume and force it through the pipeline.

Control center: The operations center of a pipeline, usually located in a terminal or main pipeline facility. All the control consoles and computer-operated systems are located in the control center.

Custody transfer: The point at which a change of ownership or responsibility for the natural gas in a pipeline takes place.

Derating: The reduction of the rating of a device to improve reliability or to permit operation at high ambient temperatures.

Diffuser: A duct, chamber, or section in which a high-velocity, low-pressure stream of air or other gas is converted into a low-velocity, high-pressure flow.

Discharge: A term that refers to the area where the gas leaves a compressor. It is sometimes called the outlet.

Discharge pressure: The total gas pressure at the discharge flange of the compressor.

Discharge temperature: The total gas temperature at the discharge flange of the compressor.

Distribution line: A pipeline that transports gas from a common source of supply to (1) a customer's meter or the connection to a customer's piping, whichever is furthest downstream or (2) the connection to a customer's piping if there is no customer meter (*49 CFR 192.3*).

Downstream: Any point farther beyond the origination than a given point.

Gathering line: A pipeline that transports gas from a current production facility to a transmission line (*49 CFR 192.3*).

Lockout/tagout: A formal procedure that consists of two steps: locking a piece of equipment that must be serviced and placing a tag on that equipment to alert employees that the equipment must not be used. Only qualified employees may lock out and tag out equipment, certify that equipment is ready for use, and remove the locking device and tag.

Material safety data sheet (MSDS): A form containing information on a potentially hazardous material, including variations on the material's name, contents, test results, characteristics, potential hazards, and handling procedures.

Maximum allowable operating pressure (MAOP): The highest level of internal gauge pressure allowed for a particular pipeline or pipeline segment, as set by the U.S. Department of Transportation.

Nominal: A closely approximate measurement rather than an actual measurement. For example, a 2 × 4 (the nominal size) piece of wood actually measures 1⅝ × 3⅝ inches.

Odorant: A liquid added to natural gas (which is odorless) as a safety feature to make it smell like rotten eggs.

Poppet: A valve that rises straight up to or from its seat.

Seat: The nonmoving part of a valve upon which the moving part rests to form a seal and close the valve.

Shut in: To isolate a segment of the gas pipeline or a pipeline facility associated with that segment in response to an abnormal operating condition (AOC) or to make repairs.

Solenoid: A coil of wire shaped into a long cylinder. A moveable core is drawn into the coil when current flows.

Specified minimum yield strength (SMYS): The specified minimum yield strength of steel pipe, expressed in pounds per square inch (psi) as defined in Department of Transportation regulations.

Stuffing box: A device that prevents leaks along a moving part, such as a piston rod.

Suction: A term that refers to the area where the gas enters a compressor. It is sometimes called the inlet.

Suction pressure: The total pressure of a gas or fluid entering the suction valve of a compressor.

Torque: A force that produces rotation, turning, or twisting.

Transmission line: A pipeline that transports gas from a gathering line or storage facility to a distribution center, storage facility, or large-volume customer that is not downstream from a distribution center (*49 CFR 192.3*).

Upstream: Any point located closer to the origination than a given point.

Valve: A device used to control the rate of flow in a pipeline, open or shut off a pipeline, or serve as a safety device.

APPENDIX

Pipe Wall Thickness

Dimensions of Welded and Seamless Pipe Carbon, Alloy and Stainless Steel

ASME B36.10, B36.19

Nominal Pipe Size (in inches)	Outside Diameter	Sch 5S	Sch 10S	Sch 10	Sch 20	Sch 30	Sch 40S	STD	Sch 40	Sch 60	Sch 80S	XS	Sch 80	Sch 100	Sch 120	Sch 140	Sch 160	XXS
⅛	10.29	–	1.24	–	–	–	1.73	1.73	1.73	–	2.41	2.41	2.41	–	–	–	–	–
¼	13.72	–	1.65	–	–	–	1.73	1.73	1.73	–	3.02	3.02	3.02	–	–	–	–	–
⅜	17.14	–	1.65	–	–	–	2.31	2.31	2.31	–	3.20	3.20	3.20	–	–	–	–	–
½	21.34	1.65	2.11	–	–	–	2.77	2.77	2.77	–	3.73	3.73	3.73	–	–	–	4.75	7.47
¾	26.67	1.65	2.11	2.11	–	–	2.87	2.87	2.87	–	3.91	3.91	3.91	–	–	–	5.54	7.82
1	33.40	1.65	2.77	2.77	–	–	3.38	3.38	3.38	–	4.55	4.55	4.55	–	–	–	6.35	9.09
1¼	42.16	1.65	2.77	2.77	–	–	3.56	3.56	3.56	–	4.85	4.85	4.85	–	–	–	6.35	9.70
1½	48.26	1.65	2.77	2.77	–	–	3.68	3.68	3.68	–	5.08	5.08	5.08	–	–	–	7.14	10.16
2	60.32	1.65	2.77	2.77	–	–	3.91	3.91	3.91	–	5.54	5.54	5.54	–	–	–	8.71	11.07
2½	73.02	2.11	3.05	3.05	–	–	5.16	5.16	5.16	–	7.01	7.01	7.01	–	–	–	9.52	14.02
3	88.90	2.11	3.05	3.05	–	–	5.49	5.49	5.49	–	7.62	7.62	7.62	–	–	–	11.13	15.24
3½	101.60	2.11	3.05	3.05	–	–	5.74	5.74	5.74	–	8.08	8.08	8.08	–	–	–	–	16.15
4	114.30	2.11	3.05	3.05	–	–	6.02	6.02	6.02	–	8.56	8.56	8.56	–	11.13	–	13.49	17.12
5	141.30	2.77	3.40	3.40	–	–	6.55	6.55	6.55	–	9.53	9.53	9.53	–	12.70	–	15.88	19.05
6	168.28	2.77	3.40	–	–	–	7.11	7.11	7.11	–	10.97	10.97	10.97	–	14.27	–	18.24	21.95
8	219.08	2.77	3.76	–	6.35	7.04	8.18	8.18	8.18	10.31	12.70	12.70	12.70	15.06	18.26	20.62	23.01	22.22
10	273.05	3.40	4.19	–	6.35	7.80	9.27	9.27	9.27	12.70	12.70	12.70	15.06	18.26	21.44	25.40	28.58	25.40
12	323.85	3.96	4.57	–	6.35	8.38	9.52	9.52	10.31	14.27	12.70	12.70	17.48	21.44	25.40	28.58	33.32	25.40
14	355.60	3.96	4.78	6.35	7.92	9.52	–	9.52	11.13	15.06		12.70	19.05	23.83	27.79	31.75	35.71	–
16	406.40	4.19	4.78	6.35	7.92	9.52	–	9.52	12.70	16.66		12.70	21.44	26.19	30.96	36.52	40.46	–
18	457.20	4.19	4.78	6.35	7.92	11.12	–	9.52	14.27	19.05		12.70	23.82	29.36	34.92	39.67	45.24	–
20	508.00	4.78	5.54	6.35	9.52	12.70	–	9.52	15.06	20.62		12.70	26.19	32.54	38.10	44.45	49.99	–
22	558.8	4.78	5.54	6.35	9.52	12.70	–	9.52	15.87	22.22		12.70	28.58	34.92	41.28	47.62	53.98	–
24	609.4	5.54	6.35	6.35	9.52	14.27	–	9.52	17.48	24.61		12.70	30.93	38.89	46.02	52.37	59.51	–
26	660.4	–	–			–	–	9.52	–	–		12.70	–	–	–	–	–	–
28	711.2	–	–	7.92	12.70	15.88	–	9.52	–	–		12.70	–	–	–	–	–	–
30	762.0	6.35	7.92	7.92	12.70	15.88	–	9.52	–	–		12.70	–	–	–	–	–	–
32	812.8	–	–	7.92	12.70	15.88	–	9.52	17.48	–		12.70	–	–	–	–	–	–
34	863.6	–	–	7.92	12.70	15.88	–	9.52	17.48	–		12.70	–	–	–	–	–	–
36	914.4	–	–	7.92	12.70	15.88	–	9.52	19.05	–		12.70	–	–	–	–	–	–
38	965.2	–	–	–	–	–	–	9.52	–	–		12.70	–	–	–	–	–	–
40	1,016.0	–	–	–	–	–	–	9.52	–	–		12.70	–	–	–	–	–	–
42	1,066.8	–	–	–	–	–	–	9.52	–	–		12.70	–	–	–	–	–	–
44	1,117.6	–	–	–	–	–	–	9.52	–	–		12.70	–	–	–	–	–	–
46	1,168.4	–	–	–	–	–	–	9.52	–	–		12.70	–	–	–	–	–	–
48	1,219.2	–	–	–	–	–	–	9.52	–	–		12.70	–	–	–	–	–	–

102A01.EPS

Figure Credits

Instromet, Inc.	102F28
Emerson Process Management, Daniel Division	102F29, 102F30, 102F31
Fisher Controls International, LLC	102F32
The American Society of Mechanical Engineers	Appendix A

NCCER CURRICULA — USER UPDATE

NCCER makes every effort to keep its textbooks up-to-date and free of technical errors. We appreciate your help in this process. If you find an error, a typographical mistake, or an inaccuracy in NCCER's curricula, please fill out this form (or a photocopy), or complete the online form at **www.nccer.org/olf**. Be sure to include the exact module ID number, page number, a detailed description, and your recommended correction. Your input will be brought to the attention of the Authoring Team. Thank you for your assistance.

Instructors – If you have an idea for improving this textbook, or have found that additional materials were necessary to teach this module effectively, please let us know so that we may present your suggestions to the Authoring Team.

NCCER Product Development and Revision

13614 Progress Blvd., Alachua, FL 32615

Email: curriculum@nccer.org
Online: www.nccer.org/olf

❑ Trainee Guide ❑ Lesson Plans ❑ Exam ❑ PowerPoints Other ______________________

Craft / Level: Copyright Date:

Module ID Number / Title:

Section Number(s):

Description:

Recommended Correction:

Your Name:

Address:

Email: Phone:

Module 60102-02

Basic Pipeline Hydraulics and Equipment

Introduction to the Pipeline Operator Qualification Rule and NCCER

It could be said that petroleum products keep America running. Petroleum products like oil and natural gas are the principal sources of energy in the United States and are used in many manufacturing and chemical processes. Petroleum products are primarily transported by pipeline. There are more than a million miles of liquid and natural gas pipeline in this country, ranging in diameter from two inches to four feet. These pipelines bring crude oil and natural gas from production fields to processing plants and refineries—and ultimately to the consumer.

Pipeline Operator Qualification

Petroleum products are considered hazardous, so pipeline personnel must take extra precautions and follow specific procedures when working with them. To help ensure the safe handling and transportation of petroleum products, the United States Department of Transportation (DOT) mandates safe practices as federal regulations.

One such regulation, the Operator Qualification (OQ) rule, requires pipeline operators to develop a written qualification plan to evaluate their personnel and their contract personnel (see *49 CFR Part 192 Subpart N—Gas*, and *49 CFR Part 195 Subpart G—Liquid*). The evaluation hinges on the operator's ability to perform certain covered tasks and to be able to recognize and respond to abnormal operating conditions that may be encountered while performing these covered tasks.

Covered Tasks

A pipeline performance task is deemed covered if the activity passes the following four-part test:

1. Is it performed on a pipeline facility?
2. Is it an operations or maintenance task?
3. Is it performed as a requirement of *49 CFR Part 195 (Liquid)* or *49 CFR Part 192 (Gas)*?
4. Does it affect the operation or integrity of the pipeline facility?

If the answer to all four questions is yes, then the activity is considered covered and must figure into the evaluation process of pipeline personnel.

Consortium on Operator Qualification

The Consortium on Operator Qualification (COOQ) became the guiding force for the identification and development of covered tasks for the liquid pipeline industry. Their work product, API publication 1161, *Guidance Document for the Qualification of Liquid Pipeline Personnel*, formed the basis for the resulting Pipeliner Training and Assessment Program.

The covered task list created in API publication 1161 has now evolved to include gas, as well as construction covered tasks. NCCER maintains a version of the covered task list, which can be accessed at **www.nccer.org**.

NCCER

With the guidance and support of the Consortium on Operator Qualification, NCCER was able to design and implement the Pipeliner Training and Assessment Program.

NCCER is a not-for-profit 501(c)(3) education foundation established in 1995 by the world's largest and most progressive construction and maintenance contractors and national contractor associations. NCCER was founded to address the severe workforce shortage facing the construction and maintenance industries and to develop a standardized training process and curricula. Today, NCCER is supported by hundreds of leading construction and maintenance companies, manufacturers, and national associations and developed more than 70 construction and maintenance titles.

NCCER also maintains a Registry, which contains the records of trainees and instructors who have received NCCER training and certification. All completion information for every trainee is sent to NCCER and filed within the Registry. The Registry can then confirm training and skills for workers as they move from company to company, state to state, or even within their own company.

NCCER Standardized Curricula

NCCER standardized curricula are developed using the expertise of industry subject matter experts (SMEs). The use of SMEs as an integral part of every new development or revision ensures that the curriculum is current, industry-driven,

MODULE 60102-02 CONTENTS

Figures

Tables

MODULE 60102-02

Basic Pipeline Hydraulics and Equipment

Objectives

When you have completed this module, you will be able to do the following:

1. Explain basic hydraulic systems.
2. Describe hydraulics as applied to pipeline operations.
3. Identify and explain basic types of hydraulic pumps and valves.
4. Describe the operation of pumps used in pipelines.
5. Identify and explain basic types of valves used in pipelines.

Prerequisites

Before you begin this module, it is recommended that you successfully complete the following: Core Curriculum; Pipeline Core Modules 66101 and 66102.

Required Trainee Materials

1. Pencil and paper
2. Appropriate personal protective equipment

1.0.0 ◆ INTRODUCTION

Hydraulics is a study of the behavior of fluids at rest and in motion. Basic principles that apply to machine hydraulics, which are used to accomplish work of various kinds, are the same as those used to understand pipeline hydraulics. Pipeline hydraulics is a science of transportation, involving the movement of a vast volume of fluid over a great distance. No resultant work is intended or accomplished except this transportation.

The Romans discovered the basic principles of hydraulics centuries ago and used them to build aquaducts; however, these hydraulics principles were not developed further until modern times. With the coming of the machine age and the development of the internal combustion engine came an increasing demand for huge volumes of petroleum. Locating and transporting it in an efficient, safe, and economical way became a necessity. Pipeline hydraulics were developed as the answer to that need.

Today, pipelines carry vast volumes of products for thousands of miles, safely, efficiently, and economically. Product quality is assured by the way that pipeliners operate their hydraulic systems.

Pipeline safety receives major attention. Products must be highly pressurized to accomplish the necessary movement. Due to the volumes, pressure, and hazardous products involved on the pipeline, a release or break in the line is a serious problem. If not handled appropriately, such occurrences can quickly become a disaster, affecting personnel, the public, and the environment. Operations personnel must be specialists in the field of hydraulics to manage the operation safely. Pipelines depend upon trained and qualified personnel for every facet of their operation.

This module covers the following principles and subjects:

- Pipeline hydraulic safety
- Pipeline hydraulic principles
- Hydraulic principles of petroleum products
- Pipeline valves
- Pipeline pumps
- Pipeline products
- Pipeline design factors

2.0.0 ◆ PIPELINE HYDRAULIC SAFETY

Pipeline construction in this country is strictly controlled by U.S. Government standards and regulations under the jurisdiction of the Department of Transportation (DOT). Other standards included as law are incorporated by reference such as those set by the American National Standards Institute (ANSI) and the National Fire Prevention Association (NFPA). If there is a conflict between *CFR 49* and *ANSI/NFPA 58* and *59*, the ANSI/NFPA standard always prevails.

To prevent personal injury when working with hydraulic systems, take the following safety precautions:

- Do not operate hydraulic equipment without pump and motor guards in place.
- Wear gloves to prevent burns from hot lines and components.
- Wear safety goggles when opening any hydraulic system.
- Ensure that the system has been shut down and that all pressure has been relieved before opening the system.
- Do not touch or work on operating equipment.
- Warn personnel in the area before starting any equipment.
- Observe all company safety rules.

3.0.0 ◆ HYDRAULIC SYSTEM PRINCIPLES

Many types of hydraulic systems are used in industry today. Although the systems differ in many ways, the basic principles of hydraulics remain the same. To understand how a hydraulic system works, you must first understand several basic principles, which include the following:

- Pressure
- Work
- Energy
- Pascal's law
- Fluid flow
- **Cavitation**
- Bernoulli's principle

3.1.0 Pressure

Hydraulics refers to the movement of fluid in an enclosed space. For a hydraulic system to operate, a pump must supply enough force to push the fluid through the system. When fluid is forced through an opening, pressure is created. This pressure is a form of energy that is used to perform work.

Pressure is the amount of force exerted against a specific area. Pressure is normally measured in pounds per square inch (psi). The amount of force in pounds divided by the number of square inches that the force is in contact with equals the pressure in psi. For example, a 1,000-pound block of metal exerting force in a downward direction against a surface of 80 square inches creates 12.5 psi of pressure. *Figure 1* illustrates this example.

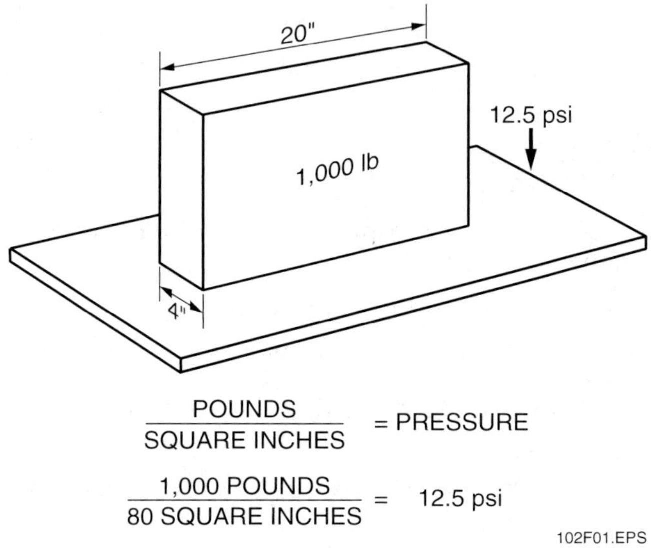

Figure 1 ◆ Pressure.

3.2.0 Work

Work is the movement of force, in pounds, over a distance. The amount of work is measured in foot-pounds (ft-lb). Multiplying the number of pounds moved by the number of feet that they are moved equals work in foot-pounds. For example, if a 100-pound object is raised 10 feet, the work accomplished is 1,000 ft-lb. In hydraulic systems, the pressure applied to the surface area of the piston multiplied by the distance that the piston travels equals foot-pounds of work.

3.3.0 Energy

Energy is the ability to do work. It is measured in the same manner as work. To operate properly, a hydraulic system must use energy to perform the work and to overcome the friction in the system. Three types of energy that are present in hydraulic systems are potential energy, **kinetic energy**, and **heat energy**.

3.4.0 Pascal's Law

Pascal's law is a basic principle of hydraulic operation on the pipeline. Pascal's law states that a confined fluid transmits externally applied pressure in all directions without change in magnitude (*Figure 2*).

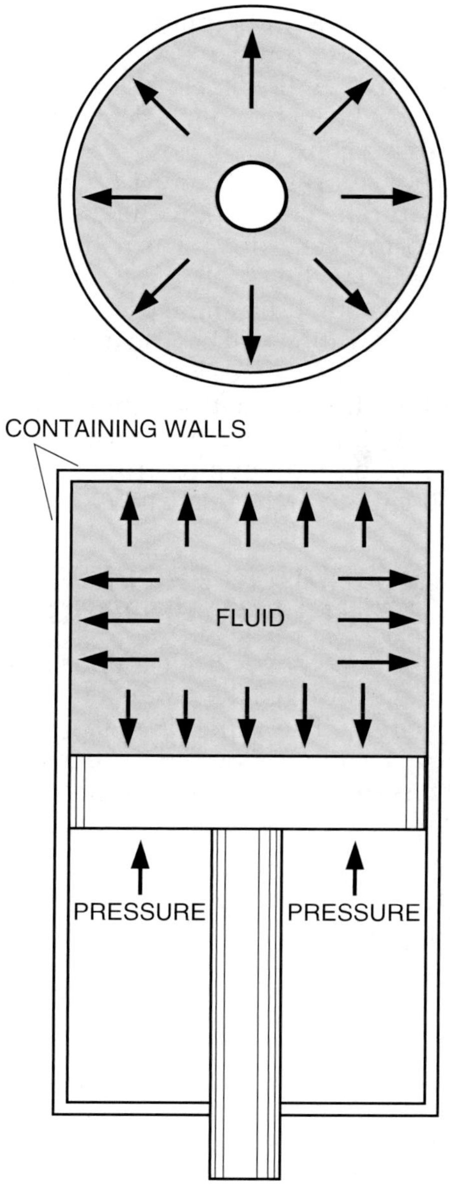

Figure 2 ◆ Pascal's law.

Basic pipeline design includes this principle in every piece of equipment because pressure is the basic operating force that moves product through the pipe. The basic result is that when pumps are started, pressure rises. Whatever pressure is applied by the pump is equally applied to every inch of the outside wall of the pipeline over its complete length, with adjustments for friction loss and other minor factors. The same pressure is also applied to any internal parts that may be encompassed by the product.

Every part of the system must be designed to withstand the **maximum operating pressure** (**MOP**) that will be applied by the pumps. The same standard applies to all valves and fittings that are installed in the pipeline. For example: consider an installation of pipeline pumps designed to pump at a pressure of 1,000 pounds per square inch. Any section of pipeline pipe not strong enough to withstand that pressure will rupture, even if it is 50 miles away from the pumps.

3.5.0 Fluid Flow

The pump in a hydraulic system pushes fluid and creates flow. Flow is measured by velocity and flow rate. Velocity is the speed at which the fluid moves past a given point. Flow rate is the amount of fluid that moves past a given point in a given amount of time. The two types of flow in a hydraulic system are laminar flow and turbulent flow.

3.5.1 Laminar Flow

Laminar flow, also known as *streamline flow*, occurs when the fluid moves in straight, parallel flow paths (*Figure 3*). This is the best type of flow in a hydraulic system because it creates little friction. Gradual changes in cross section and fluid direction do not interfere with laminar flow. Laminar flow best describes the operation of a pipeline operation.

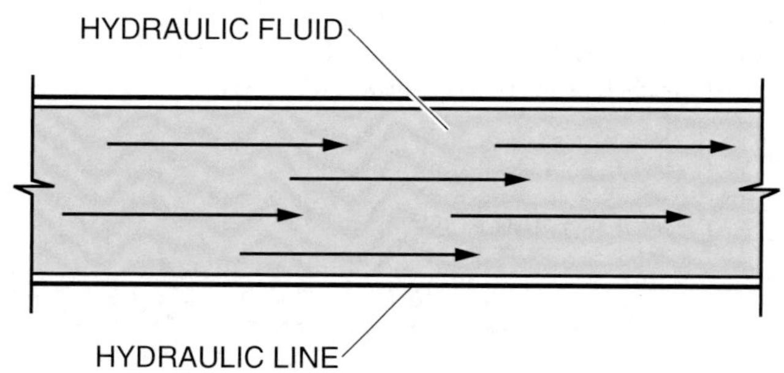

Figure 3 ◆ Laminar flow.

3.5.2 Turbulent Flow

Turbulent flow occurs when crosscurrents develop in fluid flow and the fluid no longer moves in smooth, parallel paths (*Figure 4*). It can be caused by an abrupt change in cross section or direction or by piping that is too small for the desired flow. Turbulent flow increases friction and wastes energy.

3.6.0 Cavitation

Cavitation is the formation and collapse of bubbles, or *voids*, in a liquid. These voids can be caused by turbulent flow or by trapped air caused by other mechanical problems. Sometimes the voids are caused by boiling, which occurs when the fluid reaches a very low absolute pressure. These voids in the liquid reduce the lubrication and increase wear.

As the voids are exposed to the high pressure at the pump outlet, the walls of the voids collapse and generate tons of force per square inch. This release of energy erodes the metal surfaces and can damage components in the system. The primary symptoms of cavitation are a decreased flow rate and a loud shrieking sound caused by the high-amplitude vibrations transmitted throughout the system. If cavitation is extreme, it can stop a hydraulic system from operating.

3.7.0 Bernoulli's Principle

Bernoulli's principle is a statement describing the conservation of energy in the steady flow of an incompressible, inviscid fluid. Generally stated, the quantity of liquid remains constant along any streamline (pipeline) where the pressure, flow, and mass density is constant. The energy in the system is equal at all points along the flow as long as every other factor remains constant. Variations of this rule are applied to pipelines in order to plan shipments and set up pumping orders. Calculations are made to determine how much pumping pressure must be applied to a particular product of known temperature and specific gravity, expressed in degrees API, to move the stated volume to the point of delivery in the specified time.

The principle is also applied to the setup of relief valves and relief tanks, and throttling the flow with other valves. When adjustments to pipeline pressure and flow are necessary, this principle comes into use. If pressure begins to rise above a stated point, volume is relieved into relief tanks so that the flow remains the same in the pipeline because the pressure is held to the correct level. If pressure begins to fall, valves and pumps are set to throttle points to maintain proper operating pressure and prevent cavitation. If the pipeline size changes, the difference must be included in calculations so that flow can remain constant.

The influence of gravity must be included in areas where product flows downhill because gravity tends to increase flow. Throttling must be applied to the main line so that flow remains more constant. In the other direction, product that must be lifted over elevation will have to be calculated in a different way because the elevation will add extra mass (**head**) to the load that the pumps must push over the elevation. Pump horsepower must be increased and/or extra pumps brought on line to help with the lift to higher elevations.

4.0.0 ◆ HYDRAULIC PROPERTIES OF PETROLEUM PRODUCTS

Petroleum products have certain unique properties that determine how well the product will flow in a pipeline system. The pipeline system must be designed to accommodate the properties of the fluid in the system. The petroleum product properties that need to be considered when designing a pipeline system include the following:

- Viscosity
- Viscosity index

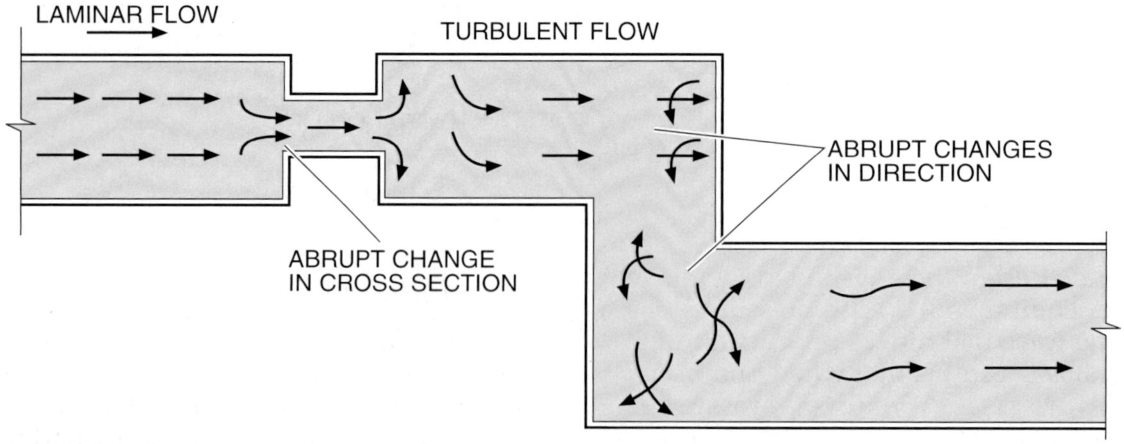

Figure 4 ◆ Turbulent flow.

- Pour point
- Flash point
- Fire point
- **Oxidation** resistance
- Emulsion resistance

4.1.0 Viscosity

The measure of the internal friction of a fluid is known as the *viscosity*. This friction is created when a layer of fluid is forced to move in relation to another layer. The greater the friction, the greater the force required to cause this movement, which is called *shear*. The shear rate is the measurement of the speed at which the intermediate layers move with respect to each other. The shear strength is the force measured to resist the shearing action. Shearing occurs whenever the fluid is physically moved, such as when it is poured, pumped, or flushed through a gearbox.

More force is needed to move highly viscous fluids, such as crude oil, than is needed to move less viscous fluids, such as gasoline. Viscosity is the thickness of a liquid or the ability of the liquid to flow at a specific temperature. Fluids may have low or high viscosity. Low-viscosity fluids are light oils that flow freely when poured. High-viscosity fluids are heavy oils that flow slowly when poured. Fluid viscosity changes with the temperature. When the temperature rises, fluids flow more freely and the viscosity is lower. When the temperature drops, the fluids flow more slowly and the viscosity is higher. Oils with different viscosities are used in different temperature conditions.

4.2.0 Viscosity Index

The viscosity index (VI) refers to the viscosity-temperature relationship of a given fluid because the viscosities of some fluids change when they are subjected to changes in temperature. All fluids do not change viscosity at the same rate when subjected to the same temperature changes. The viscosity index measures the rate at which the viscosity changes as the temperature changes. If a fluid has a low VI, the viscosity of the fluid changes rapidly with changes in the temperature. A high VI represents a lower change of viscosity with temperature changes. The temperature changes of a system must always be considered before shipping a product. When the system operates at a stable temperature, the VI is not a major consideration.

4.3.0 Pour Point

The pour point of a fluid refers to the lowest temperature at which a fluid will flow freely. This must be considered in cold weather applications and in refrigeration systems. A low pour point indicates that the fluid flows freely at very low temperatures. A high pour point indicates that the fluid stops flowing freely at low temperatures.

4.4.0 Flash Point

The flash point of a fluid is the temperature at which the vapor or steam of the fluid will flash into flame. This temperature is not high enough for the fluid to support combustion. When the flash point is reached, the fluid film between the mating surfaces is destroyed, and scoring of the metal can occur.

4.5.0 Fire Point

The fire point of a fluid refers to the highest temperature that a fluid can withstand before the vapors in the fluid will support combustion and burn steadily. In most product applications, neither the flash point nor the fire point are major concerns. These factors must be considered only in high-temperature applications.

4.6.0 Oxidation Resistance

Most pipeline products are a hydrogen-oxygen mixture known as a hydrocarbon. As hydrocarbons are exposed to air and heat, they take on oxygen from the air. This process is known as *oxidation*. Oxidation resistance is one of the most important qualities of a product because all products tend to oxidize over time. Oxidation breaks down the chemical structure of the product and eventually destroys the lubricating qualities of the product. The oxidation process is normally slow, but if the product splashes around excessively or is exposed to high temperatures, the oxidation process speeds up.

4.7.0 Emulsion Resistance

An emulsion is a combination of substances that do not readily mix, such as water and oil. Because water and oil do not readily mix, an emulsion is formed when the water is broken down into tiny droplets and carried around by the oil. This process is known as emulsification. Emulsion is unwanted in product because it decreases the lubricating qualities of the product and holds particles of dirt and foreign contaminants that can damage and cause excessive wear to bearings and other components in the system.

5.0.0 ◆ PIPELINE VALVES

The correct fluid pressure must be maintained in a pipeline system for the system to function properly and to prevent damage to the system components. Pressure-control valves limit and regulate the pressure in the system. The following are some common types of pressure-control valves:

- Relief
- Sequence
- Counterbalance
- Control
- Block

5.1.0 Relief Valves

Relief valves protect the hydraulic system from excessive pressure buildup. When pressure in the system reaches the design pressure of the relief valve, the valve opens to relieve the excess pressure. When the pressure has been adequately relieved and drops back to the proper range, the valve automatically closes. *Figure 5* shows relief valve operation.

5.2.0 Sequence Valves

Sequence valves determine the order in which fluid flows to different parts of the system. A sequence valve connected to two lines directs fluid only to the primary line when the valve is closed. When pressure in the primary line reaches a given point, the valve opens. When the valve is open, it allows fluid to flow to both the primary and secondary lines (*Figure 6*). When the pressure in the primary line falls back to the proper range, the valve closes and again directs fluid to the primary line only.

5.3.0 Counterbalance Valves

Counterbalance valves work as check valves do by allowing fluid to flow in one direction and resisting flow in the other. When a load is raised, counterbalance valves maintain pressure on the hydraulic system to prevent the load from falling. A counterbalance valve has an adjustable spring that adjusts the amount of pressure the valve can hold. *Figure 7* shows counterbalance valve operation.

5.4.0 Control Valves

Control valves are commonly used on the pipeline to regulate pressure and flow; they are also used to relieve pressure in some applications. They may or may not provide complete shutoff. Port sizes and flow patterns through these valves must be continuously adjusted to provide control by throttling action, increasing their tendency toward damaging vibration, erosion, and corrosion. These valves are used principally at pump stations, although they may be needed at other points in the line.

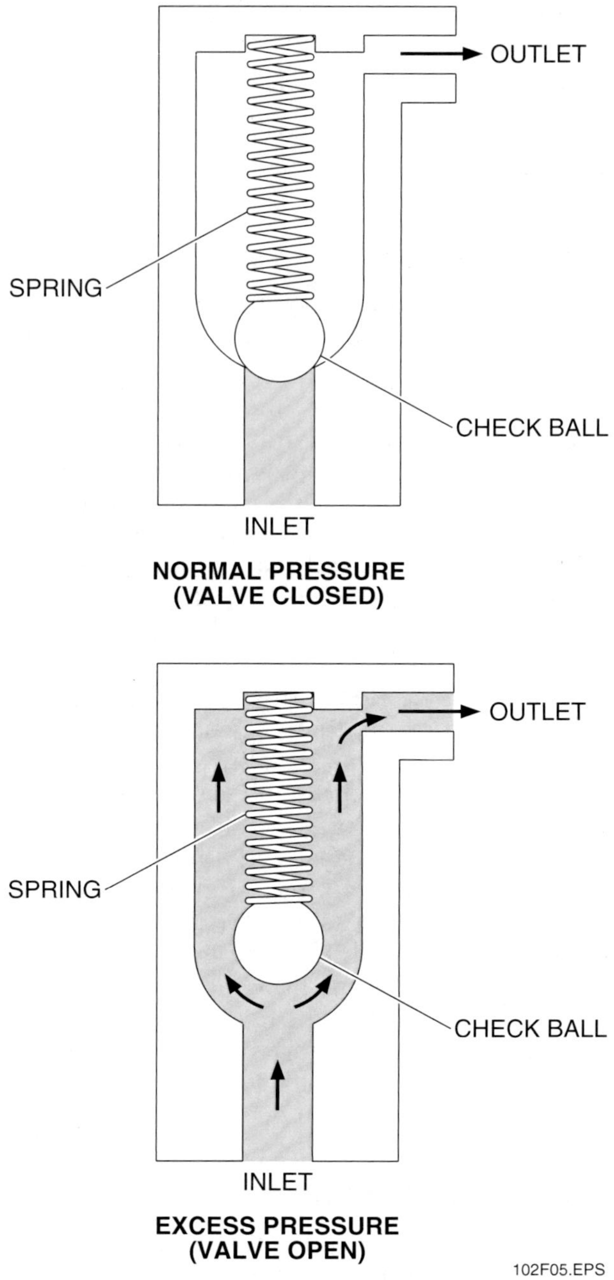

Figure 5 ◆ Relief valve operation.

5.5.0 Block Valves

Main-line block valves are large valves installed in the pipeline that completely shut off all flow on the pipeline. They are used for isolating a particular section during an emergency or for maintenance.

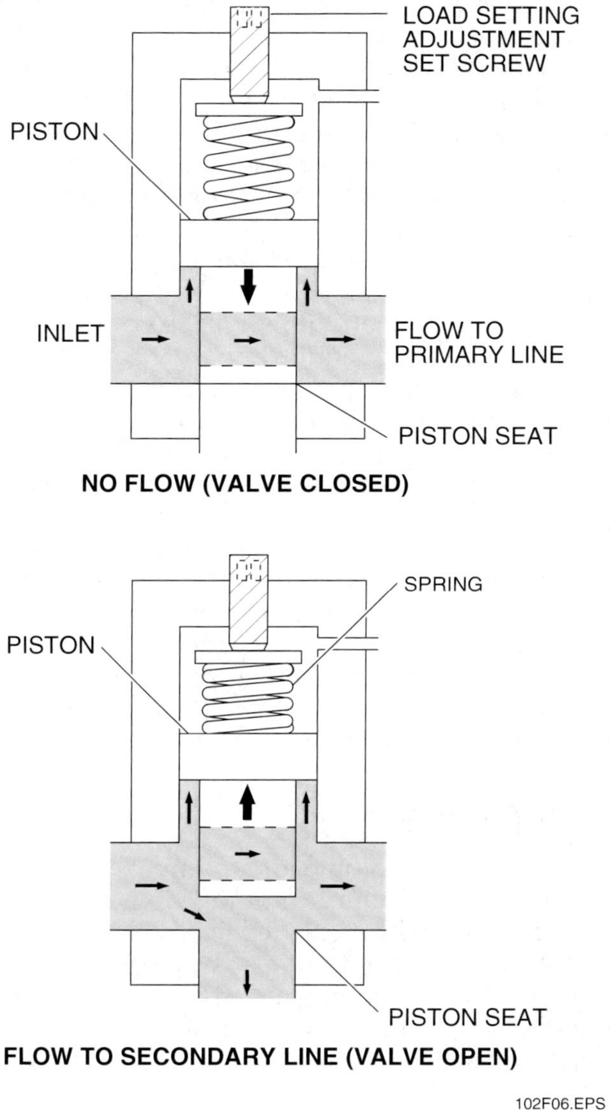

Figure 6 ◆ Sequence valve operation.

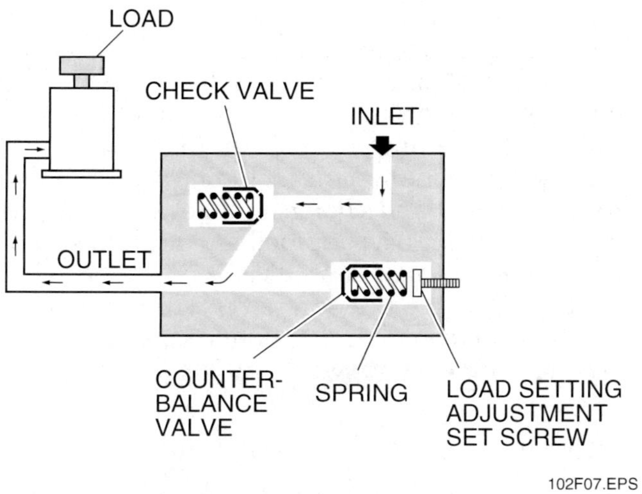

Figure 7 ◆ Counterbalance valve operation.

6.0.0 ◆ PIPELINE SYSTEM PUMPS

Pumps in piping systems move product through the system. The two things that must be considered when selecting a pump are the pump classification and the pump type. The type of pump driver and pump combinations should also be considered when selecting a pump.

6.1.0 Pump Classification

Pumps are classified according to displacement, capacity, and pressure rating. Factors to consider when rating and selecting hydraulic pumps include the following:

- Displacement
- Differential
- Capacity
- Pressure
- Drive speed
- Efficiency
- Fluid pumped
- Reliability and size
- Discharge control

6.1.1 Displacement

All pipeline pumps create pressure. The action of a pump taking product in (called *suction*) and moving it out the other side (called *discharge*) is called *displacement*. Displacement is the amount of product the pump moves with one single revolution of the impeller. Displacement is determined by the number and size of chambers within the pump. For example, a pump designed with 12 chambers, each holding a volume of 1 cubic inch, produces a displacement of 12 cubic inches with each revolution.

Pumps are classified in two ways: positive-displacement and nonpositive-displacement. Positive-displacement pumps are designed with very little clearance so that a seal is created between the inlet and outlet sides of the pump. Product that passes the seal cannot return or backflow, allowing the pump to force the product into the system so that pressure can build. Pressure is necessary to overcome resistance to flow and push the product through the pipeline. Greater distances and elevation require greater pressure and therefore require more pumping power. Most pipeline pumps are of the positive-displacement type.

Nonpositive-displacement pumps only move product when there is no load. These have no seal between the inlet and outlet and cannot prevent backflow when pressure builds in the system. When there is no resistance to flow, these will move product. If the line becomes blocked, the movement stops, although the pump may continue to run. Such pumps might be used to drain product from one level to another.

6.1.2 Differential

Pump differential is a term denoting the difference in pressure from the suction side of the pump to the discharge side of the pump. If pump differential pressure is too great, damage is likely to occur. Some of the results of a high pump differential include cavitation, excessive pump noise, excessive wear, overheating, and automatic pump shutdown.

Simply stated, a sufficient volume of product must be supplied to the suction side of the pump to equal the volume that the pump is trying to discharge. If the line pressure is not high enough to supply a proper amount of suction to the pump, the amount of pressure and volume will be insufficient near the eye of the impeller. When this happens, cavitation occurs in sufficient magnitude to seriously damage the pump. In a worst-case scenario, where the pump is pulling suction from the line, a valve being closed would cause the product in the pipeline to literally be pulled apart. The vacuum in the pipeline would cause vapor to form. Such vapor or cavitation causes a noticeable increase in pump noise. Pumping must cease.

Most main-line pumps are set to shut down on low pressure. Companies fit pumps with differential pressure switches and temperature switches to protect the equipment from damage caused by low pressure, no pressure, no flow, and to avoid exceeding MOP.

> **CAUTION**
>
> If a unit shuts down because of low suction, the pipeline operator should review the system's hydraulics, investigate the flow path, and review the tank level. Do not attempt to restart the pump until these issues have been resolved.

On the pipeline, the control center calculates the rate of discharge and the rate of suction needed to supply the flow rate. Pumps are then set at the best levels for the particular shipment and product. Suction and discharge are monitored, and if a change develops, an adjustment is made on the control center console. If a change continues, the reason must be found.

On the other side, the pump may be operating at very high pressure when a valve is closed that should not be closed. It is possible that a valve can be inadvertently closed from the control center console due to misinformation or just human error. The pressure can rise suddenly on the discharge side of the pump and overheat the pump or cause a line rupture. The pipeline will build up backpressure very quickly, and flow will cease if the situation is not dealt with quickly.

A similar situation could be caused by a hung **pig** shutting off flow and causing a sudden pressure differential. The problem must be investigated immediately and a solution found to open the pipeline. If not, the pipeline will have to be shut down until the problem is solved. Trained, experienced control center personnel may be able to make a judgment call as to whether the pipeline can be kept running. If, in the judgement of the control center personnel, there appears to be enough time to make corrections, the pipeline may be kept running while adjustments are made.

6.1.3 Capacity

The capacity, or volume, of a pump is the amount of fluid that the pump can move in a given time. Capacity is usually measured in barrels per hour or barrels per day, but it can also be measured in gallons per minute. Pump capacity is determined by the displacement, the number of revolutions that the pump makes in one minute, and the amount of resistance to flow.

6.1.4 Pressure

The purpose of a pump is to create pressure. The pressure rating of a pump is the amount of pressure it can withstand before it is damaged. A pressure rating determines the amount of product the pump can effectively move through the pipeline. Pressure created by the pump overcomes the resistance to flow in the pipeline system, including such things as product density and changes in elevation. Operations personnel must set pipeline pumps at speeds high enough to maintain efficient operation while maintaining a safe pressure level. For that reason, operations personnel must monitor pressure readings on the pipeline very closely.

6.1.5 Drive Speed

The drive speed is the speed at which the pump rotates. Drive speed is measured in revolutions per minute (rpm). The drive speed is one factor in determining the amount of flow, or volume, to the system. The higher the drive speed, the higher the volume.

6.1.6 Efficiency

The efficiency of a pump is a comparison of how well the pump actually works with how well it should work under ideal conditions. Efficiency can be decreased by internal leakage and by wear on the working parts of the pump.

Operations personnel should become completely familiar with indicators of pump condition such as flow rate, differential pressure, and power consumption (amps) because they reveal the level of pump efficiency. On constant-speed pumps, flow rate is the only indicator of efficiency. On variable-speed pumps, operations personnel should learn to recognize pressure and power changes that are not due to changes in flow rate or product gravity. If a change cannot be explained, it should be reported and investigated because it suggests a reduction in pump efficiency. Such response by operations personnel can often lead to large savings in power costs because efficiency problems are soon recognized and eliminated.

6.1.7 Fluid Pumped

The pump in a piping system must be appropriate for the type and viscosity of the fluid that the pump is expected to handle. Most pumps work well with petroleum-based fluids. Other types of fluids require a pump that can handle them without damaging seals and other pump parts.

6.1.8 Reliability and Size

The reliability of a pump is determined by how well it performs under the system conditions and how much maintenance it requires. To be reliable, a pump must be closely matched to the requirements of the system and be properly operated and maintained. A reliable pump can significantly reduce costs and downtime because it requires little maintenance and repair.

The size and weight of a pump are determined by the requirements of the system and the space available for the pump. With many pump types and sizes available, there is usually no problem matching the pump size to the system.

6.1.9 Discharge Control

The discharge of a pump must be carefully controlled to meet the requirements of the system. The discharge of some pumps is controlled by varying the speed of the pump. Pumps are equipped with internal mechanisms, such as impellers, that vary the output of the pump.

6.1.10 Temperature

A pump's motor speed affects temperature. When being operated at lower speeds, the ability to dissipate heat is reduced due to slower cooling fan speed. Motor speed is a factor that operations personnel must consider when setting the horsepower ratings for pressure and flow on main-line units. If a motor is set too slow or slows down for some reason, operations personnel must adjust the speed quickly to keep the motor from overheating and possibly burning the motor's winding. Operations personnel may need to switch to another main-line unit to prevent a failure.

6.2.0 Pump Types

There are many types of designs of hydraulic pumps. The pump used in a particular system is usually determined by the system designer or engineer. Centrifugal pumps are used most often. Some of the most common types of pumps include the following:

- Centrifugal
- External gear
- Internal gear
- Screw
- Cycloidal
- Balanced vane
- Unbalanced vane
- Radial piston
- Axial piston
- Turbine

6.3.0 Pump Drivers

A driver is the device that activates the pump. In general, most drivers are electric motors of the proper horsepower and speed rating. Other drivers are gasoline, natural gas, or diesel engines. In some applications, a generator may be included.

Pumps requiring electricity may use a commercial source or employ a diesel, gasoline, or natural gas engine that will operate a generator in the case of the failure or absence of electricity. Engines can also be a direct source of power by coupling the engine directly to the pump, using a drive shaft and gear box. *Figure 8* shows a pump with an electric motor.

102F08.EPS

Figure 8 ◆ Pump with an electric motor.

6.4.0 Pump Combinations

Pump combinations are selected based on product type, distance, and elevation on the pipeline. Usually one to two days before shipment begins, control center personnel will calculate the amount of pump power required for moving the product to the delivery point. Generally, the amount of pump power is expressed in brake horsepower. As brake horsepower is increased, the flow rate is increased. When main-line pumps are started downstream, the flow rate will increase. If main-line pumps are idled downstream, there will be a drop in flow at every location. It is up to control center personnel to schedule the number and setting of all pumps so that shipment moves on schedule.

Horsepower settings for reasonably level ground are not sufficient to push the product over a mountain range or to raise the product to a higher plateau. Calculations are made to determine the amount of extra horsepower necessary to do the work. A downstream station may need to bring several main-line units on line to supply enough pressure to do the job. *Figure 9* shows different horsepower settings.

102F09.EPS

Figure 9 ◆ Different horsepower settings.

Changes in pressure or flow rate may occur because other pumps have been brought on or taken off line without the control center operator's knowledge. Schedules may have been incorrect or misunderstood. Human error does occur. The possibility of such an occurrence increases as the distance of the shipment increases. Pressure and flow rates are monitored closely to prevent such mistakes. When they do occur, the pressure and flow should be adjusted as quickly as possible.

Other pumping combinations are used to achieve the rate of flow. Booster pumps are often required to supply increased suction pressure to main-line pumps running at high settings. If the main-line unit does not receive enough supply to the suction side of the pump, the pump may be seriously damaged from cavitation. In some cases, the level of product in the tank may supply enough pressure to feed the suction of the pump. If the tank level does not supply enough pressure, a booster pump is brought on line to supply suction to the main-line pump. A batch change may require booster pumps to increase suction for the difference in **API gravity** or viscosity.

7.0.0 ◆ PRODUCTS IN THE PIPE

The pipe is the first thing to be considered when we talk about hydraulics because the characteristics of the pipe determine the way it can and must be operated.

Pipe is selected for the material's compatibility with the products to be pumped. Design factors are prescribed by documents such as *49CFR* (Code of Federal Regulations) and accepted engineering practices.

Pipelines are by nature long distance runs. Many run, by necessity, through less-than-friendly terrain. Some pipelines are hundreds of miles long and cannot be visually observed every minute of the day, yet the system is expected to perform perfectly for the wellbeing of both the customer and the environment. Success is achieved by the quality of the system design and by the dedicated people who are responsible for operations.

7.1.0 Properties of Products

Every product in the pipe has a unique set of properties with different operating characteristics that must be considered when each operation is planned and carried out. Calculations must be made on the basis of weight, viscosity, pressure and **compressibility**, and environment.

The pipeline is affected by the weight of the product being moved because each different liquid has specific characteristics of its own.

Materials of low viscosity, such as gasoline, are much easier to pump than are materials of high viscosity, such as pure crude oil. It is much harder to pump heavier liquids uphill than lighter liquids.

Pipeline **linefill** is the amount of product held in a segment of the pipeline (*Figure 10*). Linefill, which is the pipeline inventory, is a function of the size of the pipeline and the properties of the fluid contained in the pipeline. Linefill is affected by barometric pressure, temperature, and pumping pressure applied. For a product to reach the purchaser, the entire volume of the pipeline must be displaced with more of the same or another product. In reality, a pipeline is physically a long tank lying on a horizontal plane. It must be filled as if it were a tank, displacing product before the product can be transported and delivered to the customer.

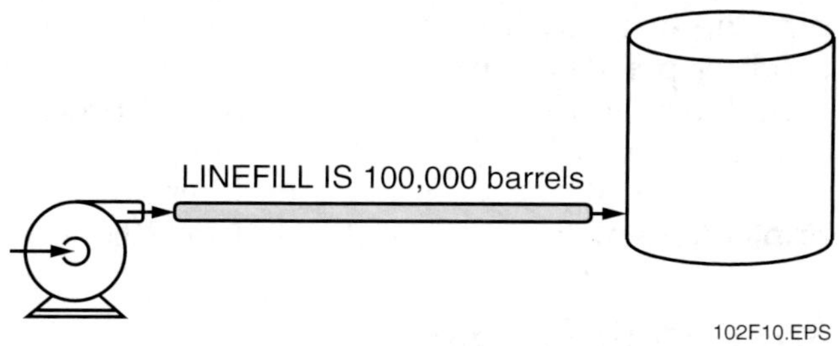

Figure 10 ◆ Linefill.

7.2.0 Products Under Pressure

Pressure is the most fundamental concept in the pipeline industry. Pressure affects most of the decisions made. Pressure is defined as the force acting on a surface area. It is described or measured in units of pounds per square inch (psi) and pounds of force per square inch (lbf/in^2).

Every successful operation depends completely on pressure to move the product. Some products, like propane, are in a liquid state only because they are kept under high pressure. The pressure must be held constant to keep vapor from forming.

Pressure is created and applied to the pipeline by means of pumps. Line pack is a result of compressibility of the product within the pipeline. Because most products are at least somewhat compressible, every operational decision must be based on **lag time**.

Lag time, or *hydraulic wave*, is an expression used to describe the time difference between starting the **upstream** pumps and when pressure begins to reside at the **downstream** end of the pipeline. Lag time is created by a combination of compressibility and pressure. The more compressible the product, the longer the time differential. It will take a natural gas liquid (NGL) a lot longer to pack than a crude, so it takes longer to see the effects of an operational change.

In a location where many miles are involved in a segment of pipeline, lag time might be as much as 15 minutes. In order to prevent waves and surges or water hammer effects, pumps must be started at slower rates or one at a time in sequence. Although this may take more time to reach a normal operation level, it is a safer procedure.

7.3.0 Surge or Water Hammer

Pipelines experience an effect called *surge*, or water hammer. A surge is caused by a sudden increase in pressure. This increase in pressure is the result of the velocity of the product in the pipe changing suddenly. Occurrences that cause this effect include:

- Start or stop of a pump
- Change in the speed of a pump
- Valve opened suddenly
- Valve closed suddenly
- Sudden line blockage (stuck pig or other)

7.3.1 Causes of Surge

When the kinetic energy of the mass of moving product suddenly changes into pressure energy, the result is a serious increase in pressure that can cause major damage to the pipeline and equipment. This increase of pressure causes surge. Heavier products such as gasoline or oil moving at high flow rates are more likely to cause serious damage. When the line is stopped suddenly, pumps continuing to work cause pressure to reach maximum pressure quickly. The more suddenly the incident occurs, the more likely a dangerous situation, possibly a rupture, will take place. Higher flow rates at such times create greater force and increase the rise in pressure.

For example, a closed valve or blocked line stops product from moving. When product stops moving in the pipeline unexpectedly, the pumps overload and overheat, and damage may occur. Control center personnel must take quick and decisive action. Pipelines that are operating with higher pressure and flow or heavy products require that action be taken more quickly. *Figure 11* shows how water hammer or surge occurs.

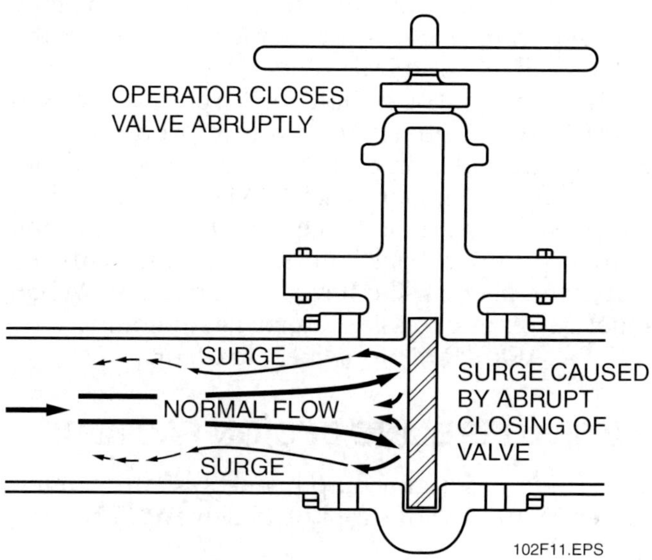

Figure 11 ◆ How water hammer or surge occurs.

7.3.2 Controlling Surge

Water hammer, or surge, can be controlled by several procedures. The first priority is to protect the pipeline. Protecting the pipeline protects personnel, equipment, and the environment. Pumps should be shut down, and the blockage must be found as quickly as possible. If it is due to a closed valve, the problem should be easily remedied. If it is a hung pig, the problem may require major intervention. The pig will have to be found and then a plan worked out to free it. Only after the problem is found and solved can the pipeline be returned to normal operation.

Water hammer or surge can be controlled in most instances by managing shipments more carefully. Valves should be opened and closed according to a schedule and closely monitored. Relief tanks and related relief valve settings should be closely monitored and changed as necessary. Pump settings should be checked and changed as required.

> **NOTE**
> Centrifugal pumps are often used in pipeline systems because they allow some slippage (backflow), which prevents water hammer.

7.3.3 Relief

Pipeline relief refers to the practice of diverting a stream of product into a tank to take presure off of the pipeline. Relief is a normal activity on the pipeline. When this condition is experienced, pipelines employ surge tanks fed from relief valves. A surge tank is constructed with pressure/vacuum vents around the top that will leak if overfilled. Tank roofs are constructed so that they will separate if overfill becomes more than the valves will allow to leak.

Under normal operation, the surge tank receives extra flow as relief valves allow product to be relieved from the pipeline to keep pressure within normal operating parameters. When the high flow has passed, the product relieved into the surge tank is then pumped back into the pipeline with the shipment, keeping the flow at a steady rate. When relief becomes excessive, abnormal operating conditions (AOCs) are more likely to occur.

8.0.0 ◆ PIPELINE DESIGN FACTORS

The original design of a pipeline system is determined by the law of supply and demand. The term *supply* includes the factors of product availability and geographical location. The term *demand* includes the factors of the customers and their location geographically.

Designing the pipe involves planning a system big enough to carry the product required with a small percentage of extra capacity. Building a pipeline system with a capacity of 5,000 barrels of product per day when a customer's maximum expected requirement is 1,000 barrels of product per day would not be reasonable. Pipes are selected on the basis of acceptable engineering principles, which are laid out in *49 CFR 195*.

8.1.0 Friction Loss

Friction is a basic design factor that must be considered in every operation. It is inherent to the material used in the construction of the pipeline and the product being pumped.

Friction limits operations by reducing the efficiency of the operation, much as extra load affects the operation of a truck.

Friction is mainly affected by:

- *Pipe roughness* – A result of construction material because some materials are smoother than others. The greater the roughness, the greater the friction factor in the pipe. Drawn tubing has the least roughness, and concrete has the most. Over time, buildups can occur inside the pipe and must be broken up by mechanical devices called pigs that are passed through the lines. Product flow over a period of time acts like sandpaper along the inner walls of the pipe as it moves along the surface, and eventually there is enough wear to adversely effect the thickness of the pipe.
- *Pipe diameter* – Determined by the demand for product. The amount of friction is reduced in relation to the increase in the size of the pipe. A small pipe results in higher friction loss than a large pipe.
- *Pipe length* – A result of demand for product, with longer pipe resulting in higher friction loss.
- *Product density and viscosity* – Affect friction loss and are a result of the temperature of the product.
- *Flow rate* – Adversely affects friction. As with most actions, the faster we go, the more power it takes.
- *Minor obstacles* – Obstructions such as valves and bends in the line affect the friction loss somewhat, but their influence on the straight-through flow should be so minimal that they are not a great consideration.

8.2.0 Maximum Operating Pressure (MOP)

MOP is determined by the characteristics of the pipeline, including the thickness of the pipe wall and the ratings of the valves used. The pipeline designers prescribe the types and sizes of materials used in construction. Everything has limits as to the amount of pressure that can be withstood before failure.

> **NOTE**
> The pipeline is only as strong as its weakest member.

MOP is set by the DOT to regulate the range of safe operations for a pipeline. The MOP will only be set as high as the weakest part of the line will stand. For example, a pump station might be able to generate 10,000 pounds of pressure on the line, but due to age and corrosion factors, the safety level on the line is calculated to be 2,100 psi. The MOP is the lessor of the two.

The MOP of a given line might be subject to change from time to time because of other reasons, including age, corrosion, wear, damage from natural forces, and changes in federal and state regulations.

The MOP is calculated by using Barlow's equation:

$$\text{MOP} = 72\% \times 2 \times \text{Ys} \times \text{WT} \div \text{OD}$$

Where:

Ys = Yield Strength (psi)

WT = Wall Thickness (in.)

OD = Outer Diameter

72% = Design Factor

The MOP can be increased by:

- The use of higher grades ANSI steel
- The use of a larger wall thickness, which is more massive and heavier to transport
- The use of a smaller diameter pipe, which will mean less flow and more friction

However, the actual pipe is not the only component that determines the MOP. Every valve, flange, fitting, and gasket in the system is taken into account. In such case, one would upgrade a weak member or members to a higher ANSI rating.

The MOP of flanges and valves is determined by the ANSI. ANSI is a private organization that works to establish consensus among qualified groups for standards in industry, including pipeline fittings and valves. *Table 1* shows the MOP codes.

Table 1 MOP Codes

ANSI	MOP (psi)
150	275
300	720
400	960
600	1,440
900	2,160
1,500	3,600

Again, the actual MOP of the pipeline is determined by its weakest member. For example, consider the following pipe system:

1 mile of 12 inch X-42 pipe with a wall thickness of 0.312 inches
ANSI 600 control valve
ANSI 900 check valve
What is the MOP?

12 inch, X-42 pipe means the OD = 12.75 inch
Ys = 42,000 psi

MOP pipe = 72% × 2 × 42,000 psi × 0.312 ÷ 12.75 in. = 1,480 psi

MOP control valve ANSI 600 rating = 1,440 psi

MOP check valve ANSI 900 rating = 2,160 psi

MOP for the system = 1,440 psi

Why is the MOP for the system only 1,440 psi? Because the ANSI 600 control valve will fail first, so it limits the MOP of the entire system.

8.3.0 Elevation Profile

Elevation plays an important role in the hydraulics of a pipeline because product that requires two pumps to push it uphill may not require any pumps to flow downhill. Elevation profile can be discussed in terms of head and gravity.

8.3.1 Head

Elevation affects pressure and is described in terms of feet above sea level. In any operation involving any increase in elevation, feet of head must be calculated correctly for the operation to succeed. Operations personnel must be able to accurately calculate the feet of head to calculate the pressure necessary to accomplish the work.

Pressure is calculated by the equation:

$$P = H \times SG \div 2.31$$

Where:

P = Change in pressure (psi)

H = Head (feet)

SG = Specific gravity

2.31 = NPSH (net pump suction head, which is a pump factor)

9.2.2 Line Numbers

Line numbers are used on P&IDs to indicate the size of pipe, service of line, piping material specification group number, number of the pipe in a particular service, and insulation required. In other P&IDs or other types of drawings of the same system, line numbers are used to identify lines and to transfer pipelines from one drawing to another.

Line numbers are normally enclosed in an oval-shaped symbol with an arrow at one end. The arrow indicates the direction of flow through the pipeline and should be used as often as necessary to maintain continuity. The line number can be shown directly on the line or above the line, with an arrow pointing to the line (see *Figure 16*).

In *Figure 16*, 6" indicates the nominal pipe size; LW is the fluid symbol and represents white liquor in this case; the number 20 indicates the piping material specification; the number 12 indicates the particular line number in the white liquor system; and the letter A indicates the insulation group numbers from the specifications. If no insulation letter is given, insulation is not required.

Line numbers vary between different engineering companies, and you must always check the legend of the P&ID to determine the system for reading the line numbers.

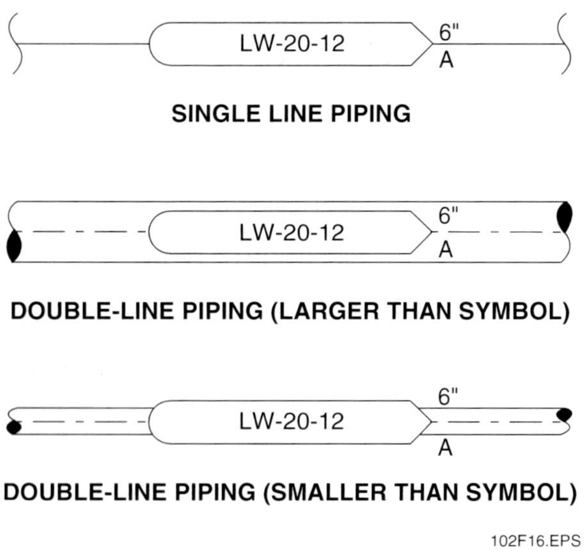

Figure 16 ◆ Typical line numbers.

9.2.3 Match Lines

Reading a P&ID is a simple matter of tracing a system from one location or piece of equipment to another and interpreting the symbols and abbreviations on the diagram. Sometimes, an entire system is shown on a single sheet, which makes it very easy to trace the system. Larger systems may require the use of two or more sheets to show the entire system, and the system must be traced from one P&ID to another. Match lines help you do this.

A numbered match line is provided at the end of the P&ID to show where the system leaves one sheet and starts again on another sheet. The system is traced to the numbered match line on one sheet and continues at the same number on the match line of the next sheet. The system always continues on the next P&ID in the numbering sequence of P&ID sheets unless otherwise noted on the match line. If the system continues on a P&ID sheet other than the next sheet in sequence, a note to the effect appears in a box on the system line at the match line (see *Figure 17*).

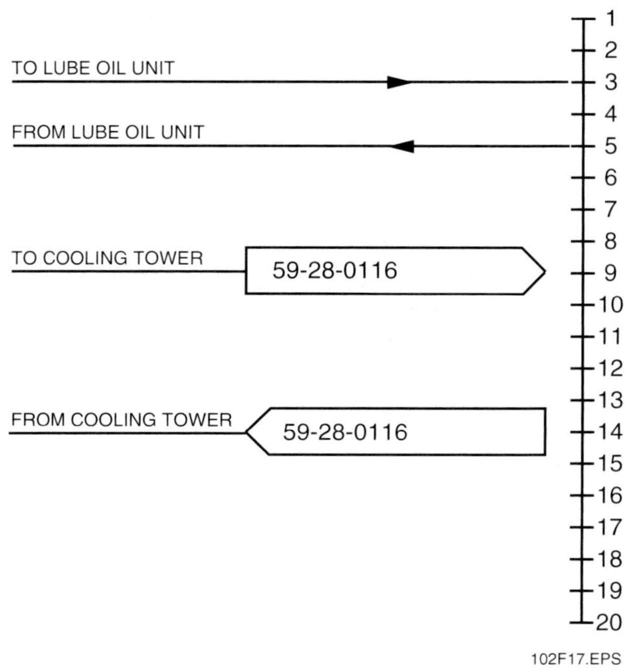

Figure 17 ◆ Match lines.

9.3.0 Piping Components

P&IDs show piping system components, such as valves, strainers, expansion joints, clean-outs, flexible connections, dampers, desuperheaters, drains, and vents. P&IDs use symbols and abbreviations to describe these components. Always check the legend on the drawings to determine exactly what the symbols stand for. These symbols differ between engineering companies. *Figure 18* shows some commonly used piping component symbols.

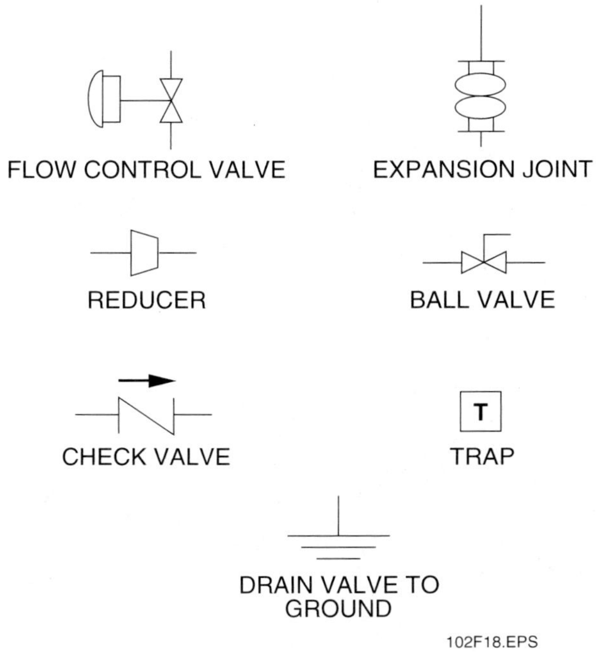

Figure 18 ◆ Some commonly used piping component symbols.

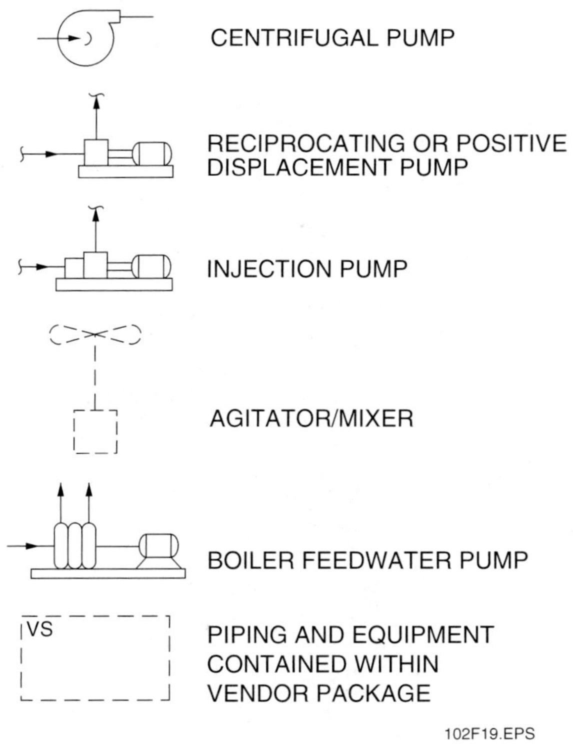

Figure 19 ◆ Some commonly used equipment symbols.

9.4.0 Pipeline Facility

P&IDs show pipeline facility equipment, such as pumps, vessels, motors, turbines, condensers, and other major equipment in a system. P&IDs also give information about the size, capacity, horsepower, and operating parameters of the equipment. *Figure 19* shows commonly used equipment symbols.

9.5.0 Instrumentation

P&IDs show system instrumentation, such as control valve operators and analytical, flow, pressure, level, and temperature instruments. P&IDs use symbols and abbreviations to describe the instruments and the functions in a system. *Figure 20* shows typical instrumentation symbols and the meaning of each.

9.6.0 Isometric Drawings

An isometric drawing, or ISO, combines the plan and elevation views of a system or part of a system into one drawing. Usually, only one pipeline is shown on an isometric drawing, although for simple or duplicate pipelines, more than one line may be shown. An isometric pipe drawing shows the details needed to fabricate and install a section of a piping system (see *Figure 21*). An ISO usually includes a bill of material, which lists the types of pipe, flanges, valves, bolts, and gaskets required for the pipe run.

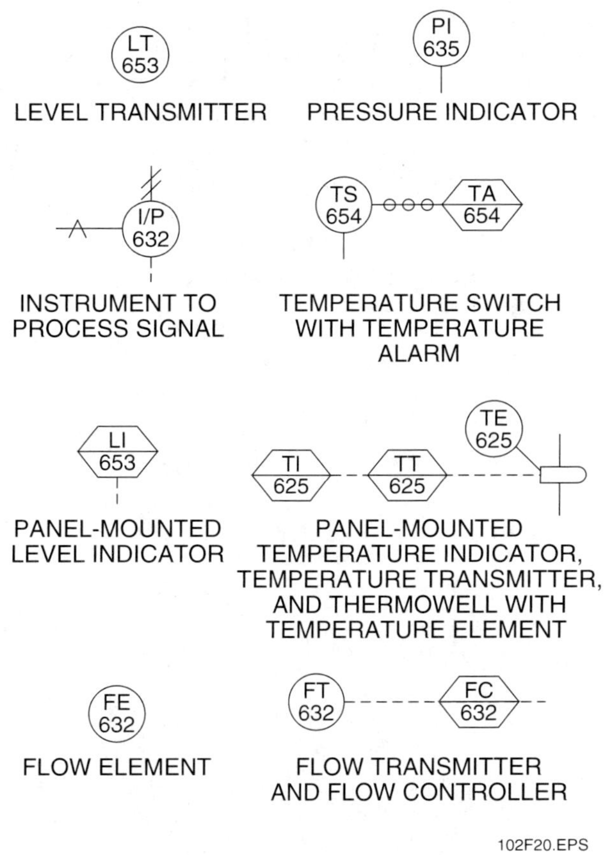

Figure 20 ◆ Typical instrumentation symbols.

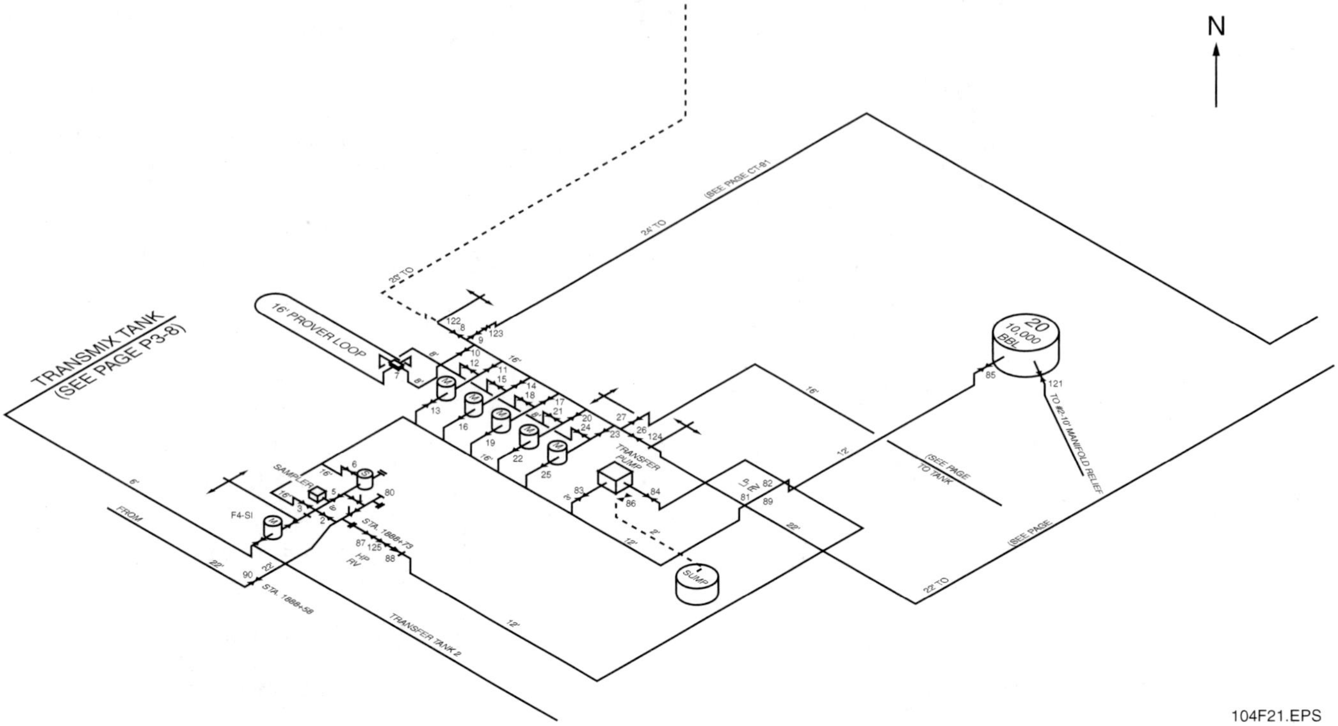

Figure 21 ◆ Piping isometric drawing.

Isometric piping drawings use the same symbols as P&IDs or any other piping drawings. The ISO has arrows that show the direction of flow. To read an ISO, begin at the end of the pipe run nearest the source and work in the direction of flow. Trace the line from the first fitting, and identify all the fittings and components by their symbols and abbreviations. Also, note the dimensions from one fitting or component to the other. The dimensions for valves and other components are shown on the ISO. The takeout for fittings and flanges must be calculated from the dimensions given on the iso. The ISO also provides pipeline elevations for installation purposes.

Summary

The world of pipeline hydraulics is very different from closed-loop hydraulics. The purpose of pipeline hydraulics is transportation. Devising the hydraulic parameters involved to enable such movement of vast volumes of hazardous products over thousands of miles in a safe and efficient manner is the job of the pipeliner. Pipeliners carry out these tasks daily with success.

Review Questions

1. Pipeline construction in this country is strictly controlled by the U.S. Government standards and regulations under the jurisdiction of the _____.
   a. Federal Aviation Administration
   b. Department of Transportation (DOT)
   c. Federal Bureau of Investigation
   d. Alcohol, Tobacco, and Firearms

2. Hydraulics refers to the movement of fluid in a(n) _____.
   a. enclosed space
   b. open space
   c. tank
   d. straight line

3. Work is the movement of force, in pounds, over _____.
   a. time
   b. an enclosed space
   c. a distance
   d. a graduated increment

4. Laminar flow, also known as streamline flow, occurs when the fluid moves in a _____.
   a. turbulent motion
   b. bidirectional flow path
   c. swirling motion
   d. straight flow path

5. The measure of the internal friction of a fluid is known as _____.
   a. viscosity
   b. flash point
   c. gravity
   d. oxidation resistance

6. The lowest temperature at which a lubricant will flow freely is known as the _____.
   a. viscosity
   b. flash point
   c. pour point
   d. emulsion resistance

7. All pipeline pumps create _____.
   a. viscosity
   b. pressure
   c. efficiency
   d. equilibrium

8. The difference in pressure from the suction side of the pump to the discharge side of the pump is known as pump _____.
   a. displacement
   b. differential
   c. drive speed
   d. efficiency

9. The amount of product held in a segment of pipeline is known as _____.
   a. linefill
   b. linepack
   c. MOP
   d. batch fill

10. The additional pressure that is required to push fluid up an elevation is known as _____.
    a. gravity
    b. profile
    c. head
    d. MOP

GLOSSARY

Trade Terms Introduced in This Module

API gravity: Measure of the density or gravity of liquid petroleum products in the United States, derived from relative density by a formula; API gravity is expressed in degrees—10° API is the equivalent of 1.0, the specific gravity of water.

Cavitation: Gas- or vapor-filled cavities are formed within liquids by mechanical forces. In pipelines, it specifically refers to formation of vapors from the motion of the fluid that causes the formation and collapse of bubbles due to turbulent flow or trapped air.

Compressibility: The unit of change in volume that will occur in a product by the application of a unit of pressure at a given temperature.

Downstream: In pipeline operations, downstream refers to any point beyond the origination.

Head: The height of a column of liquid required to produce a specific pressure; more properly referred to as *hydraulic head*. In a pipeline, it is the height of a theoretical column of fluid extending from the lowest level of a line or tank to a given point of elevation, usually measured in feet.

Heat energy: Energy that is produced by friction or the resistance of the fluid flow. Introduced in the first law of thermodynamics.

Kinetic energy: The energy that a body possesses because of its motion. In pipeline operations, it is energy produced by the movement of fluid.

Lag time: Time differential between the start of pumps on one end of the pipeline and the increase in pressure on the other end of that line. Operational decisions must include accurate lag time calculations.

Linefill: The volume of the fluid required to completely fill the inside of the pipeline.

Maximum operating pressure (MOP): The maximum internal gauge pressure that a hazardous liquid or carbon dioxide pipeline system or segment of such may be operated. Highest level of pressure allowed for a particular pipeline as set by the Department of Transportation to regulate the range of safe operations for that pipeline. Broadly speaking, it is set at 80 percent of the original design pressure, considering all components or the highest pressure the pipeline was operated at during the last five years.

Oxidation: A chemical reaction that increases the oxygen content of a compound. In pipeline operations, it is a chemical reaction that occurs when oxygen mixes with fluids to form harmful chemicals.

Pigs: Mechanical devices designed to travel with the flow of product within a pipeline for a variety of purposes including scraping, cleaning, and inspecting the line.

Upstream: In pipeline operations, upstream is a relative term that refers to any point before your current location.

NCCER CURRICULA — USER UPDATE

NCCER makes every effort to keep its textbooks up-to-date and free of technical errors. We appreciate your help in this process. If you find an error, a typographical mistake, or an inaccuracy in NCCER's curricula, please fill out this form (or a photocopy), or complete the online form at **www.nccer.org/olf**. Be sure to include the exact module ID number, page number, a detailed description, and your recommended correction. Your input will be brought to the attention of the Authoring Team. Thank you for your assistance.

Instructors – If you have an idea for improving this textbook, or have found that additional materials were necessary to teach this module effectively, please let us know so that we may present your suggestions to the Authoring Team.

NCCER Product Development and Revision

13614 Progress Blvd., Alachua, FL 32615

Email: curriculum@nccer.org
Online: www.nccer.org/olf

❑ Trainee Guide ❑ Lesson Plans ❑ Exam ❑ PowerPoints Other ______________________

Craft / Level: Copyright Date:

Module ID Number / Title:

Section Number(s):

Description:

Recommended Correction:

Your Name:

Address:

Email: Phone:

Module 60103-02

Pipeline Communications

Introduction to the Pipeline Operator Qualification Rule, API, and NCCER

It could be said that petroleum products keep America running. Petroleum products like oil and natural gas are the principal sources of energy in the United States and are used in many manufacturing and chemical processes. Petroleum products are primarily transported by pipeline. There are more than a million miles of liquid and natural gas pipeline in this country, ranging in diameter from two inches to four feet. These pipelines bring crude oil and natural gas from production fields to processing plants and refineries—and ultimately to the consumer.

Pipeline Operator Qualification

Petroleum products are considered hazardous, so pipeline personnel must take extra precautions and follow specific procedures when working with them. To help ensure the safe handling and transportation of petroleum products, the United States Department of Transportation (DOT) mandates safe practices as federal regulations.

One such regulation, the Operator Qualification (OQ) rule, requires pipeline operators to develop a written qualification plan to evaluate their personnel and their contract personnel (see *49 CFR Part 192 Subpart N—Gas,* and *49 CFR Part 195 Subpart G—Liquid*). The evaluation hinges on the operator's ability to perform certain covered tasks and to be able to recognize and respond to abnormal operating conditions that may be encountered while performing these covered tasks.

Covered Tasks

A pipeline performance task is deemed *covered* if the activity passes the following four-part test:

1. Is it performed on a pipeline facility?
2. Is it an operations or maintenance task?
3. Is it performed as a requirement of *49 CFR Part 195 (Liquid)* or *49 CFR Part 192 (Gas)*?
4. Does it affect the operation or integrity of the pipeline facility?

If the answer to all four questions is *yes*, then the activity is considered *covered* and must figure into the evaluation process of pipeline personnel.

American Petroleum Institute

The OQ rule and its related covered tasks provided the catalyst to develop a unified pipeline training and assessment program—so that pipeline operators and owners have ready access to the assessment and training tools they need. This effort was the direct result of the vision and leadership provided by The American Petroleum Institute (API) and API members.

Efforts to unify the industry are nothing new to API. Indeed, the impetus for forming the Institute in 1919 was the need to standardize engineering specifications for drilling and production equipment. Today, the most pressing issues revolve around public perceptions and government policies toward the industry. Speaking with one voice on these issues has become as essential as having interchangeable parts in the field. API's members determine the petroleum industry's positions on public policy as well as the standards that govern its day-to-day operations worldwide.

Consortium on Operator Qualification

API assembled a focus group whose primary purpose was to identify covered tasks and provide compliance recommendations to pipeline operators.

The Consortium on Operator Qualification (COOQ) became the guiding force for the identification and development of covered tasks for the liquid pipeline industry. Their work product, API publication 1161, *Guidance Document for the Qualification of Liquid Pipeline Personnel,* formed the basis for the resulting Pipeliner Training and Assessment Program.

The covered task list created in API publication 1161 has now evolved to include gas covered tasks and is maintained by NCCER. The most current covered task list can be downloaded at www.nccer.org.

National Center for Construction Education and Research

With the guidance and support of API, its members, and the Consortium on Operator Qualification, the National Center for Construction Education and Research (NCCER) was able to design and implement the Pipeliner Training and Assessment Program.

NCCER is a not-for-profit 501(c)(3) education foundation established in 1995 by the world's largest and most progressive construction and maintenance contractors and national contractor associations. NCCER was founded to address the severe workforce shortage facing the construction and maintenance industries and to develop a standardized training process and curricula. Today, NCCER is supported by hundreds of leading construction and maintenance companies, manufacturers, and national associations and developed more than 35 construction and maintenance titles.

NCCER also maintains a National Registry, which contains the records of trainees and instructors who have received NCCER training. All completion information for every trainee is sent to the NCCER and filed within the Registry. The Registry can then confirm training and skills for workers as they move from company to company, state to state, or even within their own company.

Standardized Curriculum (NCCER's Contren™ Learning Series)

NCCER curricula are developed using the expertise of industry subject matter experts (SMEs). The use of SMEs as an integral part of every new development or revision ensures that the curriculum is current, industry-driven, and indicative of industry practices. The development and/or revision process is facilitated from start to finish by NCCER project managers, who work closely with the SMEs and a select team of technical writer(s) on the development and presentation of the curriculum.

Pipeline: Five Training Paths

The pipeline curriculum focuses on essential skills and competencies within five skill areas:

- Corrosion Control
- Electrical and Instrumentation
- Maintenance
- Mechanical
- Operations (Field and Control Center)

SMEs from the pipeline industry have determined the skills and competencies needed based on API's covered tasks and the consensus of the group/industry. Specific company information is generally not incorporated into the curriculum unless the SMEs determine that the information is the industry standard.

Training Modules and Covered Tasks

The smallest unit of instruction in the curriculum is a *module*. Each curriculum is composed of modules (the equivalent of stand-alone chapters) that are logically sequenced into complete *levels* of instruction. Every module is task-oriented and requires the successful completion of a written (knowledge-based) examination and a performance (hands-on) test. Modules may be individually selected and/or combined for efficient task training. Modules in the pipeline curricula, to the extent possible, will align with API-developed covered tasks. While not all modules will relate to covered tasks, all covered tasks will relate to at least one module.

Benefits of Training to the Pipeline Marketplace

Applying NCCER's standardized training process allows pipeline owners and contractors in the United States to establish the benchmark for acceptable skill and competency levels within the pipeline industry. This process will assist training organizations in developing and/or improving standardized pipeline training programs that effectively match their resources and operating conditions while responding to regulatory demands and to the needs of the industry. The process will also recognize training organizations that meet the minimum standards of competence, warranting regulatory, public, and professional confidence. This in turn will foster national unity within the pipeline industry for the mutual benefit of individual technicians, owners, contractors, and ultimately pipeline-users and society.

Additional Resources

To review the Pipeliner Training and Assessment Program Implementation and Resource Guide and the latest version of pipeline covered tasks, visit NCCER's Web site at www.nccer.org, or contact:

National Center for Construction Education and Research
PO Box 141104
Gainesville, FL 32614-1104
(888) 622-3720
Web site: www.nccer.org

For more information on API, please contact:

American Petroleum Institute
1220 L Street, Northwest
Washington, D.C. 20005
(202) 682-8000
Web site: www.api.org

To order *API Pub 1161*, contact Global Engineering Documents by calling 1-800-854-7149 or visiting their Web site at www.global.his.com.

Additional sources of information on the pipeline operator qualification regulation include:

DOT – Office of Pipeline Safety
www.ops.dot.gov

Transportation Safety Institute
www.tsi.dot.gov

COURSE MAP

This course map shows all of the modules in the Liquid Pipeline Field Operations curriculum. The suggested training order begins at the bottom and proceeds up. Skill levels increase as you advance on the course map. The local Training Program Sponsor may adjust the training order.

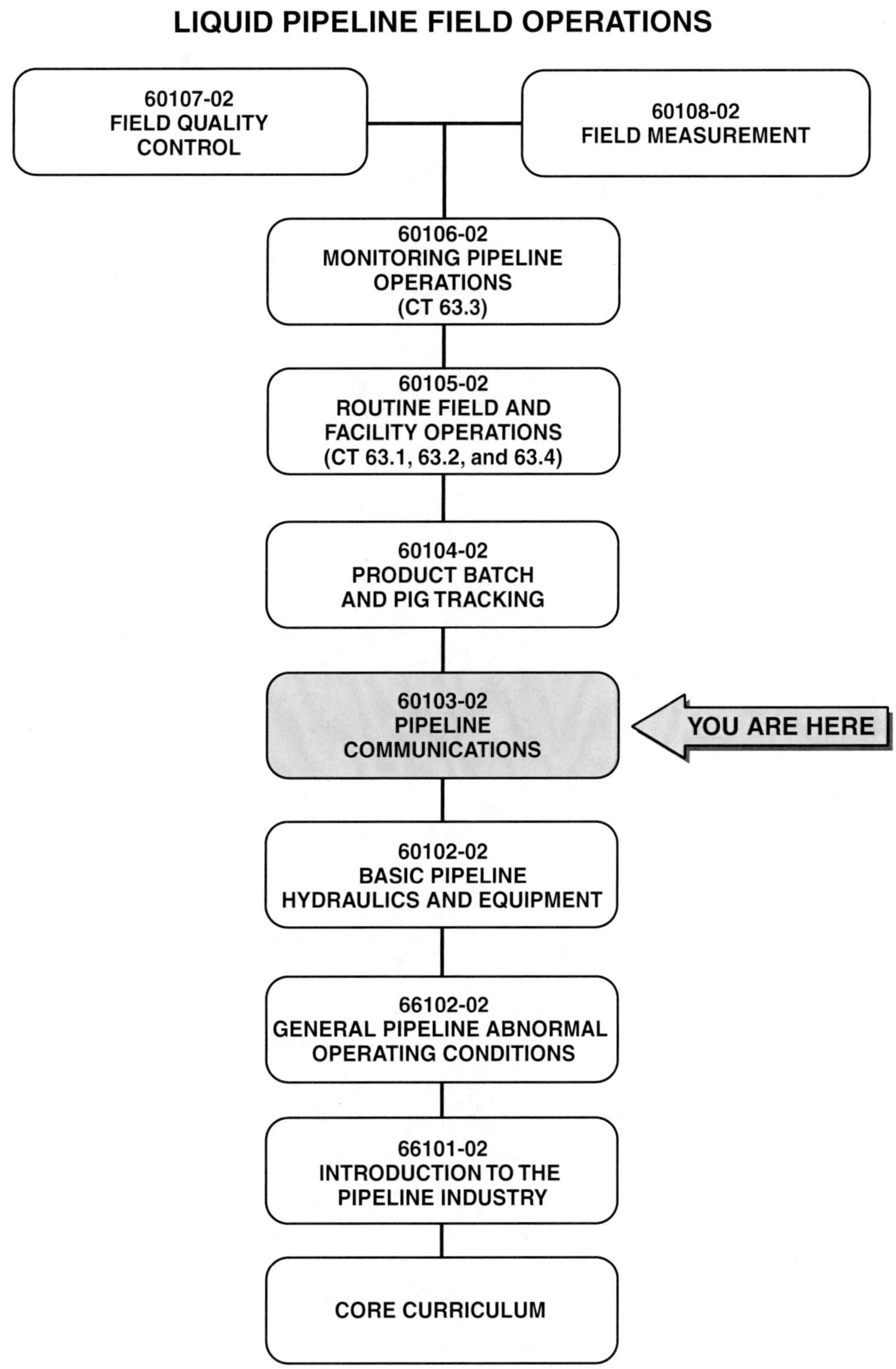

103CMAP.EPS

MODULE 60103-02 CONTENTS

Figure

Pipeline Communications

Objectives

When you have completed this module, you will be able to do the following:

1. Identify and explain the internal communications requirements that are necessary to operate the pipeline and the pipeline facility.
2. Explain the communications responsibilities of the pipeline facility scheduling department.
3. Identify and explain the external communications procedures that must be followed to inform and educate the general public of pipeline issues.

Prerequisites

Before you begin this module, it is recommended that you successfully complete the following: Core Curriculum; Pipeline Core Modules 66101 and 66102; Liquid Pipeline Field Operations, Module 60102.

Required Trainee Materials

1. Pencil and paper
2. Appropriate personal protective equipment

1.0.0 ◆ INTRODUCTION

By definition, communication is to make known or to inform. In some situations, communications within the pipline system are critical. A breakdown in communications and control in some areas could be catastrophic. By using the immense communications network, pipeline **operators** are able to keep the product flowing safely, efficiently, and productively.

Someone communicates the demand for a product. Someone else communicates supply. Pipeliners communicate the particulars of making the connection and completing the transaction. As trivial as it sounds, this is the purpose of many pipeline communications.

Pipeliners must use every means at their disposal to manage the enormous amount of information required. Much of it must be passed with meticulous correctness. Even though the latest technology in computer systems is employed to control the line, much still has to be done by other means. Communications equipment employed by pipeliners includes:

- Telephone
- Pagers
- Cell phones
- Computer network
- Microwave relay
- Two-way radio
- Closed circuit television (surveillance)
- Written documents
- Human voice

This module discusses the wide range of communication channels that must exist between pipeline operators. These communication channels are divided into internal communications, the responsibilities of the scheduling department, and external communications.

2.0.0 ◆ INTERNAL COMMUNICATIONS

Internal communications that are constantly required to operate a pipeline facility include the following:

- Communicating critical operations information to the next shift
- Communicating with the **control center**
- Coordinating with maintenance technicians
- Initiating **call-outs**
- Generating **work orders**

2.1.0 Critical Communications Between Shifts

The pipeline system operates on a rotating schedule. Communication between one shift and the next is an absolute necessity. It is critical to ensure safe and effective operation of the pipeline system. It must be specific and meticulous. Errors and misunderstandings between shifts can cause serious problems in **terminal** and pipeline operations as well as with customers.

It is the responsibility of shift personnel to effectively communicate the details and operating conditions presently in effect to the relieving shift personnel.

2.1.1 Preparations for the Next Shift

Shift personnel should make as many preparations as possible for the next shift. These preparations should be communicated both verbally and by entry into the log to the relieving shift.

2.1.2 Work Authorizations

Work authorization permits should be shared according to company policy. Any information or notification of any work that may have an impact on the following shift must be communicated to that shift both verbally and in writing per company policy.

2.2.0 Control Center Communications

Pipeline operations must be closely coordinated between **field**, scheduling, and control center. The **controller** and field personnel must make contact with each other periodically in order to remain aware of any event that might affect operations and to collect information that is not available from the **Supervisory Control and Data Acquisition (SCADA)** system. The controller must receive information from all field locations that do not show on the SCADA screens, according to company policy. One example of a situation that requires meticulous communication between the control center and the field personnel includes the request for **pipeline injections**.

Schedule orders involving another company's facilities should be confirmed prior to delivery as per company policy. Both control centers must be in constant contact with each other to ensure success.

Communications between the field and the control center must exist to prevent product contamination, to maintain the pipeline within proper operating limits, and to monitor for abnormal operating conditions.

2.3.0 Maintenance Coordinations

When pipeline maintenance is required, the field maintenance supervisor will inform, in advance, the **Scheduler** and the control center of the nature, location, and duration of the work.

Because of the potential impact that the maintenance activity might have on pipeline operating conditions, the controller must be especially diligent in monitoring pipeline conditions while the maintenance crew is working on the pipeline.

> **WARNING!**
> Maintenance crews must be kept advised of pipeline pressure and must be alerted if pressure exceeds one-half of maximum operating pressure (MOP).

It is the responsibility of the field maintenance personnel to coordinate with the pipeline scheduler. Upon receiving notice that maintenance is scheduled, operations may have to:

- Determine if any operating changes are needed due to the planned maintenance work.
- Initiate any appropriate changes before maintenance begins.
- Log all changes on the **daily operating sheet**.
- Personally calculate pipeline pressure at the work location, using the nearest pressure transmitters upstream and downstream.
- Notify field personnel of the pipeline pressure at the work location.
- Lower pipeline pressure at the work location as requested.
- Monitor pressures and rates throughout maintenance activities to ensure that planned pressures are not exceeded.

- Request the field maintenance supervisor to keep operations personnel advised of progress at regular intervals so the controller will not have to interrupt them for status reports.
- Notify appropriate personnel immediately if a line upset causes pressure changes that will affect their work.
- Verify with maintenance personnel that all work is complete and the pipeline is back in service prior to returning the pipeline to full operating conditions.

In many cases, several parties will be involved in the operation and maintenance of the pipeline system at one time. These might include multiple maintenance crews, injection and delivery facility operations, local field operations, and control center personnel. All parties must understand their roles and how their actions affect the operation of the pipeline.

It is important to recognize when human error has occurred and may have caused normal pipeline limits to be exceeded.

Some examples of human error include:

- Inadvertently blocking the line
- Opening the wrong valve
- Bypassing a safety device
- Improper operation of equipment
- Improper valve lineups resulting in wrong product in the tank

Pipeline operations should be coordinated with field personnel to verify that they are ready for the planned operation. Schedules should be reviewed and communicated to all parties to ensure accuracy before operations begin.

If a pipeline or segment of a pipeline has been shut down for report of a release or equipment malfunction, **operations personnel** must not begin startup until they have received notification from local pipeline personnel at the site of the problem that the problem is corrected.

2.4.0 Initiating Call-Outs

When situations occur that cannot be resolved or cannot wait until regular manned hours, a field person is needed to report to work outside normal working hours. This is referred to as a *call-out*.

Situations that often lead to a call-out include:

- Checking on particular changes made or needing to be made
- Switching tanks in some special manner
- Investigating a suspected problem at an automated trunk line station
- Helping a driver with problems such as a rack spill or an incorrect bill of lading
- Losing communications and requiring a field technician immediately

Call-outs of significant numbers of personnel and equipment would be expected in the case of emergency. Such a call-out might also include:

- Local fire departments and emergency response agencies
- Personnel from other pipeline companies (by mutual agreements)
- Department of Public Transportation and public facilities
- Aircraft and naval equipment, depending on location
- Industry mutual assistance organizations

2.5.0 Generating Work Orders

Work orders usually apply to maintenance and repair tasks and scheduled preventative maintenance that are conducted to keep the pipeline in good working order. These procedures must be planned and coordinated with full knowledge of how the type of work will affect operations. When work performed will affect pipeline safety, pressure, or flow rate, the person responsible must notify operations.

In some companies, regular maintenance is tracked, and orders are issued from computer modules. Such systems keep control of regulatory-required and non-deferrable activities. These activities, which are regulation driven and time consequential, must be completed within the required frequencies.

Some work orders for repairs are planned and issued far in advance of the date when the work actually begins and may have project-specific procedures that optimize safety, cost, or timing. Usually such a work order is of major size and has a great impact on operations.

Work orders may include the following information:

- Work order number
- Control center location
- Specific work required
- Exact location of the work
- Date and times for work to begin
- Date and time for work to be completed
- Name of supervisor or lead technician
- Name of initiator
- Permits required to engage in the work issued and in hand

- Signatures of appropriate control room supervisor
- Signature of maintenance supervisor

When the work is completed, the following information may be required:

- Control center location
- Shift times and dates for starting and ending work
- Name of the supervisor in charge
- Lead technician for the work
- Control room operators for day and night shifts
- Work item number
- Exact description of work accomplished (what was done and exact location)
- List of applicable procedures
- Work permit and/or work order number
- Name of person assigned to do actual work
- Initials of approving supervisor (usually the same as the supervisor in charge)
- Time completed
- Initials of person inspecting the work after completion
- Comments

Some special permits that might be required are:

- Confined space entry permit
- Hot work permit
- Cold work permit
- Electrical permit

When the work is finished, permits are kept on file.

3.0.0 ◆ COMMUNICATIONS WITH THE SCHEDULING DEPARTMENT

It is the scheduler's responsibility to make certain that pertinent information about product movement is supplied to shippers, appropriate internal departments, and personnel within the company. Issues that need to be communicated to the scheduler might include off-spec product, product outages, and deviations from schedule.

> **NOTE**
> If you ever notice any discrepancies on the schedule, notify the scheduler immediately.

3.1.0 Communications with the Field

Most of the information communicated between the scheduler and operations is schedule-related. Important information that must be provided to operations is included in the following four types of product orders:

- Pumping orders
- Line orders
- Delivery orders
- Receipt orders

These are sent to both the origin and delivery terminals.

3.1.1 Pumping Orders

Pumping orders inform the origin location of what to do with each **batch** and include product specifications, batch size, the time schedule for pumping, and the rate to be maintained. Some reference to shipper may be given, including accounting information.

3.1.2 Line Orders

Line orders include information about the arrangement of batches in the line. Line orders specify how much of each product type is passing and the order of products within the batch. Line orders are sent to all locations along the pipeline through which a batch is scheduled to pass.

3.1.3 Delivery Orders

Delivery orders instruct the delivery location exactly what to do with the batch received. Much of the information on a delivery order pertains to shippers. Examples of the information contained on the delivery order include the following:

- Specifying the amount of each product type allowed to pass before switching the stream into the terminal
- Accepting a certain volume
- Side-stripping at a certain point
- Specifying the particular tank for storage

3.1.4 Receipt Orders

Receipt orders inform the receipt location what to expect for a certain batch from the shipper, such as specifications of the product and what to do with the product when it arrives. Such orders include:

- Flow rate to maintain
- Switching instructions
- Tanks to be used
- Accounting information
- Quality-related information

3.2.0 Communications from Operations

Operations provide information to the scheduling department in the form of several standard reports that include the following:

- Physical inventory
- Terminal activity
- Operational status
- Batch information
- Quality issues

Operations provides this information so that the scheduler can make informed decisions about the most efficient way to move products and make specified deliveries. Schedulers must be notified of the particulars. Pump failures, tankage levels, equipment compatibility, and any number of other items require the scheduler to plan shipments and deliveries differently.

Common occurrences that require the scheduler to make adjustments in planning include:

- Requests for regular scheduled maintenance operations
- Emergency repair problems
- Requests for increase/decrease in delivery quotas
- Requests for schedule verification
- Requests for adjustments in pumping specifications
- Requests for product quality verification
- Requests for instructions in dealing with problems

Types of information the scheduler may need to request from operations in order to keep product moving smoothly include:

- Operational status of pumps, valves, lines, and tanks
- Availability of pumps, valves, lines, and tanks
- Changes in various manifold configurations
- Product previously stored in a tank
- Personal knowledge

In order to manage and respond quickly to such information, the scheduler must also develop a broad-based knowledge of all the physical facilities within the system. The kinds of items of which the scheduler must be fully aware include:

- Facilities having truck racks
- Facilities having access to other pipelines
- Tanks within the terminal that share a common manifold
- Tanks that can and cannot be pulled completely empty
- Types and sizes of tanks available
- Tanks easily accessible

Situations may also arise in which the shipper will call upon the scheduler to help find a way to get product to a specific location outside the pipeline system. The scheduler must have a working knowledge of other pipeline systems, access to them, physical connections to them, and the size and volume they handle in order to determine the volume and delivery rates possible.

3.3.0 Communications with Customers/Shippers

The scheduling department must communicate with customers/shippers on a regular basis. In some cases, the pipeline company and the shipper are separate departments of the same company, but regardless it is because of the customer (the shipper) that the pipeline exists and continues to operate. Thus, communication with the customer is of the ultimate importance. Pipeliners communicate carefully with the customer in order to provide the product desired as efficiently as possible.

Scheduling is the first order of business for the pipeline. In order to maintain an efficient level of operation, every step in the process must be coordinated between the shipper and pipeline personnel.

Suppose a shipper is moving a large volume of product to a terminal on the coast. There is a breakdown of communications in scheduling. Suddenly, it is discovered that the appropriate tanker vessel will not be in the proper berth and there is not enough tankage (a quick-fix solution to the problem) available to store the necessary volume until the tanker vessel arrives. This problem is not discovered until pumping has already begun. Delivery cannot be completed. The system must be shut down until the problem can be resolved in some other way. This shutdown will cause loss of time and loss of production.

The scheduling department is the primary point of contact with the shippers. Shippers call the scheduling department in matters relating to original **nominations** and product specifications. The standard types of communications received from shippers are:

- Nominations
- Product specifications
- **Product transfer orders**
- Last-minute batch changes
- Transaction reconfirmation
- Additional information

3.3.1 Nominations

Nominations are received from shippers on a monthly basis. Nominations can be described as a reservation of the pipeline and associated tankage for a certain volume of product at a certain time. Confirmation of nominations is necessary for the scheduler to be able to keep the line flowing at top efficiency.

3.3.2 Product Specifications

The customer will typically notify operations when the product is out of specification (spec). Operations will then notify scheduling of the off-spec product. Different companies may have different requirements, but the advance time is usually not less than twenty-four hours before receipt by the pipeline.

3.3.3 Product Transfer Orders

Product transfer orders (PTOs) are documents that instruct the pipeline to transfer product from one shipper's account to another shipper's account. Such communications require documentation for validity as they are legal and financial documents. The scheduler may be required to discuss such a matter with the shipper or refer the matter to a particular office or official of the company.

The scheduler may need to determine whether the shipper has sufficient product for such a transfer to be possible. Shippers often need to ask if transfers have occurred or request that a delayed transfer be scheduled. In all matters, it is up to the scheduler to help the shipper, either by providing information or directing the shipper to the proper office.

3.3.4 Last-Minute Batch Changes

Shippers regularly contact the scheduler to inquire about the possibility of increasing or decreasing batch sizes as market conditions, inventory requirements, and timing requirements for delivery or receipt change. Shippers often request grade changes during receipt in order to fill inventory deficiencies in storage, and it is the scheduler's responsibility to arrange for compliance if possible.

The scheduler is required to keep shippers informed of various changes that will affect them. Some items the scheduler will need to convey regularly are:

- Pipeline schedule
- Delivery status changes
- Technical information
- System oriented information
- Shipper's inventory status
- Problems with inventory
- Changes in operational status
- Any situation affecting schedule

3.3.5 Reconfirming Transactions

After a product transfer order has been transacted, written confirmation is sent to the shipper from the accounting department. The scheduler is often contacted by the shipper to reconfirm delivery, volume, grade, and arrival time.

In case of discrepancies, it is the responsibility of the scheduler to research the handling of the shipper's account to determine the cause. It may be as simple as a ticket written incorrectly, but the error must be found in order to balance accounts.

3.3.6 Providing Information

Scheduling must be able to furnish such data as the latest tariff information, published product information, or the characteristics of a particular product.

3.4.0 Communications with Quality Assurance

The quality assurance department is responsible for assuring the quality of the product delivered. If a problem is discovered or suspected with a particular batch or if a batch must receive some special treatment, the quality assurance department will issue special instructions on how to handle that particular batch of product.

For example, the operations personnel notice that part of a particular batch is of questionable appearance. Operations communicates the problem to quality assurance. After a solution is worked out, quality assurance will send instructions back to operations on exactly what to do in order to maintain quality in the shipment. Such instructions might include how to move the product and where to store it.

Types of instructions which might be issued by quality assurance include:

- Batch specifications
- Specifying a particular tank for storage
- Switching instructions
- Current **fungible specifications**
- Blending instructions

4.0.0 ◆ EXTERNAL COMMUNICATIONS

External communications that are constantly required to maintain and operate a pipeline facility include the following:

- Communication with the general public
- Public education
- Right-of-way markers
- Coordination with liaisons
- Audits
- One-call system compliance
- One-call system response

4.1.0 Communication with the General Public

An extremely important aspect of operations is communicating with the general public. These communications involve routing the inquiries and reports to the appropriate office or official of the company. Such communications may come from:

- The media (radio, television, newspaper)
- Emergency reports (local individuals)
- Notification of regulatory agencies
- Construction reports (local contractors)
- Patrol plane reports
- Federal, state, or local agencies

When such inquiries or information from outside agencies come in, the person receiving them should use a company form with designated headings to record the inquiry or information given so that if there is confusion during such a contact, the listed information can simply be filled in. This will result in obtaining more information with fewer omissions. *Figure 1* shows an example of a Call Investigation Report Sheet commonly in use.

Due to the reactive nature of the information such parties usually request or supply, it is a primary duty of the receiver to demonstrate professional courtesy. Assure the caller that the information or inquiry has been noted carefully and will be referred promptly to the appropriate company representative. Assure the person that a timely response can be expected.

4.1.1 The Media

Contacts from radio, television, or newspapers should all be handled the same way. Information obtained over the phone may be used for news broadcasts or printed. Operations personnel must be extremely cautious when making statements to the media because this information could be publicized and, therefore, must be accurate.

Most companies have set policies for information disclosure. Response to such inquiries must be handled exclusively by designated individuals who are properly prepared so that company policy may be strictly followed.

Displaying professional courtesy, the operations personnel should assure the individual making the inquiry that the importance of their inquiry or information is understood. Assure the person that if necessary, a representative of the company will respond in a short time.

The individual should try to get the best understanding about the inquiry or information possible during this first contact. While talking to the person, write down as much information as possible from the person, but especially:

- Person's name
- Professional association (if any)
- Location or address
- Stated inquiry or problem
- Telephone number where he or she can be reached
- Time they can be reached

After this information is collected, pass the inquiry on to the appropriate person, as per company policy.

4.1.2 Emergency Reports

The most important information that operations can receive from the public is the report of a possible release. Care must be used in handling such a contact to ensure the ability to properly respond. Operations personnel should obtain as much exact information as possible while the person is speaking because it may be the only chance. It is possible that the caller may refuse to give a name or any information other than the location, yet the report must be investigated with all due haste. Operations should obtain as much of the following information as possible:

- Name of contact
- Telephone number where he or she can be reached
- Exact location of the release
- What was released (if known)
- Is there ground coverage?
- Any geographical markers near the sight

> **NOTE**
> The operator should document in writing exactly what the caller stated and ALWAYS get the telephone number.

CALL INVESTIGATION REPORT

COMPANY NAME:	**DATE:**
CALLER'S NAME:	**DATE:**
CONTACT TELEPHONE NUMBERS/NAMES:	
NAME:	**NUMBER**
NAME:	**NUMBER**
AREA/LOCATION OF REPORT:	
DESCRIPTION:	
NOTIFIED NAME:	**DATE:**
CONTROLLER'S NAME:	**TIME:**

103F01.EPS

Figure 1 ◆ Call Investigation Report Sheet.

Operations should then carefully and calmly instruct the caller to exercise caution by staying a safe distance away until help arrives. If the release is significant, operations should instruct the reporting party to stop traffic in the area and keep people away from the area.

After the report has been received, company policy should be followed for exact instructions relating to the involvement of emergency management organizations.

4.1.3 Notification of Regulatory Agencies

It is the responsibility of the designated company representive to make all first notifications to the National Response Center (NRC) for all pipeline facilities. Reportable occurrences include:

- Any emergency that includes fire, explosion, or injury
- All releases on water
- Releases that exceed 50 barrels (2100 gallons) on land
- Releases of highly volatile liquids (HVLs) that exceed five barrels

The NRC should be called any time any of the above criteria are met.

4.1.4 Construction Reports

Operations will receive reports from construction contractors indicating there will be activity such as excavation on the line. This should be reported as quickly as possible to the appropriate office or company official. Personnel will then be assigned to monitor the work.

Information that operations must obtain when receiving the report includes:

- Contracting company's name
- Nature of the work
- Name and title of person reporting
- Telephone number where the person can be reached
- Geographical location of the site
- Any other information to help pinpoint exactly where the pipeline and the contractor's activities may intersect.

Finally and most importantly, operations should inform the construction contractor of exact company policy relating to such work on the pipeline.

4.1.5 Patrol Planes

Operations will receive calls from patrol planes that inspect and monitor the pipeline. When such a report is received, operations should make detailed notes of all critical information. Items that should be obtained include:

- Report number
- Date
- Exact time
- Type of product involved
- Closest milepost or exact geographical location (if known)
- Any other information useful in determining appropriate follow-up action

Finally, operations should call the area management immediately so that personnel can be dispatched to investigate the site as quickly as possible.

4.1.6 Federal, State, and Local Agencies

Federal, state, and local agencies call operations to make inquiries or furnish information. Such agency contacts should be referred to the designated company official or office. Because such calls often involve legal standards and compliance matters, they must be dealt with by individuals who are properly prepared to do so.

Operations should get as much information from the individual as is practical. A broad understanding of such agency contacts is needed because it is helpful for company management to have some sense of direction before speaking to these individuals. Included should be:

- Caller's name
- Agency represented
- Position or title
- Stated inquiry or problem
- Telephone number where the contact can be reached
- Times when the contact can be reached

Operations should assure the agency contact that the matter will be referred to the appropriate company representative as soon as possible and that a timely response can be expected. With professional courtesy, operations should restrict the conversation to obtaining the needed information only.

4.2.0 Public Education

An important part of pipeline communications has to do with the education of the general public. It is required by law under *49 CFR 195.402 (c) (12), 195.410, 195.440, 195.442 (b) (2) (i-ii)* that the general public living near pipeline facilities carrying hazardous materials be educated concerning the affairs of the pipeline and its operations. This section provides only a broad interpretation of these requirements and makes suggestions for fulfilling these legal obligations.

A company's public education program is conducted through the safety or personnel department and is distributed in English unless significant portions of local residents do not understand English. In such situations, literature must be distributed in bilingual format such as Spanish, French, or Portuguese, and meetings must be held with interpreters present.

4.2.1 Purpose of Public Education Programs

Public education programs are designed to help customers, residents along the pipeline right-of-way, residents near installations, the general public, appropriate government officials and agencies, and persons engaged in excavation-related activities to recognize and respond to emergencies.

The goals of public education programs are to educate and inform the above parties:

- How to identify the location of pipelines.
- The purpose of the company's damage prevention program.
- How to recognize and respond to a pipeline emergency.
- About the existence of and instructions for use of one-call notification systems.
- About emergency response personnel with company facilities and emergency response procedures.
- About the equipment and personnel each agency has available in emergencies involving company facilities.

General information that must be communicated by the public education program includes, but is not limited to:

- Facts about company operations
- Importance of recognizing and reporting an emergency
- How and to whom to report an emergency
- Personal conduct during an emergency
- Specific action to take when releases are detected

These programs are also designed to help prevent dig-ins and reduce encroachments by helping the public to recognize pipeline facilities and to improve advance notification of construction activities that may encroach upon pipeline rights-of-way.

4.2.2 Company Policy

Written programs for public education guide the actions of companies in educational and liaison activities to enable them to interact in a positive way with the public. The written program make certain that all areas are covered properly. It becomes a part of company policy. It furnishes a clear statement of how the company will respond to the needs of local residents and assigns personnel responsibility for particular actions and activities that must be carried out in order to provide proper educational opportunities.

Properly designed and carried out, such programs greatly aid the public in responding appropriately to emergencies involving pipeline facilities. Ultimately, such programs help decrease or eliminate accidents, loss of life, and damage to property.

4.2.3 Basic Activities

Some basic activities necessary to conduct public education programs include:

- Establishing of residential directories (lists)
- Producing and distributing printed literature
- Holding educational meetings
- Erecting right-of-way markers
- Maintaining liaisons with local emergency response agencies
- Reviewing and auditing the public education program

4.2.4 Residential Directories

In order to facilitate the education of everyone who lives near pipeline rights-of-way, the company is required by federal law (*49 CFR 195.440* and *195.442*) to create a list that identifies every residence and business within the vicinity. *49 CFR 192.614* also requires a list of those in the vicinity who engage in excavation of any kind.

These lists must be kept on a current basis or updated regularly so that no one is missed. The only successful way to affirm the integrity of such a list is to have a representative of the company visit each residence and business and verify the information and map location. This is necessary because some people may not have a telephone and would need someone who knows where they are to contact them personally in time of emergency. Others may have built new homes or businesses in the vicinity and are not on any list. The company must find ways to become specifically familiar with the area along its pipeline system.

4.2.5 Vicinity

References noted above use the word *vicinity* to describe an applicable boundary for legal purposes. The vicinity of the pipeline is subject to interpretation. It could be measured by a corridor two miles wide on either side of the pipeline right-of-way or just one-quarter mile on either side of the right-of-way, depending upon the normal product flowing in the pipeline installation and what is believed to be a safe buffer area. The company must make this decision, possibly in consultation with federal, state, and local officials.

4.2.6 Area

As pointed out earlier, some of the terms used in the Code of Federal Regulations are somewhat open to interpretation. Relating to excavators and contractors, *area* might be considered to be the entire county or some other designated geographical area where people who normally engage in excavation activity close to the pipeline would be located.

Some companies develop such a list through reasonably available means such as contractor associations and contractor awareness programs. Some decisions require investigation of the business arrangements of the locality. For example, some domestic well drillers, septic tank installers, and plumbers, often travel for miles to a job.

The basic idea is that everyone who might become involved in an emergency be notified for education and in case of pipeline emergency.

4.2.7 Production and Distribution of Printed Literature

Separate brochures must be designed for each pipeline system. Some of the information included by some companies includes:

- A brief explanation of the specific pipeline system (products transported, product end uses)
- A full-color depiction of a pipeline marker sign and a system map
- An explanation of pipeline marker signs, including the type of marker used, the information provided on the marker, and where markers are normally located
- An explanation of how to identify the location of pipelines
- A reminder to notify the appropriate **one-call notification system** prior to any digging, excavating, or plowing near the pipeline
- A listing of applicable one-call notification system system telephone numbers
- The company's emergency toll-free telephone number
- An explanation of how to recognize a pipeline release by sight, sound, and smell
- A list of appropriate actions that should be taken in the event of a pipeline release
- An explanation of the company's damage prevention program
- A request to pass the information along to family members or business associates

Companies produce items and distribute them during personal contact with residents. Some of the items include magnets, telephone stickers, pens, pencils, note pads, key rings, and bottle openers. These bear the "Call Before You Dig" logo and the one-call telephone number.

Companies document and retain documentation of contacts, mailings, and other distributions so that such activities can be verified. Often, residents are asked to initial a copy of the scheduled visit list. Mailing lists and other activities are documented by date and addressee. Random checks of document receipt by residents on mailing lists are documented as verification of compliance.

The scheduling of these efforts is important. Literature might be distributed at one time during the year, while meetings could be held at another time in order to smooth the work loads. However, scheduling is necessary to make certain these things are done in a timely fashion.

The location manager is responsible for maintaining public education programs within the area of responsibility.

4.2.8 Educational Meetings

Companies organize, publicize, and carry out educational meetings. These educational meetings are held periodically as a means of informing specific groups about encroachments, right-of-way precautions, coordinating councils, emergency procedures, and other pertinent topics. Slide and movie presentations are often a recommended part of these agendas. Such presentations are subject to company approval of the content and visual representations of facilities.

4.3.0 Right-of-Way Markers

Right-of-way markers are signs and placards that are posted along the pipeline right-of-way. *49 CFR 195.410* requires the location of such markers at every point of crossing public roads and railroads and in sufficient numbers along the remainder of each buried line so that the location of the pipeline is accurately known.

Design of such line markers must include:

- Lettering on a sharply contrasting background
- The word "WARNING," "CAUTION," or "DANGER" followed by the words "Petroleum (or the name of the hazardous liquid transported) Pipeline" or "Carbon Dioxide Pipeline"
- Lettering that is at least 1-inch high with a quarter-inch stroke
- The name of the company operating the line and a telephone number where the operator can be reached at all times

4.4.0 Coordination with Liaisons

Liaisons with appropriate fire, law enforcement, and public officials that may respond to emergencies must be established and maintained systemwide by the company. Two important liaisons are the Local Emergency Planning Committee (LEPC) and the local Emergency Management Director. You must check with your local government agencies to identify these officials in your community.

NOTE

Many areas, especially counties and municipalities, have a legally mandated chain of command in time of emergency. These are important relationships that must be built and, over time, involve some degree of personal acquaintance and trust.

The designated company personnel must make certain that they:

- Learn the responsibilities and resources of each government organization that may respond in an emergency.
- Acquaint the officials with the company's ability to respond in emergencies.
- Describe the types of emergencies in which the company will notify officials.
- Allow local fire and emergency personnel to visit and become familiar with company facilities.
- Provide facility maps and blueprints as appropriate in order to facilitate rapid, effective response in emergencies.
- Plan how the company and local personnel can engage in mutual assistance to minimize hazard to life and property.
- Personally contact all local officials or their representatives who may assist in case of emergency involving company facilities. These people should be personally contacted at least once each year.

4.5.0 Audits

The responsible person must ensure that in each operating area, effective public education programs are documented. Regular audits are carried out so that documentation and compliance are assured.

The designated department should review every facet of this program on an ongoing basis to ensure consistency with regulations and industry standards, effectiveness of the way information is being delivered, and productiveness of the particular format being used.

4.6.0 One-Call System Compliance

Federal law (*49 CFR 198*) requires that a telephone system that will notify the pipeline operator of activity in the vicinity of the pipeline that could threaten the safety of the pipeline be set up by states. Upon notification by a person planning to carry out demolition, excavation, tunneling, or construction in the vicinity of a pipeline facility, the operator of the facility is required by law to mark accurately, and in a reasonable and timely manner, the location of the pipeline facility in the vicinity of the planned work. The penalty for not using the one-call notification system or not heeding location information or markings can result in a fine, imprisonment up to five years, or both.

Every company operating a pipeline system carrying hazardous materials must be covered by a one-call system. The purpose of such a system is to help protect the pipeline and thereby the general public. Its basic purpose is to prevent people from accidentally digging into or damaging a pipeline carrying hazardous materials.

4.6.1 Excavator's Responsibility

Excavators are required to notify the designated one-call system that covers the area of each intended excavation activity at least 48 hours before digging. Provide the following information:

- Name of person notifying the system
- Name, address, and telephone number of the excavator
- Specific location
- Starting date
- Description of the intended excavation activity

In emergencies, excavators must be allowed to begin an excavation activity immediately, but they are required to notify the operational center at the earliest possible moment.

4.6.2 Failure to Use the One-Call Notification System

Federal law demands a firm penalty for not using the one-call notification system. Failure to use the system or to comply with location information or markings could result in fine, imprisonment up to five years, or both.

4.6.3 One-Call Action

The one-call system must have a dependable way of receiving and recording contacts from excavators so that the system is reliable. An attendant must be available to answer the calls when they are received. The call taker must have the maps, charts, and layouts necessary to pinpoint the exact location described and identify the pipeline system(s) that are directly involved.

One-call system personnel must immediately notify the management of the pipeline system(s) involved of the intent to excavate or encroach on their right-of-way. All information must be furnished with accuracy.

4.7.0 One-Call System Response

Operators of underground pipeline systems who are participating or are required to participate in a one-call system must respond according to guidelines laid out in *49 CFR 192.614* and *49 CFR 195.442*. A broad interpretation of participation in such a system would include:

- Identifying excavators
- Identifying the public
- Notification process
- Publicity
- Personal contacts
- Public meetings
- Handling reports
- Pipeline inspections
- Management

4.7.1 Identifying Excavators

Identifying on a current basis all people in the vicinity of an underground pipeline who engage in excavation activities (*49 CFR 192.614* and *49 CFR 195.442*) involves creating a list or directory, including addresses and phone numbers of the persons and businesses. The list must be updated by personal contact on a regular basis to maintain its reliability.

4.7.2 Identifying the Public

Identifying on a current basis the general public who lives in the vicinity of an underground pipeline right-of-way involves creating a list or directory of every individual person and family whose dwelling place is within the area of the pipeline right-of-way, including addresses, phone numbers, and map locations. The list must be updated by personal contact on a regular basis to maintain its reliability.

4.7.3 Notification Process

The people and excavators listed in the directories described above must be notified as often as necessary to make them aware of the operation of the damage prevention (one-call system) program. Such persons must be notified of the existence and purpose of the program and how to learn the exact location of underground pipelines before excavation activities begin.

4.7.4 Publicity

Notification may be accomplished by distributing brochures (in English and, if necessary, bilingual) bearing the telephone number of the one-call system prominently on the front, directly to persons on the lists. Such brochures might include magnets, phone stickers, pencils, pens, notepads, or other small gift items bearing the company logo and the one-call telephone number.

Signs and billboards might be employed in some areas where media coverage is questionable or non-existent. Free literature display racks placed in public places can be used to distribute pamphlets portraying facts about the pipeline system and giving the telephone number of the one-call system.

4.7.5 Personal Contacts

There should be some manner of personal contact regularly confirming and updating the residential and business lists. For example, personal visits by pipeline personnel would confirm both identity and map location. Publications could also be distributed at such time. Some companies do this at least every twelve to fifteen months because people move, new homes are built, and some are destroyed.

4.7.6 Public Meetings

Public meetings may be held to explain the one-call system and answer questions about the system. Meetings should be presented in the prevailing language of the area and must also be presented bilingually in some areas in order to gain practical coverage. An interpreter must be engaged for bilingual areas.

4.7.7 Handling Reports

When notifications from the one-call system are received, the operator of the pipeline system must have a program in place to properly deal with the contact. Proper channels of information for this program are laid out in company policy, as well as designated representatives to supervise the situation. Typical elements of a program to handle one-call reports include the following:

- The appropriate supervisor must be notified immediately.
- The company must send representatives to investigate and clearly mark the underground pipeline.
- The persons who give notice of their intent to excavate must then be given actual notice as to the type of temporary markings provided and a clear explanation of how to identify the temporary markings.
- The temporary markings must be provided as far in advance of the beginning of excavation as possible.
- Documentation of every call and the specifics of that call must be recorded.

4.7.8 Pipeline Inspections

If the company representative finds there is reason to believe that the pipeline could suffer damage from excavation activities, the operator must arrange for inspections of the area of the pipeline that could be damaged. These inspections must be carried out as frequently as necessary to verify the integrity of the line. In the case of blasting, inspections must include leak detection surveys.

4.7.9 Management

It is the responsibility of management to make certain that the guidelines of the one-call system are being strictly observed, including proper documentation of all calls received, actions taken, and required reports filed. The area director should carry out this task on an ongoing basis to make certain that industry regulations and industry standards are met.

The safety department of the company should retain access to the records of this program for verification and proof of work done.

Summary

Communications are at the heart of everything the pipeline operator does. Communications must be clear, exact, concise, and current so that personnel can understand exactly what conditions are and what must be done to carry on an efficient operation. This concept applies to both internal and external communications.

Review Questions

1. Initiating call-outs and generating work orders are examples of _____ communications that are constantly required to operate a pipeline facility.
   a. one-call
   b. public
   c. internal
   d. external

2. Work authorization permits should be shared according to _____ policy.
   a. DOT
   b. OSHA
   c. SCADA
   d. company

3. Opening the wrong valve is an example of _____ error.
   a. computer
   b. equipment
   c. human
   d. product

4. _____ orders usually apply to maintenance and repair tasks and scheduled preventative maintenance.
   a. Pumping
   b. Line
   c. Work
   d. Delivery

5. _____ might be described as a reservation of the pipeline and associated tankage for a certain volume of product at a certain time.
   a. Product specifications
   b. Product transfer orders
   c. Nominations
   d. Fungible specifications

6. Because they are legal and financial documents, _____ require documentation for validity.
   a. product specifications
   b. product transfer orders
   c. nominations
   d. daily operating sheets

7. An example of a reportable occurrence is the release of highly volatile liquids (HVLs) that exceed _____ barrels.
   a. 5
   b. 10
   c. 50
   d. 100

8. Regular _____ are carried out so that documentation and compliance are assured.
   a. meetings
   b. audits
   c. notifications
   d. changes

9. Failure to use the one-call notification system or to comply with location information or markings could result in fine or imprisonment up to _____ years, or both.
   a. two
   b. three
   c. four
   d. five

10. _____ must be carried out as frequently as necessary to verify the integrity of the line.
    a. Meetings
    b. Notifications
    c. Inspections
    d. Reports

GLOSSARY

Trade Terms Introduced in This Module

Batch: In pipelining, a quantity of crude or product of one specific gravity or weight that is pumped next to, in front of, or behind another quantity of different weight or type in the pipeline. The weight is expressed in degrees API.

Call-outs: A call for extra personnel, initiated by a supervisor, to report to work to resolve an immediate or urgent problem.

***Control center*:** Operations center of a pipeline, usually located in a terminal or main pipeline facility. All the control consoles and computer-operated systems are located in the control center.

Controller: The person responsible for the activities of controlling the pipeline, usually the control center operator.

Daily operating sheet: A record documenting all operating events of the day, which operators are required to keep. It is sometimes called a *log*.

Field: Area where petroleum products are produced and/or collected for shipment.

Fungible specifications: Requirements for certain pipeline products to be so similar as to be considered interchangeable. For example, putting a fungible commodity into a pipeline and later withdrawing that fungible commodity—it is not necessarily the same commodity, but it meets the same specifications.

Nomination: Amount of product a producer expects to take from a field as reported to state regulatory agencies. As applicable to pipelines, it refers to the reservation of time and volume on the pumping schedule for the producer's products.

One-call notification system: Telephone system required by federal law (*49 CRF 198*). Anyone who intends to perform excavating, digging, or any other underground activity in the vicinity of the pipeline is required to call the one-call telephone number to notify the operator of the pipeline so that the pipeline operator can mark the exact location of the pipeline.

Operations personnel: Field personnel and control center personnel who are responsible for moving the product.

Operator: Legal entity that owns or operates the pipeline facility.

Pipeline injection: A transfer of product received from outside the shipper's or intercompany facilities.

Product transfer orders: Legal and financial documents that instruct a pipeline company exactly what to do with a particular batch of product when it reaches the delivery point and is transferred from one account to another. To be valid, a PTO requires either a letter or other written communication.

Scheduler: Department of the pipeline company responsible for setting up the layout of volumes and times for various shipments to take place.

Supervisory Control and Data Acquisition (SCADA): Computer system that is used to monitor and control the operation of a pipeline.

***Terminal*:** Destination and collection point to which products are transported through a pipeline. It usually includes a tank farm and loading facilities.

***Work order*:** A company document that authorizes the execution of a specific work activity at a specific location and time. It is usually issued by the control center and posted for all operators to see and be aware of.

Module 64106-02

Pipeline Operations

Introduction to the Pipeline Operator Qualification Rule and NCCER

It could be said that petroleum products keep America running. Petroleum products like oil and natural gas are the principal sources of energy in the United States and are used in many manufacturing and chemical processes. Petroleum products are primarily transported by pipeline. There are more than a million miles of liquid and natural gas pipeline in this country, ranging in diameter from two inches to four feet. These pipelines bring crude oil and natural gas from production fields to processing plants and refineries—and ultimately to the consumer.

Pipeline Operator Qualification

Petroleum products are considered hazardous, so pipeline personnel must take extra precautions and follow specific procedures when working with them. To help ensure the safe handling and transportation of petroleum products, the United States Department of Transportation (DOT) mandates safe practices as federal regulations.

One such regulation, the Operator Qualification (OQ) rule, requires pipeline operators to develop a written qualification plan to evaluate their personnel and their contract personnel (see *49 CFR Part 192 Subpart N—Gas,* and *49 CFR Part 195 Subpart G—Liquid*). The evaluation hinges on the operator's ability to perform certain covered tasks and to be able to recognize and respond to abnormal operating conditions that may be encountered while performing these covered tasks.

Covered Tasks

A pipeline performance task is deemed covered if the activity passes the following four-part test:

1. Is it performed on a pipeline facility?
2. Is it an operations or maintenance task?
3. Is it performed as a requirement of *49 CFR Part 195 (Liquid)* or *49 CFR Part 192 (Gas)*?
4. Does it affect the operation or integrity of the pipeline facility?

If the answer to all four questions is yes, then the activity is considered covered and must figure into the evaluation process of pipeline personnel.

Consortium on Operator Qualification

The Consortium on Operator Qualification (COOQ) became the guiding force for the identification and development of covered tasks for the liquid pipeline industry. Their work product, API publication 1161, *Guidance Document for the Qualification of Liquid Pipeline Personnel,* formed the basis for the resulting Pipeliner Training and Assessment Program.

The covered task list created in API publication 1161 has now evolved to include gas, as well as construction covered tasks. NCCER maintains a version of the covered task list, which can be accessed at **www.nccer.org**.

NCCER

With the guidance and support of the Consortium on Operator Qualification, NCCER was able to design and implement the Pipeliner Training and Assessment Program.

NCCER is a not-for-profit 501(c)(3) education foundation established in 1995 by the world's largest and most progressive construction and maintenance contractors and national contractor associations. NCCER was founded to address the severe workforce shortage facing the construction and maintenance industries and to develop a standardized training process and curricula. Today, NCCER is supported by hundreds of leading construction and maintenance companies, manufacturers, and national associations and developed more than 70 construction and maintenance titles.

NCCER also maintains a Registry, which contains the records of trainees and instructors who have received NCCER training and certification. All completion information for every trainee is sent to NCCER and filed within the Registry. The Registry can then confirm training and skills for workers as they move from company to company, state to state, or even within their own company.

NCCER Standardized Curricula

NCCER standardized curricula are developed using the expertise of industry subject matter experts (SMEs). The use of SMEs as an integral part of every new development or revision ensures that the curriculum is current, industry-driven,

and indicative of industry practices. The development and/or revision process is facilitated from start to finish by NCCER project managers, who work closely with the SMEs and a select team of technical writer(s) on the development and presentation of the curriculum.

Pipeline: Five Training Paths

The pipeline curriculum focuses on essential skills and competencies within five skill areas:

- Corrosion Control
- Electrical and Instrumentation
- Maintenance
- Mechanical
- Operations (Field and Control Center)

SMEs from the pipeline industry have determined the skills and competencies needed based on API's covered tasks and the consensus of the group/industry. Specific company information is generally not incorporated into the curriculum unless the SMEs determine that the information is the industry standard.

Training Modules and Covered Tasks

The smallest unit of instruction in the curriculum is a *module*. Each curriculum is composed of modules (the equivalent of stand-alone chapters) that are logically sequenced into complete *levels* of instruction. Every module is task-oriented and requires the successful completion of a written (knowledge-based) examination and a performance (hands-on) test. Modules may be individually selected and/or combined for efficient task training. Modules in the pipeline curricula, to the extent possible, align with industry-developed covered tasks. While not all modules relate to covered tasks, all covered tasks relate to at least one module.

Benefits of Training to the Pipeline Marketplace

Applying NCCER's standardized training process allows pipeline owners and contractors in the United States to establish the benchmark for acceptable skill and competency levels within the pipeline industry. This process will assist training organizations in developing and/or improving standardized pipeline training programs that effectively match their resources and operating conditions while responding to regulatory demands and to the needs of the industry. The process also recognizes training organizations that meet the minimum standards of competence, warranting regulatory, public, and professional confidence. This, in turn, fosters national unity within the pipeline industry for the mutual benefit of individual technicians, owners, contractors, and ultimately pipeline-users and society.

Additional Resources

To review the Pipeliner Training and Assessment Program Implementation and Resource Guide and the latest version of pipeline covered tasks, visit NCCER's website at **www.nccer.org,** or contact:

NCCER
13614 Progress Boulevard
Alachua, FL 32615
(888) 622-3720
Website: **www.nccer.org**

Additional sources of information on the pipeline operator qualification regulation include:

DOT – Office of Pipeline Safety
www.ops.dot.gov

Transportation Safety Institute
www.tsi.dot.gov

COURSE MAP

This course map shows all of the modules in the first level of the Pipeline Electrical and Instrumentation curriculum. The suggested training order begins at the bottom and proceeds up. Skill levels increase as you advance on the course map. The local Training Program Sponsor may adjust the training order.

PIPELINE ELECTRICAL AND INSTRUMENTATION LEVEL ONE

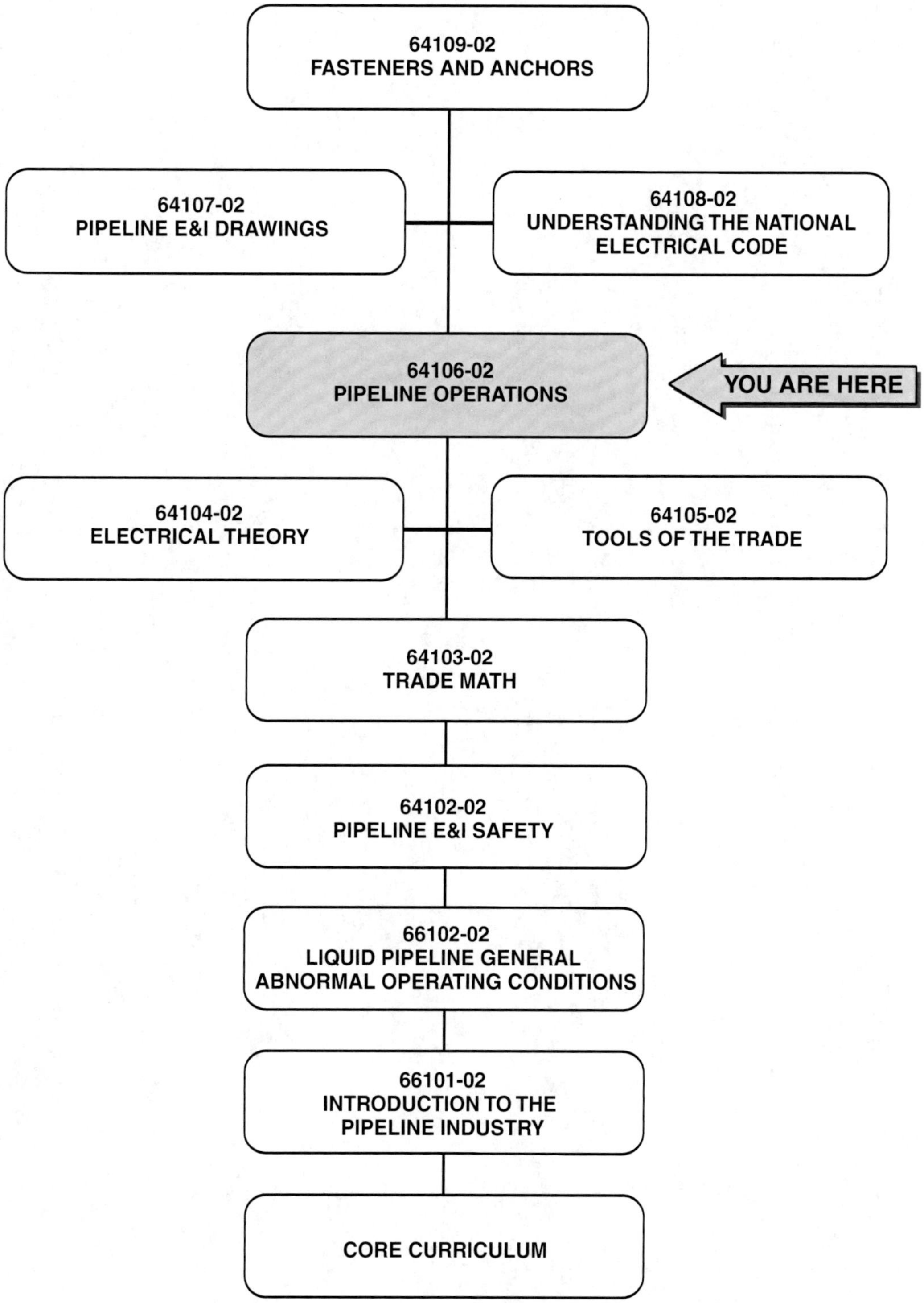

106CMAP.EPS

MODULE 64106-02 CONTENTS

Figures

Tables

MODULE 64106-02

Pipeline Operations

Objectives

When you have completed this module, you will be able to do the following:

1. Explain pipeline system hydraulics.
2. Explain ANSI ratings and standards.
3. Explain station control systems.
4. Recognize abnormal operating conditions.
5. Explain pigging operations.
6. Explain proving process meters.

Prerequisites

Before you begin this module, it is recommended that you successfully complete the following: Core Curriculum; Pipeline Core, Modules 66101-02 and 66102-02; Pipeline Electrical and Instrumentation Level One, Modules 64102-02 through 64105-02.

Required Trainee Materials

1. Pencil and paper
2. Calculator

1.0.0 ◆ INTRODUCTION

Early pipelines depended on manual control and the use of tanks to maintain a balanced system. Relying on manual control over long distances had the inherent risk of human error and delays.

Today, most pipeline operations are automated with provisions for manual control or local control as needed. The systems are programmed to operate automatically and to provide accurate data and documentation that may not have been available before.

Instrumentation and control equipment, as well as computers and microprocessors, have developed rapidly in recent years to provide the pipelines with a wide range of control systems from which to choose. They provide the means to measure and control the flow of fluids more accurately and effectively as the cost of these fluids and the price of moving them has risen.

This module focuses on control and instrumentation methods and is divided into the following sections:

- Pipeline system hydraulics
- ANSI ratings and standards
- Station control systems
- Abnormal operating conditions
- Pigging operations
- **Proving** process meters

2.0.0 ◆ PIPELINE SYSTEM HYDRAULICS

The operation of any pipeline is built on the principles of hydraulics, which is the science of the behavior of fluids at rest or in motion. These principles are, in turn, built on the properties and characteristics of fluids. Fluids include all liquids and gases. Water at ordinary temperatures is a liquid fluid, at high temperatures, a gas fluid. Air is a gas fluid. All fluids can be greatly affected by changes in temperature and pressure.

2.1.0 Fluids

Properties of fluid can determine how the fluid reacts to forces. Properties can be changed by temperature and pressure, and some are interrelated. Though there are two main properties, **viscosity** and **density**, that must be considered when working with hydraulics, there are many properties that are key to understanding the behavior of fluid. Some of these properties include:

- Viscosity
- Density
- **Elasticity**
- Inertia
- **Compressibility**
- Homogeneity
- Others

2.1.1 Viscosity

A perfect fluid is frictionless. That is, it offers no resistance to flow except that of its inertia, or its tendency to maintain status quo. However, no fluid is perfect and frictionless, and different fluids have different levels of friction, or resistance, to flow. This is called *viscosity*.

A fluid reacts differently than a solid to an applied force. Unless the force is applied to a fluid at a right angle to its surface, the force will merely cause the layers of the fluid to slide over one another. It is this layering which gives fluids the ability to flow.

How well the fluid flows will depend on how much force of attraction or friction exists between the molecules and layers in the fluid. A greater force will resist flow more than a smaller force. The resistance to flow is viscosity. Viscosity is measured in units of centipoise (cP).

2.1.2 Density

Some fluids are thicker than others. This is called *density*, and a fluid's density can be a large factor in its viscosity. It can also help to determine its compressibility, or its ability to be squeezed into a smaller mass. The density of a fluid is a measure of how tightly packed its molecules are, or how much of its matter exists in a given unit of volume. A hydrometer can be used to measure density in a liquid. Measuring the density of a gas is more difficult.

The **specific gravity** of a substance is derived from its density. Specific gravity is the ratio of the density, or mass, of a given volume of the substance to the density, or mass, of the same volume of water, both at a given temperature and pressure. API gravity is a measure developed specifically for petroleum products; it is calculated from the specific gravity.

2.1.3 Elasticity

A fluid is a substance that flows, and some fluids flow more easily than others. A fluid also is a substance that changes form when there is a slight shift in force or pressure, but it is elastic, meaning it will try to return to its previous form when the pressure is removed. *Elasticity* is a measure of the fluid's resistance to change in shape, and some fluids are more elastic than others.

2.1.4 Inertia

Inertia is the property of matter that is defined as the resistance to any change in momentum. If matter is motionless, its inertia will try to keep it from moving. If it is moving, its inertia will try to keep it from changing speed or direction. In both cases, inertia must be overcome by an outside force to effect a change.

2.1.5 Compressibility

Compressibility is a measure of how easily a fluid can be forced into a smaller volume, or amount of space, which increases its density. Pressure differences generally do not affect the density of an incompressible fluid. In practice, however, no fluid is completely incompressible.

2.1.6 Homogeneity

A homogeneous fluid has the same properties and characteristics throughout. Ordinary water would be considered homogeneous, though it could have varying layers if present in large depths. A combination of different fluids may separate into layers because of different properties, but each layer would be homogeneous. An *emulsion* is a mixture of fluids in which one fluid is scattered in the form of globules throughout the other fluid.

2.1.7 Others

Other properties of fluids, particularly those found in pipeline fluids, include volatility and flammability. *Volatility* is the tendency of a fluid to readily vaporize. Sudden vaporization can cause a violent reaction or explosion. *Flammability* is a measure of how readily a fluid catches fire when in contact with heat or sparks. The volatility and flammability of specific pipeline products is described in another section of this module. *Table 1* shows some fluid properties.

Table 1 Fluid Properties

Fluid	API Gravity @ 60°F	Specific Gravity @ 60°F	Vapor Pressure @ 100°F	Absolute Viscosity	Compressibility @ 1000 psi @ 60°F
EP Mix 80/20	214	0.41	550 psi	0.11 cP	7.3%
NGLS	146	0.51	325 psi	0.15 cP	3.8%
Raw Feed Propane	146	0.51	150 psi	0.15 cP	3.8%
Butane	112	0.58	88 psi	0.18 cP	2.0%
Anhydrous Ammonia	97	0.62	200 psi	0.25 cP	1.6%
Gasoline	63	0.66	2 psi	0.70 cP	0.8%
Jet Fuel #1	42	0.82	2 psi	1.4 cP	0.8%
Diesel Fuel #2	33	0.86	1 psi	4.7 cP	0.5%
Crudes	13-50	0.78-0.98	> 1 psi	6-400 cP	> 1.0%
Water	10	1		1.12 cP	0.3%

106T01.EPS

2.2.0 Basic Hydraulics

Some laws of hydraulics apply to gases as well as liquids but only under certain conditions. In general, the following pertains to liquid fluids. Hydraulics is usually divided into two branches: **hydrostatics** for liquids at rest and **hydrodynamics** for liquids in motion.

2.2.1 Hydrostatics

Hydrostatic principles are fundamental to the design of many machines and instruments. The design of hydraulic presses and the operation of instruments such as manometers and barometers depend on these principles.

A simple hydraulic press consists of a liquid-filled cylinder that contains two pistons, one smaller than the other. When a force is applied to the smaller piston, the force travels through the liquid to the larger piston, producing a force larger than the initial one. The increase in force is proportional to the ratio of the larger piston to the smaller one. *Figure 1* shows a simple hydraulic press.

The laws of hydrostatics are:

- The amount of pressure caused by the weight of a column of liquid is determined by the height of the column.
- An object placed in a liquid seems to lose an amount of weight equal to the weight of the liquid it displaces.
- A fluid in a container exerts pressure equally in all directions.

2.2.2 Hydrodynamics

Hydrodynamic principles are basic to hydraulic engineering, to the design of certain machines,

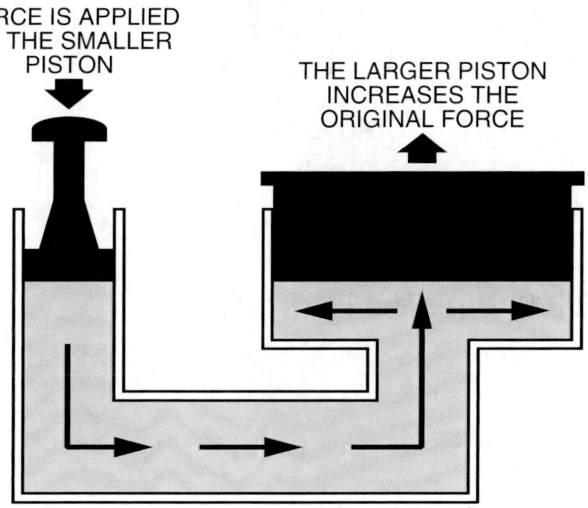

106F01.EPS

Figure 1 ◆ Simple hydraulic press.

and to pipeline engineering. Hydrodynamics define the plans for water systems, canals, and irrigation systems, as well as airplane design and hydraulic machines like water turbines.

Pipelines for fluids have to take into account the pipe size and the fluid's friction on the pipe to determine fluid velocity through the pipe. Liquid flow can be steady or unsteady. An unsteady flow results from changes in the velocity, temperature, or pressure of a liquid or its movements around obstacles. The three basic laws of hydrodynamics apply only to steadily flowing liquids. The laws of hydrodynamics are:

- The velocity of a liquid flowing through a pipe increases as the area of the pipe decreases and vice versa.
- The pressure of a liquid increases as its velocity decreases and vice versa.

- The velocity with which a liquid flows through an opening in a container equals the velocity of a body falling from the surface of the liquid to the opening.

Figure 2 shows liquid velocity and pressure relationships.

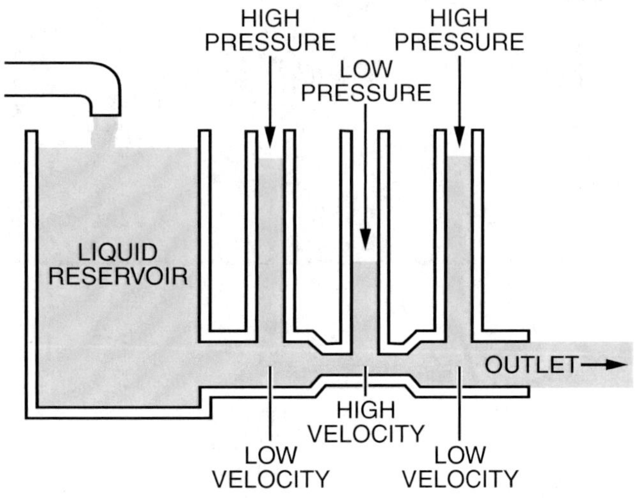

Figure 2 ◆ Liquid velocity and pressure relationships.

2.3.0 Pipeline Hydraulics

The primary factors in pipeline hydraulic design are:

- Pressure
- Friction
- Gradient
- Temperature

2.3.1 Pressure

Pressure is referenced in several ways in the pipeline industry. The liquid pipeline industry uses the term maximum operating pressure (MOP), which is defined as the maximum internal pressure expected during the operation of a pipeline. The gas pipeline industry uses the term maximum allowable operating pressure (MAOP), which is defined as the maximum internal pressure permitted during the operation of a pipeline. These pipeline industry pressure standards are defined in the Code of Federal Regulations (CFR). The liquid pipeline MOP regulations are referenced in *49 CRF 195.406 – Maximum Operating Pressure*. The gas pipeline MAOP regulations can be referenced in *49 CFR 192.619 – Maximum Allowable Operating Pressure: Steel or Plastic Pipelines, 49 CFR 192.621 – Maximum Allowable Operating Pressure: High-Pressure Distribution Systems*, and *49 CFR 192.23 – Maximum and Minimum Allowable Operating Pressure: Low-Pressure Distribution Systems.*

NOTE

Some companies use both terms even though *49 CFR Part 195* refers only to MOP. Individuals should consult the pipeline operator's maintenance and operations manual to determine how these terms are defined by the operator.

Although fluids are defined by their ability to flow, something has to start the flow and overcome the fluids' inertia and friction. A force, such as difference in pressure, gravity, or an external force physically pushing the fluid, can do this. Stopping the flow once it is started will require another force, such as resistance (friction), or a stationary object, such as a closed valve.

Using gravity to move a fluid is dependent on topography. When the fluid is traveling uphill, pumps must be used to overcome the fluid's inertia and resistance and the force of gravity. When the flow in the pipe is downhill, gravity may be all that is needed.

2.3.2 Friction

The resistance, or friction, of the fluid against the pipe will be constantly trying to slow the fluid. Most pumps work by providing a difference in pressure and must continue providing that force all along the line to overcome the friction. However, the pressure in the system must be kept under the maximum limits.

Friction is also affected by the roughness of the pipe interior, the pipe diameter and length, the product density and viscosity, the flow rate, and obstacles, such as valves and bends. Pipeline design can help reduce some of these, but higher fluid viscosity and density will increase the friction.

2.3.3 Gradient

Formulas are used to calculate pressure drop and flow rates for pipelines. In order to use these formulas, the characteristics of the fluid to be moved must first be determined. The characteristics used are primarily the specific gravity, viscosity, and either the flow rate or the allowable pressure drop.

The elevation of the terrain over which the pipeline travels must also be determined. The elevation and the pipe size are used, along with the pressure drop data, to determine the hydraulic gradient. The slope of the hydraulic gradient will represent the amount of pressure drop per unit of linear distance. As long as there is no change in pipe size or specific gravity, the pressure drop per unit will be uniform, and the gradient will be a straight line with a slope proportional to the system flow rate. The gradient is plotted in feet rather than pressure because the elevation profile is in feet. Pressure at the origin and endpoints is converted to feet using the specific gravity of the fluid. The endpoint is the end of the pipeline if no booster stations are on the line. Otherwise, each booster station will be an end point for the last gradient and a starting point for the new gradient. *Figure 3* shows a typical elevation profile plot for two gradients with no elevation change, one with a booster station and one without.

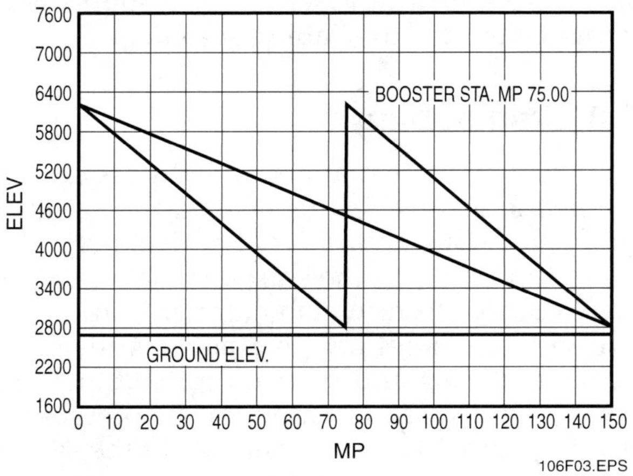

Figure 3 ◆ Typical elevation profile plot for two gradients with no elevation change.

2.3.4 Temperature

Another factor in the flow of pipeline fluids is the temperature. The higher the temperature, the lower the viscosity and vice versa. When the fluid is hot, it flows better. Heaters can be included in the pipeline equipment to keep the fluid at the optimum temperature.

3.0.0 ◆ ANSI RATINGS AND STANDARDS

The American National Standards Institute (ANSI) is a private, non-profit organization that administers and coordinates the U.S. voluntary standardization and conformity assessment system. The Institute's mission is to enhance both the global competition of U.S. business and quality of life by promoting and facilitating voluntary consensus standards and conformity assessment systems and by safeguarding their integrity.

In cooperation with its federated membership, the ANSI coordinates hundreds of standardizing organizations in the U.S. The federation consists of about 900 companies and some 200 trade, technical, and other organizations. It helps govern the International Organization for Standardization (ISO) and the International Electrotechnical Commission (IEC) and provides the only recognized system in this country for establishing standards. ANSI works closely with the American Society for Testing and Materials (ASTM), an organization that develops many standards.

ANSI standards apply to a number of pipeline components, including valves, pipe, flanges, vessels, and hydraulic systems. A standard might provide testing requirements, as shown in the following:

- *ANSI/ASME B16.37 (1980) – Hydrostatic Testing of Control Valves* — This standard establishes requirements for standard hydrostatic testing of control valves to prove the structural integrity and leak tightness of the valve-pressure retaining parts, including any closure parts but excluding moving seals.

Another might specify the types of components to choose for specific applications, such as the following:

- *ANSI/ASME B16.5 (1988) – Pipe Flanges and Flanged Fittings* — This standard covers pressure-temperature ratings, materials, dimension, tolerances, marking, testing, and methods of designating openings for pipe flanges and flanged fittings in sizes NPS ½ through NPS 24 in rating Classes 150, 300, 400, 600, 900, 1500, 2500.
- *ANSI/ASME B16.10 (1986) – Face-to-Face and End-to-End Dimensions of Valves* — This standard covers face-to-face and end-to-end dimensions of straightway valves and center-to-face and center-to-end dimensions of angle valves. It assures installation interchangeability for valves of a given material, type, size, rating class, and end connection.

It is important that ANSI ratings be verified when designing and installing systems to ensure that pressure and other ratings are appropriate. Testing of valves is addressed in the following:

- *ASME/ANSI PTC 25.3 (1988) – Safety and Relief Valves: Performance Test Codes* — This code establishes rules for conducting tests of valves to determine under specified conditions of back pressure—before opening, while relieving, and at closing—the following characteristics: simmer-warn pressure, opening pressure, etc.

ANSI also provides standards for graphic symbols to be used in diagrams and other drawings:

- *ANSI Y32.2.3 (1988) – Pipe Fittings, Valves, and Piping: Graphic Symbols* — This standard covers graphic symbols used for pipefitting, valves, piping, and associated equipment.

4.0.0 ◆ STATION CONTROL SYSTEMS

Controlling the fast-paced, continuous, and complex operation of the pipeline is a critical job that is done largely by the scheduler and controllers with the help of other personnel and systems like the Supervisory Control and Data Acquisition (SCADA) system, programmable logic controllers (PLC), and remote terminal units (RTU). Computerized control systems can gather, analyze, and act upon data in the shortest possible time. Most operators now have some form of computer-based monitoring system to run the pipeline operations. There are many functions that can be handled by these systems:

- Leak detection and location
- Batch tracking
- Pig tracking
- On-line flow compensation
- Modeling
- Instrument data and malfunction checking
- Inventory and material balance

4.1.0 Scheduling

Before a shipment of crude reaches the pipeline and comes under the control of the control center, it has to be accumulated and scheduled with the pipeline scheduler. Pipeline shipping is a wholesale-type operation; it is based on large shipments, high throughput, and efficient operation to reduce costs for the benefit of shippers and consumers. Because of this, a minimum volume is required for each single shipment. Shipments, or batches, must also be arranged according to their grade or chemical characteristics.

A shipment may be accumulated over a period of days and held in tanks until the required volume is ready. A single pipeline may carry several grades of oil for each of several shippers. The shipment is scheduled by the pipeline scheduler to leave a stated point of origin on a stated date and to be delivered to a destination point on a stated date. The shipper must have tank space to receive the oil at the destination.

To receive the oil at the downstream end, an equal amount must be entered in the upstream end. Therefore, the shipping process is a continuous one. Each shipper will always be receiving oil from and putting more oil into the pipeline.

In addition, regulatory bodies decree the allowable production rates for the wells in many states. The purpose of these allowables is to achieve maximum efficient recovery.

All of this requires careful control, monitoring, and scheduling. Computers are used to process the information on how much oil is to be produced and delivered, the locations of production and delivery, and ownership of the oil being moved. Computer programs determine the pumping stations to be used and the pumping unit combinations needed to obtain the desired rate of flow at minimum cost. These optimization programs consider the type of oil, the station power costs, batch sizes, and pump capabilities. Other factors include operating personnel, storage capacities, and available equipment.

4.2.0 Dispatching

The execution of the planned schedule is done by the pipeline dispatcher, also called the *operations control center (OCC) controller*. The dispatching job is a 24-hour one, and the shift dispatcher sits at a control console from which communication may occur with any point on the pipeline system. The controls can start and stop pump stations hundreds of miles away and monitor the pressures of lines, the rates of flow into and out of the system, and the gravities of the various oil streams. The control station may also have the capability of opening tanks to the line suction and the switching ability at a destination to change the stream from one delivery point to another. Detailed records are kept to show how, when, and what quantities were delivered, as well as characteristics of the delivered products.

Monitoring of the line also ensures security because a change in pressure or flow rate can mean leaks, breaks, or other problems. From the control center, flow in the pipeline can be stopped immediately if a problem is discovered.

Communication between all points on the pipeline system is critical for good control. Good communication verifies information to avoid errors and confusion. Communication is provided through telephone, radio, and satellite transmissions and allows the dispatcher to operate the pipeline.

4.3.0 Control Systems

A variety of control systems is available, and control systems should be chosen and customized to the particular pipeline application. *Figure 4* shows a typical control system for a single-unit pipeline booster station. A minimum control system will include a suction pressure and discharge pressure controller. Other controllers, such as a motor current controller and a ramp generator, may be connected into the control loop. *Figure 5* shows a control system for a multi-unit booster station. The control system for single and multi-unit stations is identical except, in the latter, the pressure transmitters are located so that one control system will suffice regardless of the number of units. In this system, one control valve serves both units.

This section includes:

- SCADA
- Sequencing
- Pressure control
- Level control
- Flow control

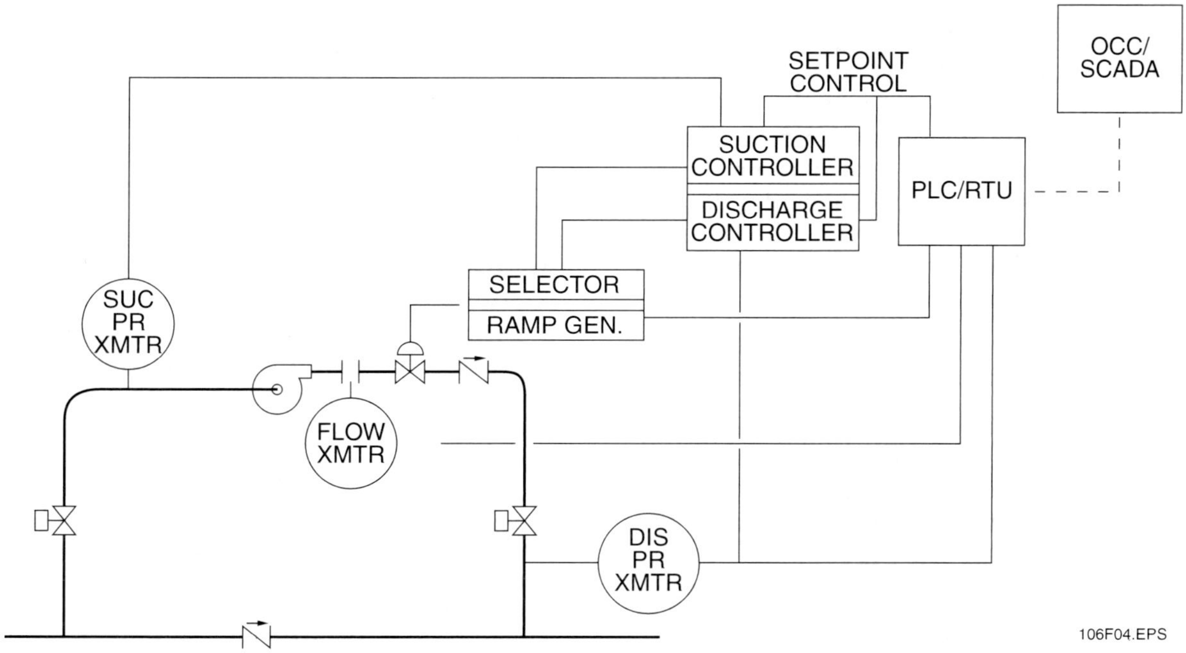

Figure 4 ◆ Control system for a single-unit pipeline booster station.

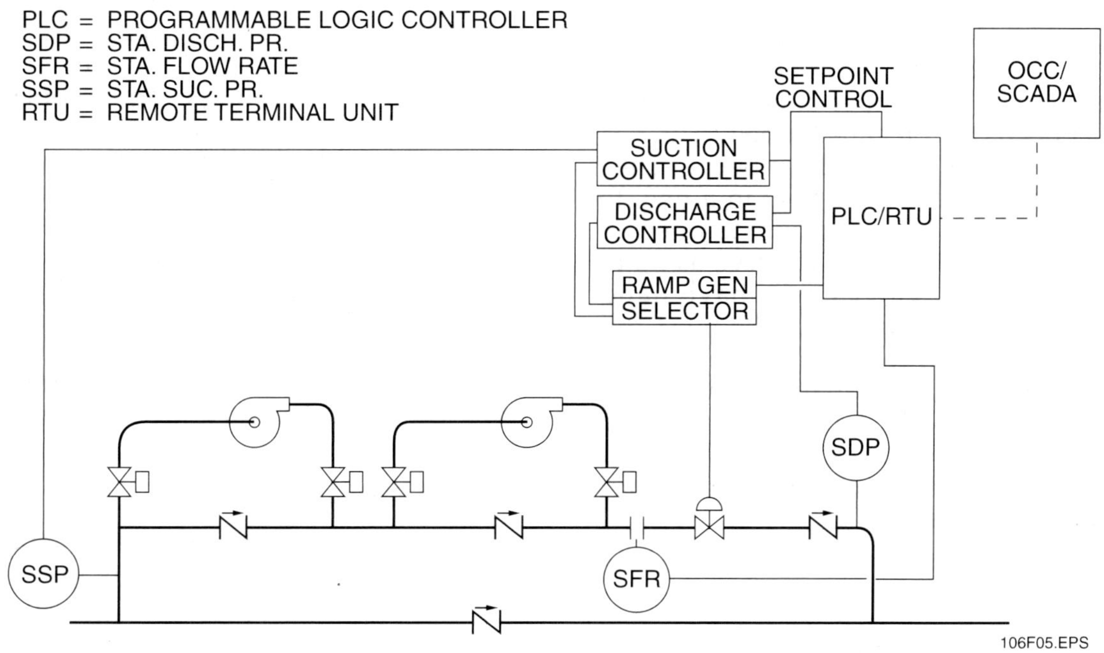

Figure 5 ◆ Control system for a multi-unit booster station.

reducing uneven wear on the sliding elements and adding to the noise that can indicate scraper position. Curved scraper blades are provided to scrape deposits from the pipe surface. These scrapers are less likely to become clogged with wax than the type that uses brushes. The scraper shaft is often fitted with a universal joint to allow the device to go through pipe bends. Tough, oil-resistant rubberlike materials form the discs or cup-shaped propulsion members that block and allow little bypassing of the oil stream. *Figure 6* shows a traditional scraper.

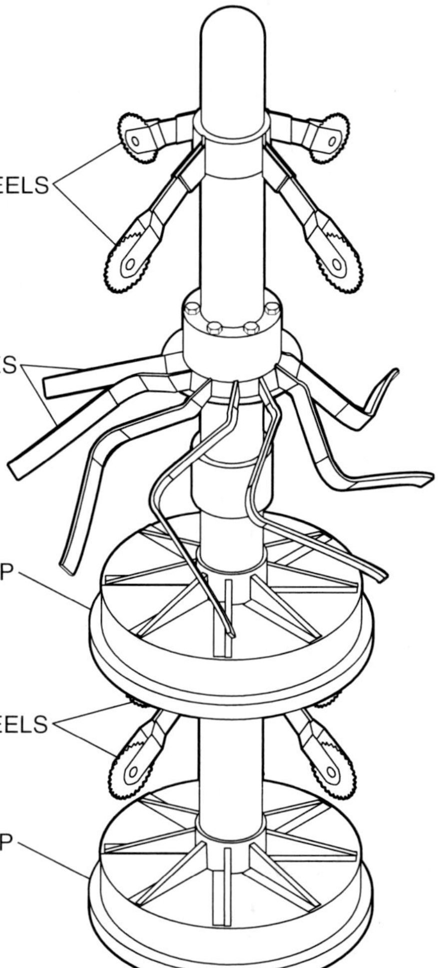

Figure 6 ◆ Traditional scraper.

Scrapers for large pipe are heavy and tend to wear on one side, allowing the stream to bypass them. Various methods are used to minimize this wear, such as attaching centering springs and making the scraper more buoyant.

6.1.2 Batch Separators

Spherically shaped pigs are made of oil-resistant plastic and are inflated with liquid. The size of the spheres can be adjusted by changing the content and pressure of the liquid. These can be distorted to some degree and can pass through fittings that other pigs cannot. The spheres are used mainly as batch separators, but they still sweep water and solids along.

Barrel-shaped pigs of plastic foam are also used as batch separators. The foam is the open-cell type that becomes saturated with oil and retains its shape under pressure. Usually the front and back are sealed against oil to provide propulsion. These pigs are compressible and can fit through smoothly contoured fittings of reduced area and in pipelines of varying size.

6.2.0 Receivers and Launchers

Receiver and launcher traps and associated MOVs are located at specific station facilities along the pipeline system to permit main pipeline pigging and scraping operations. Stations without receiver/launcher traps are equipped with a mainline bypass valving network that allows the passage of a pig/scraper through the mainline pipe.

The piping arrangement for launching and receiving a scraper includes an incoming scraper trap and an outgoing scraper trap. The arrangement and valve control prevents pigs from going through the station pumps. *Figure 7* shows a typical receiver/launcher arrangement.

6.2.1 Valve Operation—Receiving

When an incoming pig is expected, valve A and C (refer to *Figure 7*) are opened to receive the pig and valve B is closed. When the pig is in the trap, valve B is then opened, C is closed, and A is closed. This allows for removal of the pig from the trap while returning to normal flow through the station. Other valves on the trap barrel drain any remaining fluid from the trap and vent the trap before the trap door is opened.

6.2.2 Valve Operation—Launching

To insert the scraper on the outgoing scraper trap, valves G and H must be closed and valve I must be opened. Drain valves and vent valves are opened to relieve pressure in the scraper barrel. The trap door may then be opened and the scraper inserted into the barrel. Valves on the barrel are then closed. To start the scraper, valve H is cracked, and air is bled through the vent valve. Valve G is opened, valve H is opened, and valve I is partially closed. To return to normal operations after the scraper has passed through valve G, valve I is opened and G and H are closed.

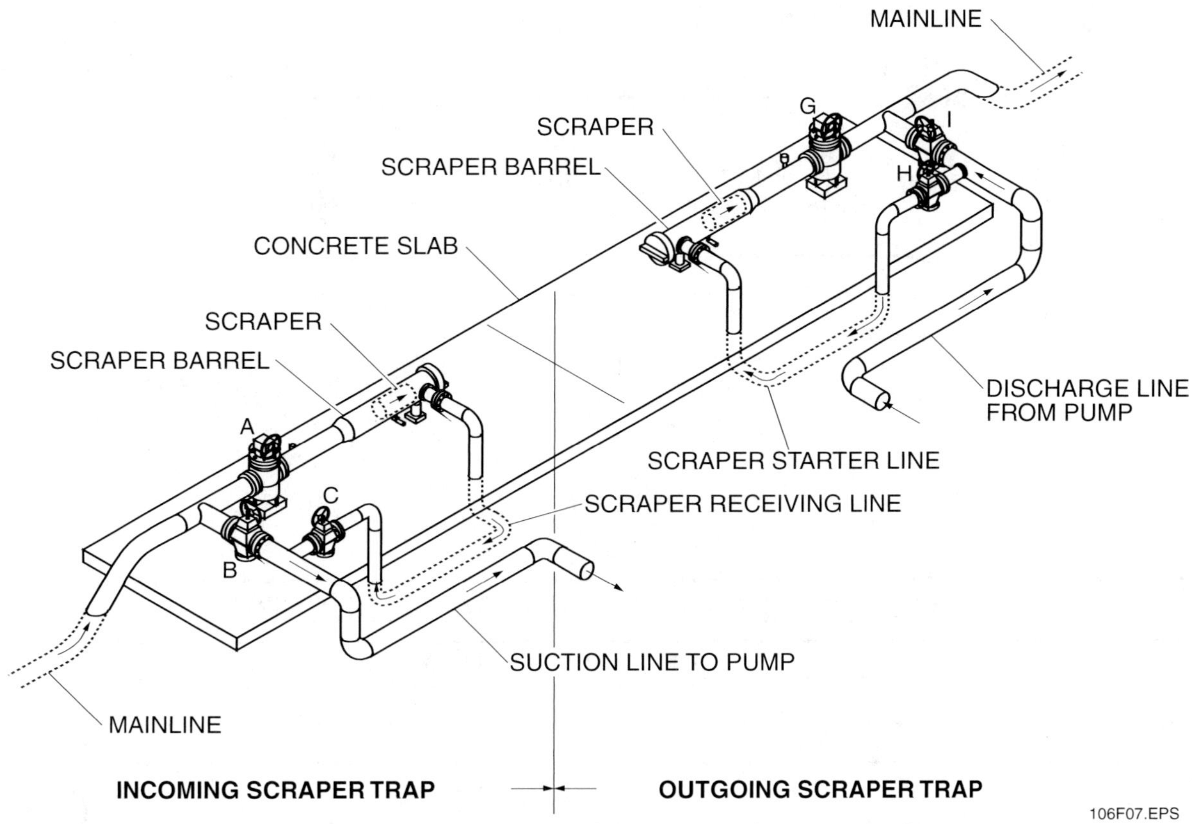

Figure 7 ◆ Typical receiver/launcher arrangement.

6.3.0 Sump Operations

During pigging operations, traps accumulate fluid along with the scraper, and this fluid must be drained from the trap to the sump. The sump also collects fluid from other operations, such as metering and analysis.

The sump, associated MOVs, sump pump, and instrumentation are designed to receive crude oil or product from the equipment and station drain-line system and to reinject this fluid into the main pipeline. If the drain rate into the sump system exceeds the evacuation rate, a high-level alarm is generated, which may result in shutdown.

6.4.0 Flare Operations

Flare operations can be either an automatic or manual task, depending on the site operation. Waste gases in a pipeline system that are accumulated or stored must be safely disposed, and flaring is a safe and practical method of disposal. Flare operations involve disposal of waste gases from the pipeline through the use of combustion. This is typically accomplished by burning the waste gases through an elevated high-temperature flare stack. These flare stacks are equipped with an igniter and/or burner that is used to burn off the gases as they pass through. Using an elevated flare stack, waste gases flow up through the stack and are incinerated, and the combustion flame is carried out the top of the stack.

6.5.0 Mid-Stations/Booster Stations

Methods have to be in place at booster stations to deal with pig passage because pigs cannot go through station pumps and equipment. They would damage pumps by becoming stuck in piping elbows and tees and by introducing debris that could damage systems. To prevent this, passage indicators, automatic bypass sequencing, and station shutdown are used.

6.5.1 Passage Indicators

Pig passage indicators are used to indicate the location of a pig. These indicators may trigger an automatic bypass system or a receiver/launcher sequence.

6.5.2 Automatic Bypass Sequencing

Some stations have an automatic bypass feature in their station piping system. *Figure 8* shows a bypass schematic diagram.

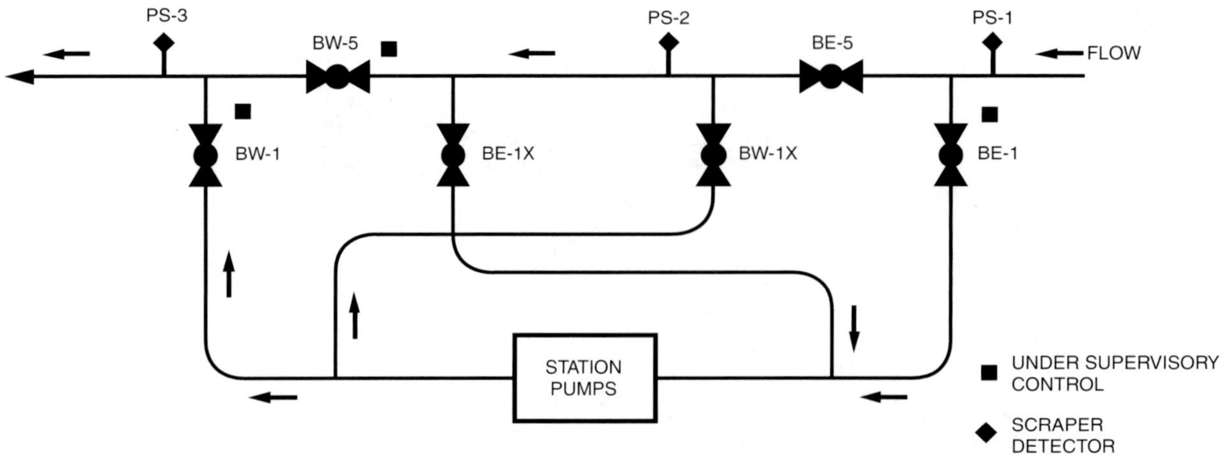

FLOW DIAGRAM

AUTOMATIC SCRAPER BYPASS VALVE SEQUENCE

	VALVE POSITIONS & OPERATIONS					
	BE-5	BW-5	BE-1	BW-1	BE-1X	BW-1X
ORIGINAL	O	C	O	O	O	C
PS-1 CLOSES (BE-5 & BE-1X OPEN, BW-5 CLOSED)	–	–	CLOSES ½	–	–	–
PS-2 CLOSES	–	–	OPENS	–	–	–
BE-1 OPEN	CLOSES	–	O	–	CLOSES	–
BE-5 & BE-1X CLOSED, (BW-1 OPEN)	C	OPENS	–	O	C	–
BE-5 OPEN	–	O	–	–	–	OPENS
BW-1X OPEN	–	–	–	CLOSES ½	–	O
PS-3 CLOSES	–	–	–	OPENS	–	–
BW-1 OPEN	–	CLOSES	–	O	–	CLOSES
BW-5 & BW-1X CLOSED	OPENS	C	–	–	OPENS	C
FINAL	O	C	O	O	O	C

O = OPEN C = CLOSED PS = PIGSIG (SCRAPER DETECTOR)

106F08.EPS

Figure 8 ◆ Bypass schematic diagram.

The fast-acting gate valves respond to the pig signals or scraper detectors as the pig/scraper approaches. Normal flow through the station pipe is blocked while the pig passes through. When the scraper passes through and leaves the station, the outgoing pig detector restores the valve positions for normal flow. This bypass feature saves the trouble of manually removing the pig from the incoming line and reinserting it in the outgoing line.

6.5.3 Station Shutdown

Another method of automatic scraper bypassing entails the automatic shutdown of the pump station and closure of the station suction and discharge valves when a scraper arrives at the first pig detector. The valves open and the station restarts when the scraper has passed the downstream detector.

7.0.0 ◆ PROVING PROCESS METERS

Various meters are used along the pipeline to monitor the volume and characteristics of the fluid flowing through the pipeline. Meter readings taken before and after deliveries are utilized to determine the volume of throughput. Meters are typically equipped with transmitting, remote-reading, and printout accessories to facilitate the delivery of correct product volumes and for preparation of delivery tickets. Data is often transmitted to remote locations. Strainers are installed upstream of meters to remove any foreign material from the stream before it goes through the meter. Strainers are generally more coarse than filters.

The meter proving system provides the means of calculating a daily meter factor for each meter in a meter set. This is done by comparing the oil volume through a meter to a known volumetric displacement of oil in the prover. This value is compared to a computer-generated meter curve. The meter must prove within certain limits along the meter curve. During normal proving operations, the automatic proving sequence is initiated by an operator from the computer terminal in the control room.

Meters are proved any time a question of accuracy arises, any time changes or repairs are made to a meter, upon request by any party to the custody transfer, and at least once during a ticket period. A *ticket period* is defined as an operating day, which is from midnight to midnight. Meters are proved under conditions as close to constant flow, temperature, and pressure (or normal conditions) as can be obtained.

This section covers the following:

- Process meters
- Meter provers
- Correction factors
- Proving instrumentation and readings

7.1.0 Process Meter Types

Practically all measurements are made by turbine meters or positive-displacement meters (see *Figure 9*). This section includes the following types of meters:

- Turbine
- Positive displacement
- Ultrasonic
- Orifice

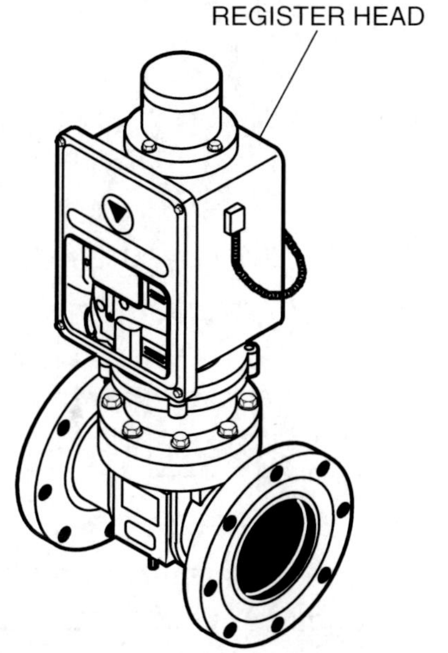

TURBINE METER (WITH REGISTER HEAD)

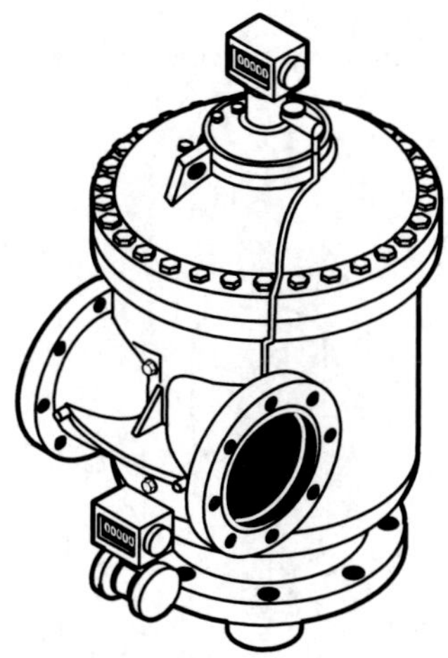

POSITIVE-DISPLACEMENT METER

106F09.EPS

Figure 9 ◆ Turbine and positive-displacement meters.

7.1.1 Turbine

Turbine meters are widely used in hydrocarbon pipeline applications. They suffer from bearing limitations, however, and require regular calibration. *Figure 10* shows a diagram of turbine meter operation.

Twin-rotor designs make the meter self-diagnostic by providing simultaneous measurements of the speed of both rotors. These meters can detect their own bearing wear and any deviations due to blade damage. Two designs have been used in pipeline systems. One has two rotors and two independent sets of bearings. The shaft and slave rotor ride on one set of bearings with the measurement rotor riding on the other. This makes it an almost frictionless flowmeter.

Another twin-rotor type is also self-checking and self-adjusting. The rotors are next to each other with the measuring rotor downstream. Deviations in the fluid leaving the first rotor are sensed by the second rotor, which triggers adjustments in speed. *Figure 11* shows a cross section of a twin-rotor meter.

7.1.2 Positive Displacement

Positive-displacement (PD) meters are mechanical devices that rotate as fluid passes through them. Some have compared the PD meter's actions to that of an old-fashioned water wheel.

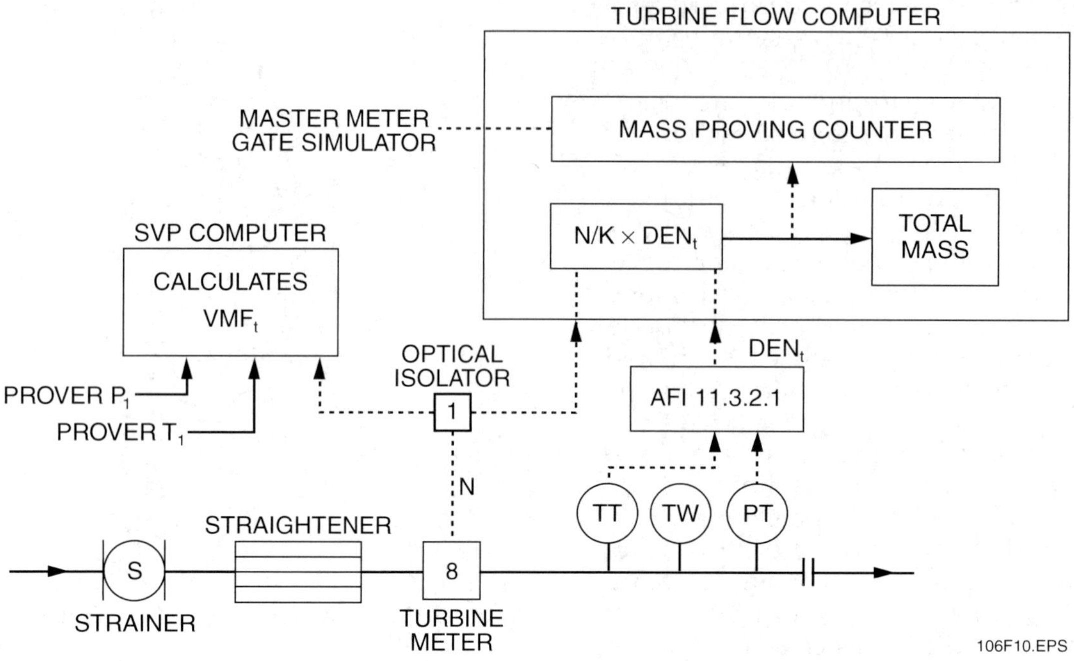

Figure 10 ◆ Diagram of turbine meter operation.

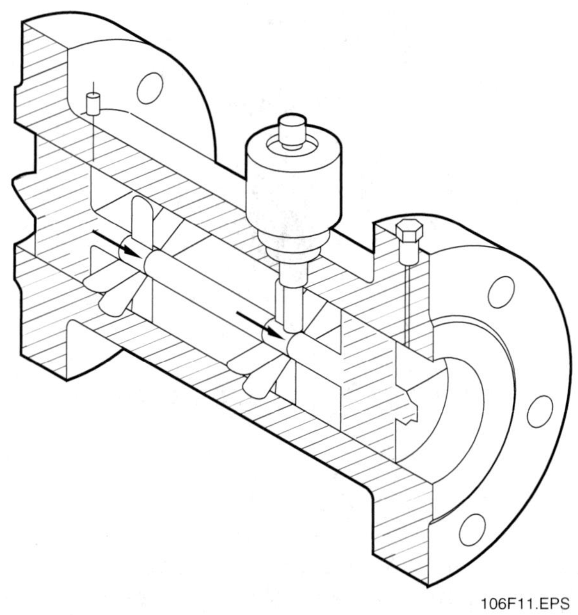

Figure 11 ◆ Cross section of a twin-rotor meter.

Measured amounts of liquid are caught in distinct segments as fluid passes through the unit. The distinct or exact segment may be an ounce or several gallons per revolution, depending on the size of the meter. A wide variety of sizes is available. While one portion is being received into the meter, another portion is being measured, and a third portion is passing out of the meter and back into the line. PD meters are designed to rotate in proportion to the amount of liquid passing through. The shaft attached to the rotor transmits the revolutions of the meter to a geared unit that, in turn, operates the meter stack. Each time the unit rotates, a sensor is activated and counts the number of rotations, somewhat like the odometer on an automobile. Electronic signaling devices can also be connected to the meter to relay information to computers, recorders, and other equipment that will record and total the operation.

One of the problems with PD meters relates to design. In most models, the blade or vane that causes the meter to turn does not actually touch the casing, so a minute amount of fluid, called *slippage,* is allowed to pass by it. The viscosity of fluids and meter wear will affect the amount of slippage in PD meters.

NOTE

Temperature, viscosity, and pressure seriously affect the accuracy of a PD meter.

7.1.3 Ultrasonic

Two types of ultrasonic meters are the Doppler and the transit time meters; both use ultrasonics to infer flow. Doppler types are rather crude flow-indicating devices, but transit time meters can indicate flow to around plus or minus 2% of reading. The advantage of ultrasonic meters is the negligible head loss and the ability to install other portable or dedicated systems without line shutdown.

7.1.4 Orifice

A simple, efficient, and reliable measurement system is the orifice flow-measuring element with an on-line density meter and microprocessor flow computer. This basic assembly provides real time-mass flow output, and the orifice is the key component.

The orifice is a static device that is generally inert to fluid conditions. Calibration is simply measurement of physical dimensions and conformance to measurement tolerances. The orifice meter is one of the few measurement devices accepted for custody transfer that does not require in-service proving.

The primary element of an orifice-type measuring station consists of the meter tube, orifice plate fitting, orifice plate, pressure taps, and straightening vanes. *Figure 12* shows a diagram of how an orifice meter works.

7.2.0 Meter Provers

A meter prover is used to calibrate meters for specific conditions of flow and product type. Meters are proved by passing a known volume of fluid through the meter at a known rate and comparing this volume with a standard. Two types of provers frequently used are the bell prover and the piston prover. Variations of provers include the following:

- Bidirectional provers
- Portable provers
- Compact provers

7.2.1 Bidirectional Provers

A bidirectional meter prover is used to calibrate meters for specific conditions of flow and product type. It is operated by having a close-fitting piston traverse the accurately calibrated interior of the pipe from one end to the other and return. A detector mounted near each end of the pipe section follows the passage of the piston, and the amount of flow metered is the volume existing in the pipe between the detectors. An electrical counter counts the pulses from the meter to be proved as the piston moves between the detectors; the count begins at one and ends at the other. *Figure 13* shows a bidirectional prover.

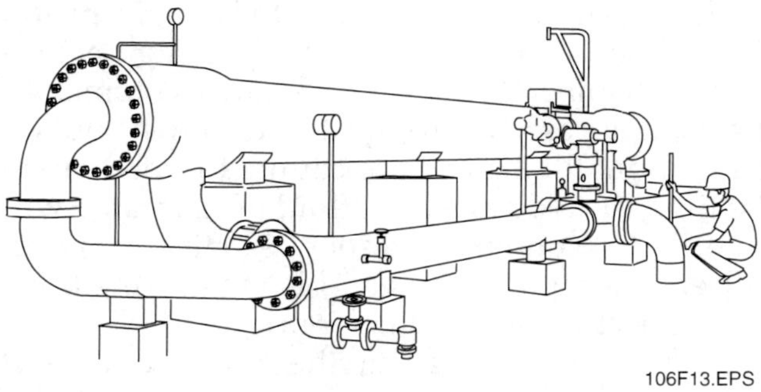

Figure 13 ◆ Bidirectional prover.

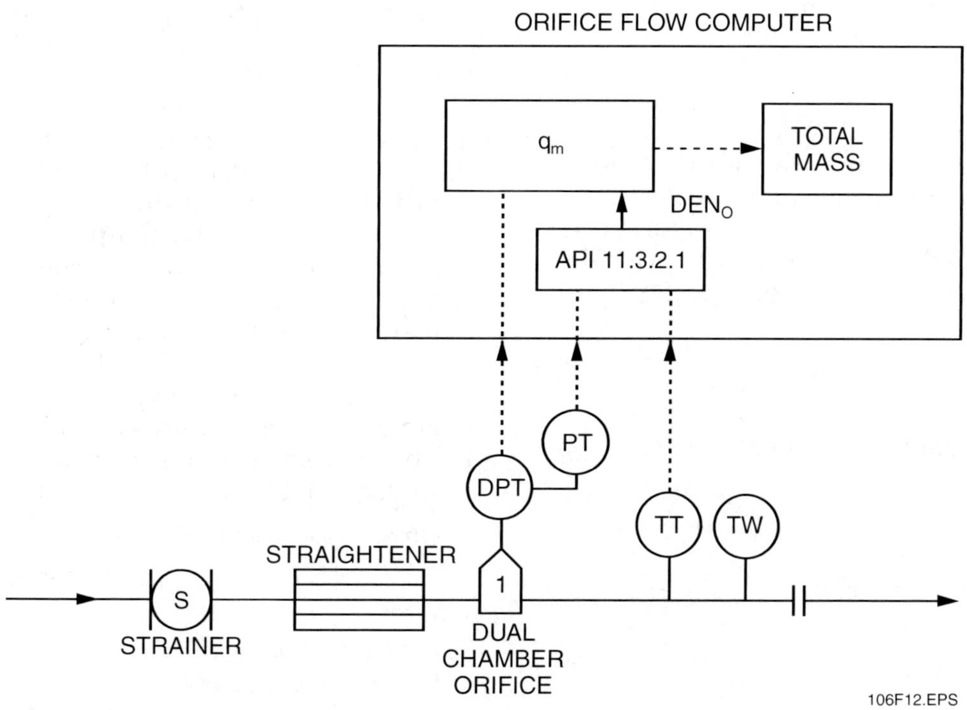

Figure 12 ◆ Diagram of how an orifice meter works.

7.2.2 Portable Provers

Portable proving equipment may be used to prove meters in the field. Such a prover might include a small volume prover (SVP), a master turbine meter, and associated instrumentation. The proving is based on the concept that a master meter can be proved and then act as a transfer standard between the primary standard SVP and another meter.

7.2.3 Compact Provers

Compact provers are essentially time/volume calibrators that eliminate the plus or minus 1 count of larger provers. It is much smaller in size but provides the same or better performance.

The prover uses pneumatics to launch a piston that is then driven by the fluid through a calibrated chamber displacing a known volume. As the piston passes through the chamber, pulses from an internal clock are counted to determine the time, A, to displace the fluid of calibrated volume, D. A second timer responds to the first input pulse from the flowmeter. This records the time, B, to collect whole flowmeter pulses, C.

During the proving run, there were $C \times A \div B$ meter pulses, and the meter factor is calculated by dividing this count by the volume displaced, or $K = A \div D \times C \div B$.

Some compact provers are unidirectional and some are bidirectional. In unidirectional types, pneumatics or hydraulics are used to launch and retrieve the piston. In bidirectional types, process fluid drives the piston. *Figure 14* shows a unidirectional compact prover.

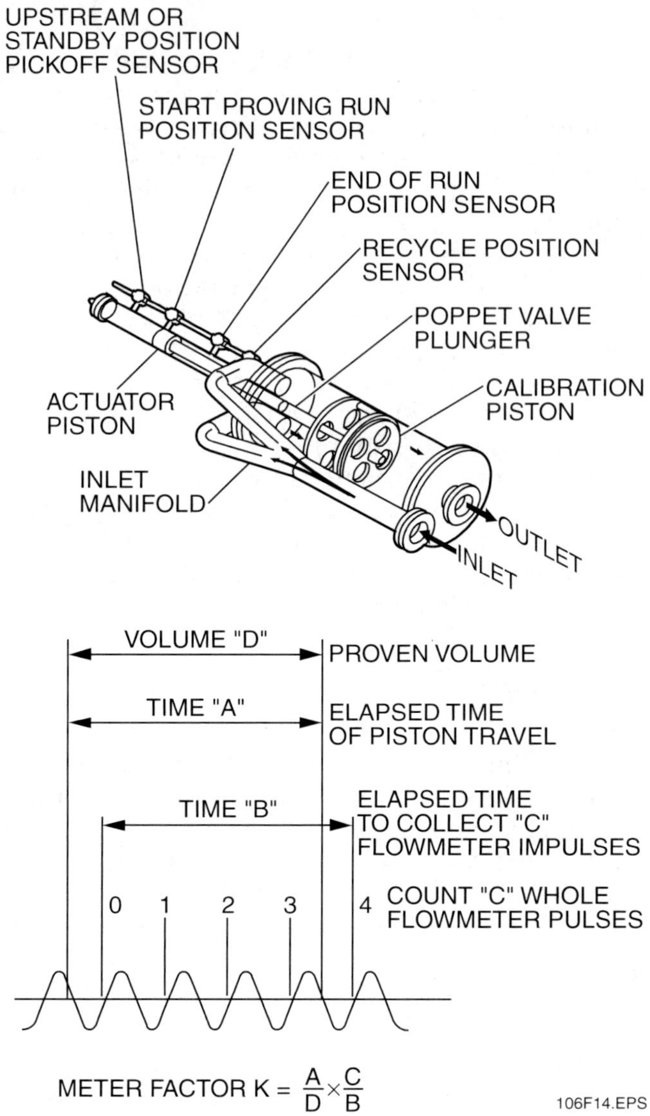

Figure 14 ◆ Unidirectional compact prover.

7.3.0 Factors

Various factors are used in proving calculations. Although modern technology allows these factors to be automatically calculated, it is important to understand the mathematics behind the calculations. This section includes the following types of factors:

- Gross meter volume
- K-Factor
- Corrections for temperature of and pressure on steel and liquid
- Meter factor
- Net volume
- Volume correction factor

7.3.1 Gross Meter Volume

Gross meter volume is obtained by making several prover runs to show the repeatability and accuracy of a prover. Each time a prover run is made, data is collected to establish the baseline conditions of operation. Prover temperature, prover pressure, meter temperature, meter pressure, and the total number of pulses are typically recorded. When several consecutive runs are in agreement, the data for these runs is then averaged to obtain the baseline data.

In order to obtain the gross meter volume, the average number of pulses is divided by the number of pulses per barrel. This is illustrated in the following equation:

$$\text{Gross meter volume} = \frac{\text{Average pulses}}{\text{pulses per barrel}}$$

Where:

Average pulses = 85,359
Pulses per barrel = 16,800
Gross meter volume = 85,359 ÷ 16,800
Gross meter volume = 5.0809

7.3.2 K-Factor

One of the correction factors calculated and used is the K-factor. The basic information needed to calculate the K-factor is available by counting the meter pulses that occur during the time the calibrated prover volume is displaced. The K-factor is determined by the ratio of total pulses counted to the calibrated volume of the prover. This is illustrated in the following equation:

$$\text{K-factor} = \frac{\text{Total pulses}}{\text{calibrated prover volume}}$$

Where:

Total pulses = 16,800

Calibrated prover volume (barrels) = 5.0991

K-factor = 16,800 ÷ 5.0991

K-factor = 3,294.6990

> **NOTE**
> The K-factor of a meter is a constant for the meter, and is typically stamped on the meter by the manufacturer.

7.3.3 Corrections for Temperature of and Pressure on Steel and Liquid

The K-factor is not the only factor taken into account. Due to the need for the highest level of precision and accuracy, effects from the environment on the equipment and product must also be considered. Since the steel that the meter and prover are made from expands and contracts due to temperature and pressure, corrections must be made for these effects. These corrections are abbreviated as CTS (corrections for temperature of steel) and CPS (corrections for pressure on steel). The liquid volume also changes based on temperature and pressure. These corrections are abbreviated as CTL (corrections for temperature of liquid) and CPL (corrections for pressure on liquid). The values for CTS, CPS, CTL, and CPL are referenced using the appropriate API tables. These figures are then multiplied together to get the corrected prover volume. This is illustrated in the following equation:

Corrected prover volume = Calibrated volume × CTS × CPS × CTL × CPL

Where:

Calibrated volume = 5.0991

CTS = 1.0003

CPS = 1.0001

CTL = 0.9940

CPL = 1.0005

Corrected prover volume = 5.0991 × 1.0003 × 1.0001 × 0.9940 × 1.0005

Corrected prover volume = 5.0731

Corrected volume for the meter can also be obtained by multiplying the meter volume by the correction factors that apply. For example, if the gross meter volume is 5.0809 temperature compensated, apply only the CPL portion of the correction. This is illustrated in the following equation:

Corrected meter volume = Gross meter volume × CPL

Where:

Gross meter volume = 5.0809

CPL = 1.0005

Corrected meter volume = 5.0809 × 1.0005

Corrected meter volume = 5.0834

7.3.4 Meter Factor

After making corrections for temperature and pressure effects for the meter and prover, this information can be used to determine the meter factor. The meter factor is calculated by dividing the corrected prover volume by the corrected meter volume. This is illustrated in the following equation:

$$\text{Meter factor} = \frac{\text{Corrected prover volume}}{\text{corrected meter volume}}$$

Where:

Corrected prover volume = 5.0731

Corrected meter volume = 5.0834

Meter factor = 5.0731 ÷ 5.0834

Meter factor = 0.9980

The meter factor obtained from these calculations allows the meter's registered volume to be corrected. For example, if the meter's registered volume is 10,000 barrels, this volume can be corrected to a more accurate volume by multiplying the registered volume by the meter factor. This is shown in the following equation:

Corrected registered volume = Registered volume × meter factor

Where:

Registered volume = 10,000 barrels

Meter factor = 0.9980

Corrected registered volume = 10,000 × 0.9980

Corrected registered volume = 9,980 barrels

7.3.5 Net Volume

Modern instrumentation has the ability to calculate the volume correction factors for the pressure and temperature of liquids and allows net proving to be used to verify proper operation of a metering system. Net proving is very similar to gross proving with the addition of some signal processing and calculations. In most cases, two pulse streams from the same meter are monitored and recorded. One of the two pulse streams represents the gross volume, which has the K-factor and meter factor applied from previous proofing runs. The other pulse stream is considered the net volume and is illustrated in the following equation:

Net volume = Gross volume $\times$ CTL for the meter $\times$ CPL for the meter

Where:

Gross volume = 5.0809

CTL meter = 1 (if temperature corrected)

CPL meter = 1.0005

Net volume = 5.0809 $\times$ 1 $\times$ 1.0005

Net volume = 5.0834

Analyzing the calculation shows that the net volume is essentially the corrected meter volume.

7.3.6 Volume Correction Factor

Where a linear meter, such as a turbine or positive displacement device is used for metering, the relationship between the K-factor and the meter factor can be used to determine the volume correction factor. This is illustrated in the following equation:

Volume = (Pulses $\div$ K-factor) $\times$ meter factor

Where:

Pulses = 16,800

K-factor = 3294.699

Meter factor = 0.9980

Volume = (16,800 $\div$ 3294.699) $\times$ 0.9980

Volume = 5.0991 $\times$ 0.9980

Volume = 5.0889

7.4.0 Proving Instrumentation and Readings

Instrumentation is required for accurate measurement and control of prover operations. Valve movement and proving operations can be controlled from the control room. Semi-automatic and manual functions are initiated at meter control panels.

7.4.1 Instruments

Daily comparisons between transmitted values and indicated values verify the prover is operating within acceptable parameters. This equipment is maintained on regular preventive maintenance (PM) cycles by E&I technicians:

- *Pressure gauges* at inlets and outlets indicate the pressure of the fluid.
- *Pressure transmitters* at inlets and outlets convert the gauged pressure to an analog signal and transmit it to the flow computer.
- *Temperature probes* at inlets and outlets show prover temperatures.
- *Temperature transmitters* at inlets and outlets transmit analog signals of temperature to the flow computer.

7.4.2 Flow Computers

The flow measurement computer is the master metering computer for each meter set. The flow computer functions as a pulse comparator, a prover counter, and a prover flow controller, as well as flow computer for its respective meters. There is one flow computer for each metering station and a metering station can contain several meters.

Summary

Modern pipeline operations depend heavily on automated systems and the instrumentation that controls them. Automatic operation of motors, valves, and pumps allows for more immediate response to pipeline conditions along the miles of pipeline. It also provides a more consistent approach to problems. Measurement systems provide accurate, immediate information for decision-making and help to determine automatic tasks to be performed.

Pipeline operations are based on the principles of hydraulics—the behavior and flow of fluids. Understanding the pipeline's hydraulics requires an understanding of the laws of hydrostatics and hydrodynamics.

Pipeline equipment and components are subjected to a variety of pressures and temperatures. The materials and design of these components are rated for certain ranges of temperature and pressure, and the integrity of the pipeline relies on using the correctly rated component in a particular application. ANSI is the main resource for ratings standards.

Control systems for pipeline operations are under the direction of the control center where the scheduler, dispatcher, and other personnel operate the systems. SCADA is one of the main systems used on pipelines to supervise operation of sequencing and control of pressure, levels, and flow.

Abnormal operating conditions occur when operating conditions exceed design limits or when events cause possible hazards. Correct response to AOCs is described in company procedures.

Two important procedures for control and maintenance of the pipeline are pigging operations and proving process meters.

Review Questions

1. Fluid properties can be changed by temperature and _____.
   a. elasticity
   b. pressure
   c. viscosity
   d. density

2. The measure of how readily a fluid catches fire when it comes in contact with heat or sparks is known as _____.
   a. homogeneity
   b. volatility
   c. flammability
   d. compressibility

3. The maximum internal pressure permitted during the operation of a pipeline is known as _____.
   a. operating pressure (OP)
   b. allowable operating pressure (AOP)
   c. maximum operating pressure (MOP)
   d. maximum allowable operating pressure (MAOP)

4. The organization that helps govern the ISO and IEC and provides the only recognized system in this country for establishing standards is _____.
   a. MOP
   b. ASTD
   c. ANSI
   d. CTS

5. Controlling the fast-paced, continuous, and complex operation of the pipeline is done with the help of systems like _____.
   a. SCADA
   b. ANSI
   c. Doppler
   d. ASTM

6. Shipments, or batches, must be arranged for shipment according to their _____.
   a. shipping temperature
   b. storage date
   c. shipping pressure
   d. grade and chemical characteristics

7. The execution of the planned shipment schedule is done by the pipeline _____, also called the OCC controller.
   a. communicator
   b. dispatcher
   c. owner
   d. field technician

8. The critical process of opening and closing valves during operations to provide the correct flow path for product is called _____.
   a. sequencing
   b. dispatching
   c. SCADA
   d. scheduling

9. A _____ can be thought of as a section of pipe with variable resistance.
   a. switch station
   b. control valve
   c. flow controller
   d. compressor

10. A(n) _____ occurs when operating design limits are exceeded or when a condition or event occurs that could cause a hazard if not corrected.
    a. abnormal operating condition
    b. pigging operation
    c. flare operation
    d. meter factor

11. Pigs used to check the pipeline for wall reductions due to dents or gouges are called _____ pigs.
    a. monitor
    b. caliper
    c. smart
    d. scraper

12. The two most common uses for pigs are cleaning out the pipe and _____.
    a. separating batches of product
    b. removing dents
    c. applying internal coatings
    d. gauge product flow

13. Fluid accumulated in traps and from other operations, such as metering and analysis, is drained to a _____.
    a. flare
    b. sump
    c. booster station
    d. mid-station

14. The disposal of waste gases through the use of combustion and an elevated high-temperature stack is called _____ operations.

a. sump
b. pigging
c. booster
d. flare

15. Meters are _____ any time a question of accuracy arises, any time changes or repairs are made to a meter, upon request by any party to the custody transfer, and at least once during a ticket period.

a. proved
b. strained
c. factored
d. aligned

16. A type of meter that is widely used in hydrocarbon pipeline applications but suffers from bearing limitations and requires regular calibration is known as a _____ meter.

a. turbine
b. positive-displacement
c. ultrasonic
d. orifice

17. A small time/volume calibrator that eliminates the plus or minus 1 count of larger provers but provides the same or better performance is called a _____ prover.

a. bidirectional
b. portable
c. compact
d. directional

18. The ratio of total pulses counted to the calibrated volume of the prover provides the _____.

a. gross meter volume
b. K-factor
c. meter factor
d. correction factor

19. The volume correction factor for the pressure on the liquid that the meter and prover are calculating is abbreviated as _____.

a. CTS
b. CPS
c. CTL
d. CPL

20. A device, which is used at inlets and outlets, that transmits analog signals of temperature to the flow computer is called a _____.

a. pressure transmitter
b. pressure gauge
c. temperature probe
d. temperature transmitter

Trade Terms Introduced in This Module

Compressibility: A measure of how easily a fluid can be forced into a smaller volume.

Density: A fluid's thickness or mass per volume.

Elasticity: A measure of the fluid's resistance to change in shape.

Hydrodynamics: The science of moving fluids.

Hydrostatics: The science of stationary fluids.

Proving: The process of calibrating the accuracy of a meter.

Specific gravity: The ratio of density of a fluid to that of water.

Viscosity: The level of a fluid's resistance to flow.

NCCER CURRICULA — USER UPDATE

NCCER makes every effort to keep its textbooks up-to-date and free of technical errors. We appreciate your help in this process. If you find an error, a typographical mistake, or an inaccuracy in NCCER's curricula, please fill out this form (or a photocopy), or complete the online form at **www.nccer.org/olf**. Be sure to include the exact module ID number, page number, a detailed description, and your recommended correction. Your input will be brought to the attention of the Authoring Team. Thank you for your assistance.

Instructors – If you have an idea for improving this textbook, or have found that additional materials were necessary to teach this module effectively, please let us know so that we may present your suggestions to the Authoring Team.

NCCER Product Development and Revision

13614 Progress Blvd., Alachua, FL 32615

Email: curriculum@nccer.org
Online: www.nccer.org/olf

❑ Trainee Guide ❑ Lesson Plans ❑ Exam ❑ PowerPoints Other ____________________

Craft / Level: Copyright Date:

Module ID Number / Title:

Section Number(s):

Description:

Recommended Correction:

Your Name:

Address:

Email: Phone:

Module 62103-02

Release Identification and Response

Introduction to the Pipeline Operator Qualification Rule and NCCER

It could be said that petroleum products keep America running. Petroleum products like oil and natural gas are the principal sources of energy in the United States and are used in many manufacturing and chemical processes. Petroleum products are primarily transported by pipeline. There are more than a million miles of liquid and natural gas pipeline in this country, ranging in diameter from two inches to four feet. These pipelines bring crude oil and natural gas from production fields to processing plants and refineries—and ultimately to the consumer.

Pipeline Operator Qualification

Petroleum products are considered hazardous, so pipeline personnel must take extra precautions and follow specific procedures when working with them. To help ensure the safe handling and transportation of petroleum products, the United States Department of Transportation (DOT) mandates safe practices as federal regulations.

One such regulation, the Operator Qualification (OQ) rule, requires pipeline operators to develop a written qualification plan to evaluate their personnel and their contract personnel (see *49 CFR Part 192 Subpart N—Gas*, and *49 CFR Part 195 Subpart G—Liquid*). The evaluation hinges on the operator's ability to perform certain covered tasks and to be able to recognize and respond to abnormal operating conditions that may be encountered while performing these covered tasks.

Covered Tasks

A pipeline performance task is deemed covered if the activity passes the following four-part test:

1. Is it performed on a pipeline facility?
2. Is it an operations or maintenance task?
3. Is it performed as a requirement of *49 CFR Part 195 (Liquid)* or *49 CFR Part 192 (Gas)*?
4. Does it affect the operation or integrity of the pipeline facility?

If the answer to all four questions is yes, then the activity is considered covered and must figure into the evaluation process of pipeline personnel.

Consortium on Operator Qualification

The Consortium on Operator Qualification (COOQ) became the guiding force for the identification and development of covered tasks for the liquid pipeline industry. Their work product, API publication 1161, *Guidance Document for the Qualification of Liquid Pipeline Personnel*, formed the basis for the resulting Pipeliner Training and Assessment Program.

The covered task list created in API publication 1161 has now evolved to include gas, as well as construction covered tasks. NCCER maintains a version of the covered task list, which can be accessed at **www.nccer.org**.

NCCER

With the guidance and support of the Consortium on Operator Qualification, NCCER was able to design and implement the Pipeliner Training and Assessment Program.

NCCER is a not-for-profit 501(c)(3) education foundation established in 1995 by the world's largest and most progressive construction and maintenance contractors and national contractor associations. NCCER was founded to address the severe workforce shortage facing the construction and maintenance industries and to develop a standardized training process and curricula. Today, NCCER is supported by hundreds of leading construction and maintenance companies, manufacturers, and national associations and developed more than 70 construction and maintenance titles.

NCCER also maintains a Registry, which contains the records of trainees and instructors who have received NCCER training and certification. All completion information for every trainee is sent to NCCER and filed within the Registry. The Registry can then confirm training and skills for workers as they move from company to company, state to state, or even within their own company.

NCCER Standardized Curricula

NCCER standardized curricula are developed using the expertise of industry subject matter experts (SMEs). The use of SMEs as an integral part of every new development or revision ensures that the curriculum is current, industry-driven,

and indicative of industry practices. The development and/or revision process is facilitated from start to finish by NCCER project managers, who work closely with the SMEs and a select team of technical writer(s) on the development and presentation of the curriculum.

Pipeline: Five Training Paths

The pipeline curriculum focuses on essential skills and competencies within five skill areas:

- Corrosion Control
- Electrical and Instrumentation
- Maintenance
- Mechanical
- Operations (Field and Control Center)

SMEs from the pipeline industry have determined the skills and competencies needed based on API's covered tasks and the consensus of the group/industry. Specific company information is generally not incorporated into the curriculum unless the SMEs determine that the information is the industry standard.

Training Modules and Covered Tasks

The smallest unit of instruction in the curriculum is a *module*. Each curriculum is composed of modules (the equivalent of stand-alone chapters) that are logically sequenced into complete *levels* of instruction. Every module is task-oriented and requires the successful completion of a written (knowledge-based) examination and a performance (hands-on) test. Modules may be individually selected and/or combined for efficient task training. Modules in the pipeline curricula, to the extent possible, align with industry-developed covered tasks. While not all modules relate to covered tasks, all covered tasks relate to at least one module.

Benefits of Training to the Pipeline Marketplace

Applying NCCER's standardized training process allows pipeline owners and contractors in the United States to establish the benchmark for acceptable skill and competency levels within the pipeline industry. This process will assist training organizations in developing and/or improving standardized pipeline training programs that effectively match their resources and operating conditions while responding to regulatory demands and to the needs of the industry. The process also recognizes training organizations that meet the minimum standards of competence, warranting regulatory, public, and professional confidence. This, in turn, fosters national unity within the pipeline industry for the mutual benefit of individual technicians, owners, contractors, and ultimately pipeline-users and society.

Additional Resources

To review the Pipeliner Training and Assessment Program Implementation and Resource Guide and the latest version of pipeline covered tasks, visit NCCER's website at **www.nccer.org,** or contact:

NCCER
13614 Progress Boulevard
Alachua, FL 32615
(888) 622-3720
Website: **www.nccer.org**

Additional sources of information on the pipeline operator qualification regulation include:

DOT – Office of Pipeline Safety
www.ops.dot.gov

Transportation Safety Institute
www.tsi.dot.gov

COURSE MAP

This course map shows all of the modules in the first level of the Pipeline Maintenance curriculum. The suggested training order begins at the bottom and proceeds up. Skill levels increase as you advance on the course map. The local Training Program Sponsor may adjust the training order.

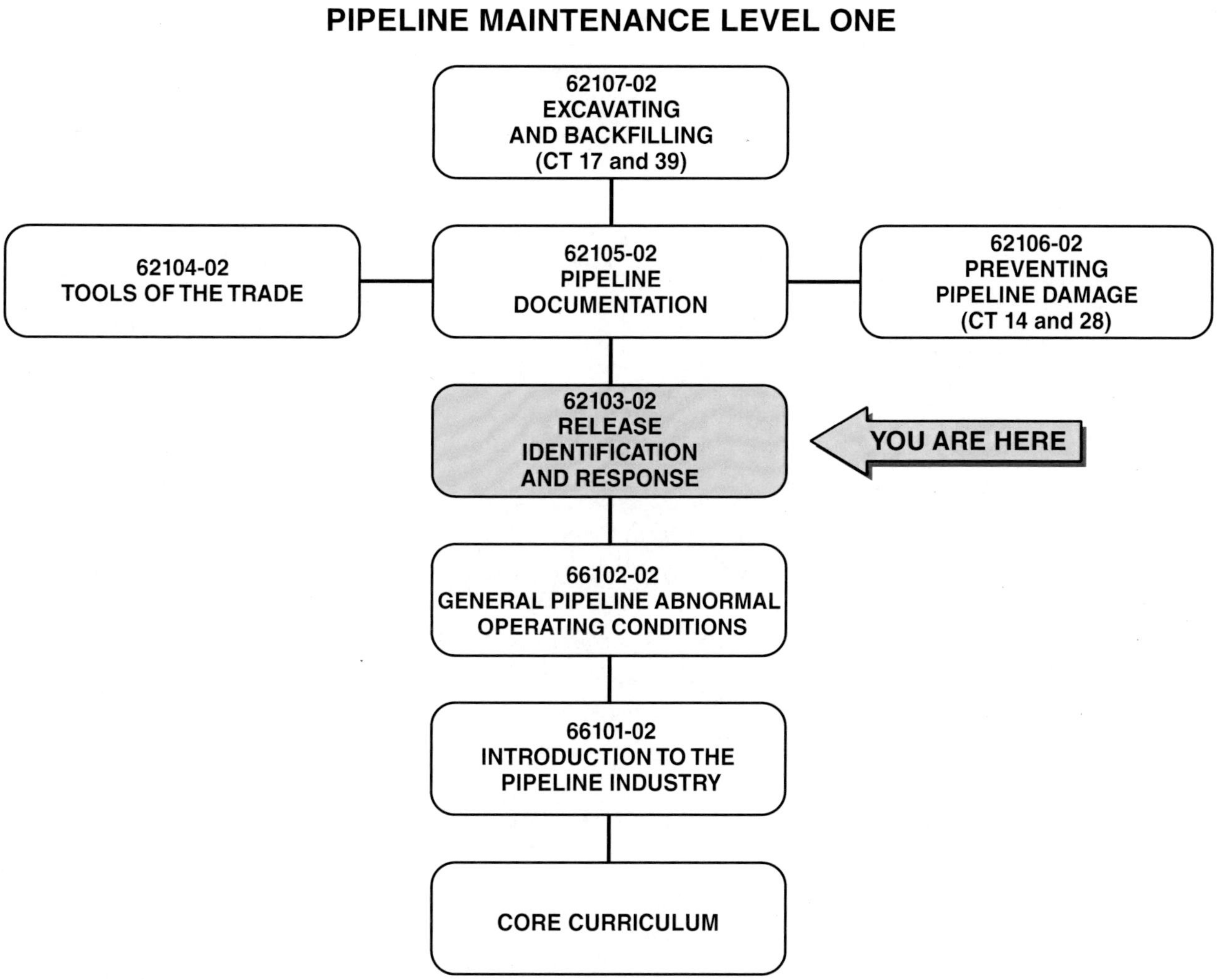

103CMAP.EPS

MODULE 62103-02 CONTENTS

Figures

MODULE 62103-02

Release Identification and Response

Objectives

When you have completed this module, you will be able to do the following:

1. Describe company environmental manuals, and regulations.
2. Describe United States Coast Guard responsibilities.
3. Describe spill prevention.
4. Recognize soil contamination.
5. Explain spill reporting and spill containment.
6. Explain proper waste handling.

Prerequisites

Before you begin this module, it is recommended that you successfully complete the following: Core Curriculum; Pipeline Core, Modules 66101 and 66102.

Required Trainee Materials

1. Pencil and paper
2. Appropriate personal protective equipment

1.0.0 ◆ INTRODUCTION

The products that pipelines carry are flammable, explosive, and/or toxic, which means that pipelines could pose a danger to people, animals, and property if these products are released to the environment. The extent of the hazard is a function not only of the characteristics of the products, but also of the pressures at which they are moved and the conditions under which they are released.

Significant quantities of natural gas from a release can ignite and/or explode. However, this gas is lighter than air, and once the release is stopped, the gas may rise and disperse. Crude oil vapors and its products are heavier than air and not so easily removed. Crude oil and petroleum products can contaminate the environment. Petroleum fuels transported in their natural liquid state also pose a fire and pollution hazard. Liquid petroleum gases, such as propane and butane, are transported as liquids under high pressure. If there is a pipeline release, these highly volatile liquids will atomize and vaporize, creating a highly flammable and explosive mixture of gas and mist. This mixture can travel long distances downwind before ignition.

Because of these dangers, federal, state, and local regulations govern nearly all aspects of the pipeline industry. Because of these dangers and because of the regulations, each company in the industry has procedures for preventing and responding to spills.

Every effort is made to prevent pipeline failure and resulting releases. If a release does occur, report it immediately. Immediate action is needed to stop the release and to contain the spill. Since most pipelines are underground, it is important to know how to recognize the signs of releases. Release identification and response are explained in the following sections:

- Company environmental manuals
- **Department of Natural Resources (DNR)** and **Environmental Protection Agency (EPA)** regulations
- National Response Center (NRC) and Coast Guard responsibilities
- Spill prevention
- Soil contamination
- Spill reporting
- Spill containment
- Waste handling

2.0.0 ◆ COMPANY ENVIRONMENTAL POLICIES

Company manuals detail the environmental hazards of the company pipeline products and the procedures to be followed for spill prevention and spill response. The company manuals include information similar to that presented in this module, but in more detail and including specific procedures for specific situations that are required to be followed.

Protecting the environment is a critical part of everyday operation. It is important that all employees study the required procedures so that they will know how to do their job in a manner that prevents emergencies and how to react if an emergency occurs.

There are many manuals relating to environmental protection, including the following:

- Oil spill response plans
- Emergency procedures

2.1.0 Oil Spill Response Plans

The Oil Spill Response Plan is a plan of action that provides decision-making criteria in the event of an oil spill. This plan satisfies the requirements of *49 CFR Part 194*, also known as the *Oil Pollution Act of 1990 (OPA 90)*. Telephone numbers and addresses of the various response centers and state and federal agencies are listed in the plan.

These plans explain in detail what will be done and by whom in the event of an oil spill. These contain lists of emergency contacts and procedures for all personnel to follow. Required notifications and reports are identified, as are the persons responsible for making the notifications and reports.

2.2.0 Emergency Procedures

Emergency procedures will address any emergency situation with the procedures to be followed, the appropriate persons to be contacted, and the proper reports to be made.

An emergency is a situation in which property, the health and safety of human life, and/or the protection of the environment are in jeopardy and prompt corrective action and the summoning of aid are essential. A pipeline emergency is an accident or incident that:

- Endangers life
- Presents a risk to the health and well-being of the public
- Presents a hazard to public and private property
- Pollutes watersheds or bodies of water
- Has the potential to damage the environment
- Endangers wildlife

Company's emergency policies' will typically cover the following topics:

- Receiving report of a potential emergency
- Identifying an emergency
- Classification of an emergency
- Immediate response locations
- Notifying appropriate personnel
- Response to impact of ground surface over the pipeline
- Response to emergencies
- Oil Movement Control Center emergency response
- Incident Command System Organization
- Emergency plan of action
- Post accident review
- Emergency records and documentation
- Analyzing failed components

The foremost consideration and objective of these policies are to take all necessary actions, including evacuation and road/railroad closure, to minimize the public's exposure to injury and potential accidental ignition, to protect the environment, and to establish and maintain the pipeline in a safe condition.

All pipeline personnel must be prepared to act quickly and efficiently when an emergency occurs, making it imperative that they have a thorough understanding of the Emergency Response Procedures Manual and the Oil Spill Response Plan. These manuals cover emergencies and do not preclude using common sense or taking reasonable and prudent actions to mitigate a problem. Employees should always respond in ways that will result in the highest degree of safety. When in doubt about a situation, employees should contact supervision immediately.

3.0.0 ◆ DNR AND EPA REGULATIONS

The EPA and appropriate state agencies work together to protect the environment and natural resources of the United States. In some states, the agency is called the Department of Natural Resources (DNR).

To provide the necessary protection, both the EPA and DNRs have established regulations. These regulations affect the pipeline industry.

3.1.0 DNR

Every state has a DNR or similar agency whose goal is to protect, maintain, and encourage wise development of state resources. The DNR in each state manages all state-owned land, water, and natural resources, except for fish and game, on behalf of the people of that state. A state's typical DNR is organized into divisions that specialize in certain areas, such as:

- Agriculture
- Forestry
- Geological and geophysical surveys
- Mining
- Land and water management
- Oil and gas
- Parks and recreation
- Support services

3.2.0 EPA

To address the potential threat posed by petroleum and non-petroleum oils, the EPA has established an oil spill program with the goals of preventing, preparing for, and responding to oil spills.

The pipeline industry, using the EPA prevention program, has reduced the number of spills from the total volume handled each year. The EPA has a long history of responding to oil spills, including several major ones, and the lessons learned have helped to improve the country's prevention and response capabilities. The program is administered through the EPA headquarters and the 10 EPA regions.

3.2.1 Prevention

To prevent oil spills, the EPA requires owners or operators of oil storage facilities to prepare and implement Spill Prevention Control and Countermeasures (SPCC) Plans that detail the facility's spill prevention and control measures. The EPA also enforces the oil spill liability and penalty provisions found in *OPA 90*, which provide incentives to facility owners/operators who take the necessary steps to prevent oil spills. The EPA also conducts on-site facility inspections to ensure that adequate measures are being taken.

3.2.2 Preparation

Two principles of the EPA's oil spill preparedness program are developing and coordinating contingency plans and conducting oil spill prevention and response training. Well-designed facility, local, area, regional, and national contingency plans assist response personnel in their efforts to contain and clean up any size spill by providing information that the teams will need before, during, and after an oil spill occurs. Training ensures that emergency responders know the what, when, and how of oil spill response.

3.2.3 Response

Although many spills are contained and cleaned up by the responsible parties, some spills require the assistance of local and state agencies and the federal government. Under the National Contingency Plan, the EPA is the lead federal response agency for spills occurring in inland waters, and the Coast Guard is the lead response agency for spills in coastal waters and deepwater ports.

The EPA tracks all reports of oil spills to United States waters. The EPA learns about the spill through responsible parties, who are required to report the spill to the federal government, or from state and local responders. Once the federal government receives the report, either through the National Response Center (NRC), the EPA, or another agency, it is recorded in the Emergency Response Notification System (ERNS). The ERNS contains historical spill information for the entire country dating from 1986.

4.0.0 ◆ NRC AND COAST GUARD RESPONSIBILITIES

The NRC and Coast Guard are closely related. The NRC is staffed by Coast Guard personnel. Both are involved in environmental or public safety emergencies.

4.1.0 NRC

The primary function of the NRC is to serve as the sole national point of contact for reporting all oil, chemical, radiological, biological, and etiological discharges into the environment anywhere in the United States and its territories. The reporting call must be made promptly. In addition to gathering and distributing spill data for Federal On-Scene Coordinators (FOSCs) and serving as the communications and operations center for the National Response Team (NRT), the NRC maintains agreements with a variety of federal entities to make additional notifications regarding incidents meeting established trigger criteria.

4.2.0 Coast Guard

The Coast Guard fields teams to respond to a variety of emergency incidents. Among these are:

- Search and rescue operations
- Law enforcement
- Oil spills
- Hazardous substance releases
- Terrorism
- Marine fires
- Multi-casualty situations

In the event of marine oil spills, the Coast Guard provides law enforcement presence, performs rescue operations as needed, supervises damaged vessel removal, and manages containment and cleanup operations.

5.0.0 ◆ SPILL PREVENTION

Spill prevention involves practically every aspect of pipeline construction, operations, and maintenance. Although spill prevention policies are required by regulations, pipeline companies are already very much interested in preventing spills. Spills mean lost production and operations shutdowns. From pipeline design to product delivery, efforts are made to prevent spills.

5.1.0 Damage Prevention

Company procedures are documented for every aspect of pipeline operations and maintenance to ensure against damage to the pipeline. Routine inspections and maintenance are targeted for keeping the pipeline operating safely.

Every company will have procedures to prevent damage due to excavation and any subsequent spills. Excavation is the primary outside force causing pipeline damage and spills. These procedures will include contractor listings and information programs, public education, One-Call systems, temporary marking protocols, and special procedures or precautions for boring and crossings by cables or other pipelines.

Specific spill prevention measures in maintenance include the following:

- Routine and thorough inspections
- Continuous observation and awareness
- Recognition of hazards and safety related conditions
- Quick and educated response to problems
- Thorough maintenance

5.1.1 Inspections

Inspections are routinely carried out along the entire pipeline. These may be done by aerial or vehicle patrol or by foot. The interval between inspections is one to three weeks. These inspections must occur at least 26 times per year. The inspections provide valuable information about what is happening on the pipeline route and may alert the company to impending problems such as encroachment, excavations, or vandalism. Inspectors may also be able to recognize signs of releases and other pipe problems.

The maintenance crew, when in the field for routine or specific maintenance tasks, also does inspections of the site. These inspections cover such things as:

- Condition of warning signs
- Condition of fences
- Soil erosion, settling, and gravel coverage
- Pipeline markers
- Paint conditions
- Chains and locks on valves, gates, and electrical panels
- Signs of releases:
  - Flanged connection
  - Valve gaskets
  - Gear boxes
  - Fittings
  - Plugs
  - Bolts
- Trash and overgrown vegetation
- Signs of vandalism
- Pipeline markers

5.1.2 Observation

Any time personnel are working along the route in the field or around equipment at facilities, it is good practice to be observant and aware of the surroundings. With experience, workers learn to recognize differences in the sounds and smells of their working environment. These can often give a clue that there is a problem. When looking at the equipment or the ground around the lines, be sure to notice any difference in soil coloring, soil consistency, or vegetation.

5.1.3 Recognition

Recognizing hazards and safety-related conditions is a way to protect the environment and everyone's safety. To learn to do this, study these and other modules associated with the hazards of pipeline products, learn the information provided by the company on environmental and health hazards, and practice using this knowledge at work.

All personnel performing operation and maintenance activities should be aware of conditions that potentially are safety related. The following are examples of conditions that, when observed, must be immediately reported to supervision:

- General corrosion
- Localized corrosion pitting
- Corrosion
- Uncontrolled movement or abnormal loading of the pipe by environmental causes
- Material defect or physical damage
- Malfunctions or operating errors
- Releases
- Events impacting the cover over the pipeline
- Conditions that could lead to imminent hazard

5.1.4 Response

Response to problems on the pipeline must be quick and done correctly. If the first person aware of the problem is not trained in how to respond, then his or her job is to inform someone who is. It is usually not a good idea, nor is it approved by the company, to make a stab at solving a problem you don't understand.

5.1.5 Maintenance

Pipeline maintenance is critical to prevent spills. Regulations require regular maintenance of critical parts. Companies, in compliance with these regulations, set schedules and provide specific maintenance procedures. The schedules are often monitored by the company computer control system, which automatically issues work orders and lists the required tasks at the appropriate times. Maintenance workers check these schedules and work orders daily.

6.0.0 ◆ SOIL CONTAMINATION

Sometimes the first sign on the surface that there is a problem with the pipeline is soil contamination. Employees working on the pipeline route must be familiar with the signs of soil contamination and be observant of conditions when on site. Some general guidelines for locating contamination from an underground release are:

- Wilted or dead vegetation
- Discolored soil
- Hydrocarbon liquid or frost on the surface
- Discolored or melted snow
- Hydrocarbon smell
- Vapor clouds or shimmering in the air above the surface
- Sounds of blowing, whistling, or trickling

Surface contamination is identified by the presence of free product on the ground, as a pool of petroleum or as a sheen or film on water. Oil and less refined products are black or dirty brown and viscous; they produce a blackish sheen on water and will form a consistent sheen on the surface, even after being disturbed. Gasoline produces a multi-colored sheen on water, representing the refined hydrocarbons and additives. Most petroleum products float on water, but some components of gasoline may mix with the water molecules and be transported with the water.

The most likely indication of ground contamination is stained and odorous soil. Odor may also be present without visible staining. Vegetation around the contamination site may be affected. In some cases of gas releases, insects are drawn to the site.

7.0.0 ◆ SPILL REPORTING

The company's environmental and spill response procedures should be reviewed carefully so that appropriate response to spills can be followed right away. Because releases may occur at any time, personnel must be alert to signs indicating a loss of oil or product. Releases may be detected by pipeline monitoring equipment, personnel surveillance, outside parties, pressure changes, or computerized volume balancing. Each emergency situation involving a spill or release may be different; therefore, all employees must be familiar with and trained to respond according to company procedures.

When a spill does occur and is discovered, it must be reported immediately to the control center. Management will initiate response to the spill and be responsible for documenting any accidents and safety-related conditions, as well as reporting to regulatory agencies.

Spills on federal lands must also be reported to the Bureau of Land Management. State and local agencies will also require reports and should be listed in the company manual. Other issues of spill reporting include the following:

- Written reports
- Telephonic notice
- Safety-related conditions reports

7.1.0 Written Reports

All reportable accidents must be reported to the DOT in writing as soon as possible, but not later than 30 days after discovery of the accident. The written report must be filed on DOT form 7000-1 or a facsimile (*Figure 1*). A company representative will collect all relevant information and make this available to the regulatory compliance personnel, who will file the written report. Supplemental reports may be filed if the information changes.

Any failure in the pipeline system in which there is a release of a hazardous liquid resulting in any of the following must also be reported to the appropriate regulatory agency:

- Explosion or fire not intentionally set by the operator
- Loss of five or more barrels of liquid
- Escape to the atmosphere of more than five barrels a day of highly volatile liquids
- Death of any person
- Bodily harm to any person resulting in loss of consciousness, having to be carried away, needing medical treatment, or disability beyond the day of occurrence
- Estimated property damage exceeding $50,000

7.2.0 Telephonic Notice

Certain accidents must be reported to the NRC at the earliest possible time – within two hours after discovery unless circumstances prevent it. Because these accidents must be reported quickly, the manager or operations or control room dispatcher will telephone it in to the NRC. To facilitate this, the NRC has a telephonic notice report form that lists information NRC will ask for when an accident is reported (see *Figure 2*).

Telephonic notification must be made for any failure or release that meets any of the following:

- Results in a death or an injury requiring hospitalization
- Results in either an explosion or a fire not intentionally set by the operator
- Results in pollution of any body of water that discolors the surface of the water or shoreline or deposits sludge or emulsion
- Is significant in the judgment of the operator

7.3.0 Safety-Related Conditions Reports

Safety-related conditions must be reported if they occur on an active pipeline or pipeline facility that is:

- Within 220 yards of any building of human occupancy or assembly
- Within the right-of-way of an active railroad, paved road, street, or highway
- Offshore
- Where a spill could reasonably be expected to pollute a body of water

The report must be filed within five working days after the condition is determined, but no less than ten days after first discovery, to the DOT and relevant state and local agencies. Reportable safety-related conditions are listed in a previous section, but should be reviewed in more detail in the company manuals.

8.0.0 ◆ SPILL CONTAINMENT

When a spill does occur, the two main areas of concern are minimizing the amount and controlling the spill.

8.1.0 Minimizing the Amount

Several actions may be taken to minimize the amount of oil spilled through pipeline releases. Often stopping the pipeline pumps stops the spill. Suction pumps can be used to assist gravity. Such pumps are especially effective where elevation differences are small compared with the total lift available. When releases occur at low points on the hydraulic profile, static pressures usually cause continued pressure and spillage.

Controls at automatic pump stations are usually set to shut down the stations and signal an alarm when changes in suction or discharge pressures occur. Block valves on one or both sides of the release are closed to stop flow. Portable pumps and bypass fittings can be used to reconnect the flow around the release to keep the pipeline operating. When internal pressure has been reduced at the site of the release, the outflow is stopped by temporarily closing the hole with clamps or pegs.

The controls and valves designed to shut down flow are required to be kept in operating condition and to be inspected regularly by the DOT. Records of inspections and tests are required to be maintained.

NOTICE: This report is required by 49 CFR Part 195. Failure to report can result in a civil penalty not to exceed $25,000 for each violation for each day that such violation persists except that the maximum civil penalty shall not exceed $500,000 as provided in 49 USC 60122

Form Approved
OMB No. 2137-0047

U.S. Department of Transportation
Research and Special Programs Administration

ACCIDENT REPORT – HAZARDOUS LIQUID PIPELINE SYSTEMS

Report Date ____________

No. ____________ (DOT Use Only)

INSTRUCTIONS

Important: *Please read the separate instructions for completing this form before you begin. They clarify the information requested and provide specific examples. If you do not have a copy of the instructions, you can obtain one from the Office Of Pipeline Safety Web Page at http://ops.dot.gov.*

PART A – GENERAL REPORT INFORMATION Check: ☐ **Original Report** ☐ **Supplemental Report** ☐ **Final Report**

1. a. Operator's OPS 5-digit Identification Number (if known) /__/__/__/__/__/
2. b. If Operator does not own the pipeline, enter Owner's OPS 5-digit Identification Number (if known) /__/__/__/__/__/
   c. Name of Operator ____________
   d. Operator street address ____________
   e. Operator address ____________ City, County, State and Zip Code

IMPORTANT: IF THE SPILL IS SMALL, THAT IS, THE AMOUNT IS AT LEAST 5 GALLONS BUT IS LESS THAN 5 BARRELS, COMPLETE THIS PAGE ONLY, UNLESS THE SPILL IS TO WATER AS DESCRIBED IN 49 CFR §195.52(A)(4) OR IS OTHERWISE REPORTABLE UNDER §195.50 AS REVISED IN CY 2001.

2. Time and date of the accident

   /__/__/__/__/ hr. /__/__/ month /__/__/ day /__/__/ year

3. Location of accident
   (If offshore, do not complete a through d. See Part C.1)

   a. Latitude: ________ Longitude: ________
   (if not available, see instructions for how to provide specific location)

   b. ____________ City, and County or Parish

   c. ____________ State and Zip Code

   d. Mile post/valve station O or survey station no. O
   (whichever gives more accurate location)

   ____

4. Telephone report

   /__/__/__/__/__/__/ NRC Report Number /__/__/ month /__/__/ day /__/__/ year

5. Losses *(Estimated)*

   Public/Community Losses reimbursed by operator:

   Public/private property damage $________
   Cost of emergency response phase $________
   Cost of environmental remediation $________
   Other Costs $________
   (describe) ____________

   Operator Losses:

   Value of product lost $________
   Value of operator property damage $________
   Other Costs $________
   (describe) ____________

   Total Costs $________

6. Commodity Spilled OYes O No
   (If Yes, complete Parts a through c where applicable)
   a. Name of commodity spilled ____________
   b. Classification of commodity spilled:
      O HVLs /other flammable or toxic fluid which is a gas at ambient conditions
      O CO_2 or other non-flammable, non-toxic fluid which is a gas at ambient conditions
      O Gasoline, diesel, fuel oil or other petroleum product which is a liquid at ambient conditions
      O Crude oil

   c. Estimated amount of commodity involved :
   O Barrels
   O Gallons (check only if spill is less than one barrel)

   Amounts:
   Spilled : ________
   Recovered: ________

CAUSES FOR SMALL SPILLS ONLY (5 gallons to under 5 barrels) : ***(For large spills [5 barrels or greater] see Part H)***

O Corrosion **O Natural Forces** **O Excavation Damage** **O Other Outside Force Damage**
O Material and/or Weld Failures **O Equipment** **O Incorrect Operation** **O Other**

PART B – PREPARER AND AUTHORIZED SIGNATURE

____________ (type or print) Preparer's Name and Title ____________ Area Code and Telephone Number

____________ Preparer's E-mail Address ____________ Area Code and Facsimile Number

____________ Authorized Signature ____________ (type or print) Name and Title ______ Date ____________ Area Code and Telephone Number

Form RSPA F 7000-1 (01-2001) *Page 1 of 4*

103F01A.PDF

Figure 1 ◆ DOT report form (page 1 of 4).

PART C – ORIGIN OF THE ACCIDENT (Check all that apply)

1. Additional location information
   a. Line segment name or ID ____________
   b. Accident on Federal land other than Outer Continental Shelf O Yes O No
   c. Is pipeline interstate? O Yes O No

Offshore: O Yes O No *(complete d if offshore)*

d. Area ____________ Block # ____________

State / / / or Outer Continental Shelf ☐

2. Location of system involved (*check all that apply*)
   ☐ Operator's Property
   ☐ Pipeline Right of Way
   ☐ High Consequence Area (HCA)? Describe HCA____________

3. Part of system involved in accident
   O Above Ground Storage Tank
   O Cavern or other below ground storage facility
   O Pump/meter station; terminal/tank farm piping and equipment, including sumps
   O Other *Specify:* ____________
   O Onshore **pipeline**, including valve sites
   O Offshore **pipeline**, including platforms

   *If failure occurred on **Pipeline**, complete items a - g:*

4. Failure occurred on
   O Body of Pipe O Pipe Seam O Scraper Trap
   O Pump O Sump O Joint
   O Component O Valve O Metering Facility
   O Repair Sleeve O Welded Fitting O Bolted Fitting
   O Girth Weld

   Other *(specify)* ____________

   Year the component that failed was installed: / / / / /

5. Maximum operating pressure *(MOP)*
   a. Estimated pressure at point and time of accident: ________ *PSIG*
   b. MOP at time of accident: ________*PSIG*
   c. Did an overpressurization occur relating to the accident? OYes O No

a. Type of leak or rupture
   OLeak: O Pinhole O Connection Failure *(complete sec. H5)*
   O Puncture, diameter *(inches)* ____________
   ORupture: O Circumferential – Separation
   O Longitudinal – Tear/Crack, length *(inches)* ________
   Propagation Length, total, both sides *(feet)* ________
   ON/A
   OOther ____________

b. Type of block valve used for isolation of immediate section:
   Upstream: ☐ Manual ☐ Automatic ☐ Remote Control ☐ Check Valve
   Downstream: ☐ Manual ☐ Automatic ☐ Remote Control ☐ Check Valve

c. Length of segment isolated ______ ft

d. Distance between valves ______ ft

e. Is segment configured for internal inspection tools? OYes O No

f. Had there been an in-line inspection device run at the point of failure? O Yes O No O Don't Know
   O Not Possible due to physical constraints in the system

g. If *Yes*, type of device run (*check all that apply*)
   ☐ High Resolution Magnetic Flux tool Year run: ______
   ☐ Low Resolution Magnetic Flux tool Year run: ______
   ☐ UT tool Year run: ______
   ☐ Geometry tool Year run: ______
   ☐ Caliper tool Year run: ______
   ☐ Crack tool Year run: ______
   ☐ Hard Spot tool Year run: ______
   ☐ Other tool Year run: ______

PART D – MATERIAL SPECIFICATION

1. Nominal pipe size *(NPS)* / / / / / in.
2. Wall thickness / / / / / in.
3. Specification ____________ SMYS / / / / / / /
4. Seam type ____________
5. Valve type ____________
6. Manufactured by ____________ in year / / / / /

PART E – ENVIRONMENT

1. Area of accident
   O Under pavement
   O Underground
   O Inside/under building
   O In open ditch
   O Above ground
   O Under water
   O Other ____________

2. Depth of cover: ____________ inches

PART F – CONSEQUENCES

1. **Consequences *(check and complete all that apply)***

a.	Fatalities	Injuries
Number of operator employees:		______
Contractor employees working for operator:		______
General public:		______
Totals:		______

b. Was pipeline/segment shutdown due to leak? OYes O No
   If Yes, how long? ______ days ______ hours ______ minutes

c. Product ignited OYes O No d. Explosion OYes O No

e. ☐ Evacuation *(general public only)* / / / / / people
   Reason for Evacuation:
   O Precautionary by company
   O Evacuation required or initiated by public official

f. Elapsed time until area was made safe: / / / hr. / / / min.

2. **Environmental Impact**
   a. Wildlife Impact: Fish/aquatic O Yes O No
      Birds O Yes O No
      Terrestrial O Yes O No
   b. Soil Contamination O Yes O No
      If Yes, estimated number of cubic yards: ____________
   c. Long term impact assessment performed: O Yes O No
   d. Anticipated remediation O Yes O No
      If Yes, check all that apply: ☐ Surface water ☐ Groundwater ☐ Soil ☐ Vegetation ☐ Wildlife
   e. Water Contamination: O Yes O No (*If Yes, provide the following*)
      Amount in water ________ barrels
      Ocean/Seawater O No O Yes
      Surface O No O Yes
      Groundwater O No O Yes
      Drinking water O No O Yes (*If Yes, check below.*)
      O Private well O Public water intake

Form RSPA F 7000-1 (01-2001) *Page 2 of 4*

103F01B.PDF

Figure 1 ◆ DOT report form (page 2 of 4).

PART G – LEAK DETECTION INFORMATION

1. Computer based leak detection capability in place? O Yes O No

2. Was the release initially detected by? (check one):
O CPM/SCADA-based system with leak detection
O Static shut-in test or other pressure or leak test
O Local operating personnel, procedures or equipment
O Remote operating personnel, including controllers
O Air patrol or ground surveillance
O A third party O Other (specify) ________________

3. Estimated leak duration days ____ hours ____

PART H – APPARENT CAUSE

Important: *There are 25 numbered causes in this Part H. Check the box corresponding to the primary cause of the accident. Check one circle in each of the supplemental categories corresponding to the cause you indicate. See the instructions for guidance.*

H1 – CORROSION

1. ☐ External Corrosion

2. ☐ Internal Corrosion

(Complete items a – e where applicable.)

a. Pipe Coating
O Bare
O Coated

b. Visual Examination
O Localized Pitting
O General Corrosion
O Other ___________________

c. Cause of Corrosion
O Galvanic O Atmospheric
O Stray Current O Microbiological
O Cathodic Protection Disrupted
O Stress Corrosion Cracking
O Selective Seam Corrosion
O Other ___________________

d. Was corroded part of pipeline considered to be under cathodic protection prior to discovering accident?
O No O Yes, Year Protection Started: /__/__/__/__/

e. Was pipe previously damaged in the area of corrosion?
O No O Yes ⇒ Estimated time prior to accident: /__/__/ *years* /__/__/ *months* *Unknown* ☐

H2 – NATURAL FORCES

3. ☐ Earth Movement ⇒ O Earthquake O Subsidence O Landslide O Other ___________________
4. ☐ Lightning
5. ☐ Heavy Rains/Floods ⇒ O Washouts O Flotation O Mudslide O Scouring O Other ______________
6. ☐ Temperature ⇒ O Thermal stress O Frost heave O Frozen components O Other ______________
7. ☐ High Winds

H3 — EXCAVATION DAMAGE

8. ☐ Operator Excavation Damage (including their contractors/Not Third Party)
9. ☐ Third Party *(complete a-f)*

a. Excavator group
O General Public O Government O Excavator other than Operator/subcontractor

b. Type: O Road Work O Pipeline O Water O Electric O Sewer O Phone/Cable
O Landowner-not farming related O Farming O Railroad
O Other liquid or gas transmission pipeline operator or their contractor
O Nautical Operations O Other ___________________________

c. Excavation was: OOpen Trench O Sub-strata (boring, directional drilling, etc…)

d. Excavation was an ongoing activity (Month or longer) OYes O No If Yes, Date of last contact /__/__/__/

e. Did operator get prior notification of excavation activity?
O Yes; Date received: /__/__/ mo. /__/__/ day /__/__/__/__/ yr. O No
Notification received from: O One Call System O Excavator O Contractor O Landowner

f. Was pipeline marked as result of location request for excavation? O No O Yes *(If Yes, check applicable items i - iv)*
i. Temporary markings: O Flags O Stakes O Paint
ii. Permanent markings: O
iii. Marks were (check one) : O Accurate O Not Accurate
iv. Were marks made within required time? O Yes O No

H4 – OTHER OUTSIDE FORCE DAMAGE

10. ☐ Fire/Explosion as primary cause of failure ⇒ Fire/Explosion cause: O Man made O Natural
11. ☐ Car, truck or other vehicle not relating to excavation activity damaging pipe
12. ☐ Rupture of Previously Damaged Pipe
13. ☐ Vandalism

Form RSPA F 7000-1 (01-2001) *Page 3 of 4*

103F01C.PDF

Figure 1 ◆ DOT report form (page 3 of 4).

H5 – MATERIAL AND/OR WELD FAILURES

Material

14. ☐ Body of Pipe ⇒ O Dent O Gouge O Bend O Arc Burn O Other ____________
15. ☐ Component ⇒ O Valve O Fitting O Vessel O Extruded Outlet O Other ____________
16. ☐ Joint ⇒ O Gasket O O-Ring O Threads O Other ____________

Weld

17. ☐ Butt ⇒ O Pipe O Fabrication O Other ____________
18. ☐ Fillet ⇒ O Branch O Hot Tap O Fitting O Repair Sleeve O Other ____________
19. ☐ Pipe Seam ⇒ O LF ERW O DSAW O Seamless O Flash Weld
O HF ERW O SAW O Spiral O Other ____________

Complete a-g if you indicate ***any*** *cause in part H5.*

a. Type of failure:
O Construction Defect ⇒ O Poor Workmanship O Procedure not followed O Poor Construction Procedures
O Material Defect

b. Was failure due to pipe damage sustained in transportation to the construction or fabrication site? O Yes O No

c. Was part which leaked pressure tested before accident occurred? O Yes, *complete d-g* O No

d. Date of test: /__/__/__/__/ yr. /__/__/ mo. /__/__/ day

e. Test medium: O Water O Inert Gas O Other ____________________

f. Time held at test pressure: /__/__/ hr.

g. Estimated test pressure at point of accident: ____________________ *PSIG*

H6 – EQUIPMENT

20. ☐ Malfunction of Control/Relief Equipment ⇒ O Control valve O Instrumentation O SCADA O Communications
O Block valve O Relief valve O Power failure O Other ____________
21. ☐ Threads Stripped, Broken Pipe Coupling ⇒ O Nipples O Valve Threads O Dresser Couplings O Other ____________
22. ☐ Seal Failure ⇒ O Gasket O O-Ring O Seal/Pump Packing O Other ____________

H7 – INCORRECT OPERATION

23. ☐ Incorrect Operation

a. Type: O Inadequate Procedures O Inadequate Safety Practices O Failure to Follow Procedures
O Other ______________________________

b. Number of employees involved who failed a post-accident test: drug test: /__/__/__/ alcohol test /__/__/__/

H8 – OTHER

24. ☐ Miscellaneous, *describe:* ______________________________
25. ☐ Unknown

O Investigation Complete O Still Under Investigation *(submit a supplemental report when investigation is complete)*

PART I – NARRATIVE DESCRIPTION OF FACTORS CONTRIBUTING TO THE EVENT *(Attach additional sheets as necessary)*

Form RSPA F 7000-1 (01-2001) *Page 4 of 4*

103F01D.PDF

Figure 1 ◆ DOT report form (page 4 of 4).

NATIONAL RESPONSE CENTER
TELEPHONIC NOTIFICATION
Response Information for Reporting

NRC Call/Case # ____________

The following format was developed from the computer screens the National Response Center (NRC) watchstanders will be using when they question you regarding your telephone notification of a reportable incident. You should have all the information below as complete as possible before making your telephone report to the NRC. You will be asked these questions. You have two (2) hours from the time the incident was discovered to obtain the information and make telephone notification to the NRC. All information reported to the NRC shall be stated as factually as possible. If all the information has not been obtained within the required notification time limit, be frank and qualify your statements as not yet available. Be sure to describe all specific values stated as preliminary and approximate pending verification.

NRC TELEPHONE REPORT INFORMATION

1. REPORTING PARTY (You)	2. SUSPECTED RESPONSIBLE PARTY
Last Name: ____________	Last Name: ____________
First Name: ____________ MI ____	First Name: ____________ MI ____
Contact Phones: 1-800-322-PIPE (7473) 1-800-664-5365	Contact Phones: ____________
Company: All American Pipeline Company	Company: ____________
Org. Type: Interstate Crude Oil Pipeline Operator	Org. Type: ____________
Position (yours): ____________	Position: ____________
Address: All American Pipeline Company 10000 Ming Ave.	Address: ____________
City: Bakersfield	City: ____________
State: California Zip: 93311	State: ____________ Zip: ____________

Was Material Released? ☐ Yes ☐ No **Are you calling for the responsible party?** ☐ Yes ☐ No

Is this telephone report to meet Federal obligation ☐ Yes ☐ No **FEDERAL CODE REQUIREMENT: 49 CFR 195.52**

Crude Oil Spill release resulting in: ***(check the box(es) that apply)***

☐ **An explosion or fire** ☐ **A death or injury requiring hospitalization** ☐ **Property damage exceeding $50,000**
☐ **Polluted water** ☐ **Judged to be a significant incident** ☐ ____________

3. INCIDENT DESCRIPTION

Source and/or Cause of Incident *(Check the box that apply)*
☐ External Source (Third Party) ☐ Material Failure ☐ Equipment Malfunction ☐ Operator Error
☐ Unknown (Describe) ____________

Discovery:	NRC Notified
Date: ________ Time: (24hr.) ________	Date: ________ Time (24hr.) ________

Incident Address/Location	
Nearest City ________	Distance from city ________
State ________	Direction from city ________
County ________	Pipeline Milepost ________
Zip ________	One Call System Area ID ________
T-S-R-Q ________	

4. MATERIAL

Classification of material released ____________ Approximate Quantity ____________ BBL(s)

5. REMEDIAL ACTION

☐ Containment ☐ Recover ☐ Re-injection into system ☐ Evaluation ☐ Clean-up ☐ Disposal
☐ Current Status

6. IMPACT

Number of Injuries ______ Were there evacuations? ☐ Yes ☐ No Number Evacuated ______
Number of Fatalities ______ Was there damage? ☐ Yes ☐ No Est. Damage $ ______

7. NOTIFICATION OF OTHER AGENCIES

☐ CA OES ☐ NM Oil Conservation Division ☐ BLM ☐ Fire Department ☐ Highway Patrol
☐ AZ DPS ☐ Railroad Commission of Texas ☐ Sheriff ☐ ____________

____________ Name of person reporting to NRC ____________ NRC Watchstander taking report ____________ Date

Retention: Life of System

103F02.EPS

Figure 2 ◆ Telephonic notice report form.

8.2.0 Controlling the Spill

If the quantity of the spill is sufficient to flood the area, measures must be taken to control the spread. Dikes or dams can be thrown up quickly by bulldozers or other earth-moving equipment. Damming small, usually dry, watercourses or ditches provides catch basins. Sometimes it is heavy equipment working near the line that causes the release, and they can be used in controlling it.

Truck or trailer-mounted pumps can be used to pick up large quantities of oil and transfer it to places for safe recovery or disposal. Pneumatic or hydraulic pumps should be used instead of gasoline pumps to avoid igniting oil vapors.

When oil is spilled into and carried on the surface of water, it can be contained by floating booms or other surface barriers. Dams with low-level outlets below the oil can be used to accommodate the water flow. In large bodies of water, floating barriers are used to encircle the oil source or to trap wind-blown oil. The entrapped oil is then picked up by pumps. Spills in the oceans are complicated by wind, waves, water currents, water depths, and the great areas involved, and these require individualized solutions.

9.0.0 ◆ WASTE HANDLING

Waste and debris around pipeline facilities, including the right-of-way, must always be properly disposed. Any waste that may have been contaminated with petroleum liquids must receive particular attention for disposal.

Facilities may have incinerators for disposing of such items. They are designed to burn small amounts of highly flammable material at a time, including oily rags, sorbent pads, and drained oil filters. They will also burn punctured aerosol cans, but not liquids. Specific instructions will be provided for loading and burning refuse in the incinerators.

Summary

An important part of any job on the pipeline is the ability to recognize hazards. The effects of the pipeline products can be harmful to people, wildlife, and vegetation.

Because of the hazards, company environmental manuals contain detailed procedures to teach workers how to help prevent hazards from becoming problems and how to react if problems occur. Each pipeline company is required to develop and maintain a SPCC to be ready to react immediately when necessary. Federal and state agencies, such as the EPA, have provided guidelines and regulations for pipeline operation and maintenance to help prevent spills and to contain any spills that do occur. The NRC and the Coast Guard are trained and ready to respond to pipeline emergencies as needed.

Every facet of pipeline construction, operation, and maintenance is concerned with preventing spills in the first place, but they do still sometimes happen. Maintenance personnel are likely to be the first to spot problems. During routine maintenance inspections or in the field on specific investigations, each person must remain alert to the signs that a problem exists or is possible.

When a spill occurs, the first steps are always to report and contain it if possible while ensuring the safety of the employees and the public. Containment procedures vary according to the situation, but are outlined and planned ahead of time to save time and prevent mistakes when needed. Employees are trained ahead of time on how to react for the same reason.

Review Questions

1. Company manuals will detail the _____ of the company pipeline products.

a. chemical make-up
b. environmental hazards
c. user applications
d. expected volumes

2. The _____ is a plan of action that provides decision-making criteria in the event of an oil spill.

a. Oil Pollution Act of 1960
b. Oil Spill Response Plan
c. company manual
d. Incident Command Organization

3. Employees should always respond in ways that will result in _____.

a. the highest degree of safety
b. meeting regulations
c. the fastest evacuation
d. the best public relations

4. The pipeline industry, using the _____ prevention program, has reduced the number of spills.

a. DNR
b. NRC
c. EPA
d. Coast Guard

5. In the event of marine oil spills, the Coast Guard provides _____.

a. cleanup crews
b. vessels
c. pipeline repairs
d. law enforcement presence

6. With experience, workers learn to recognize differences in the _____ of their working environment.

a. climate
b. sounds and smells
c. requirements
d. geography

7. Regulations require that maintenance of critical parts be done _____.

a. regularly
b. weekly
c. when repairs are needed
d. monthly

8. The most likely indication of ground contamination is _____.

a. gathering of insects
b. melted snow
c. dead wildlife
d. soil stain and/or odor

9. When releases occur at low points on the hydraulic profile, static pressures usually cause _____.

a. spillage to increase
b. continued pressure and spillage
c. pumps to stop the flow
d. controls to shut down

10. Waste and debris around pipeline facilities must always be _____.

a. buried
b. recycled
c. burned
d. properly disposed of

GLOSSARY

Trade Terms Introduced in This Module

Department of Natural Resources (DNR): Departments in state governments that supervise the use of the state's natural resources.

Environmental Protection Agency (EPA): Supervises and controls activities that affect the quality of the environment.

NCCER CURRICULA — USER UPDATE

NCCER makes every effort to keep its textbooks up-to-date and free of technical errors. We appreciate your help in this process. If you find an error, a typographical mistake, or an inaccuracy in NCCER's curricula, please fill out this form (or a photocopy), or complete the online form at **www.nccer.org/olf**. Be sure to include the exact module ID number, page number, a detailed description, and your recommended correction. Your input will be brought to the attention of the Authoring Team. Thank you for your assistance.

Instructors – If you have an idea for improving this textbook, or have found that additional materials were necessary to teach this module effectively, please let us know so that we may present your suggestions to the Authoring Team.

NCCER Product Development and Revision

13614 Progress Blvd., Alachua, FL 32615

Email: curriculum@nccer.org
Online: www.nccer.org/olf

❏ Trainee Guide ❏ Lesson Plans ❏ Exam ❏ PowerPoints Other ________________________

Craft / Level: Copyright Date:

Module ID Number / Title:

Section Number(s):

Description:

Recommended Correction:

Your Name:

Address:

Email: Phone: